BRE Building Elements

Foundations, basements and external works

Performance, diagnosis, maintenance, repair and the avoidance of defects

H W Harrison, ISO, Dip Arch, RIBA

P M Trotman

BRE
Garston
Watford
WD25 9XX

Prices for all available
BRE publications can be
obtained from:
CRC Ltd
151 Rosebery Avenue
London, EC1R 4GB
Tel: 020 7505 6622
Fax: 020 7505 6606
email:
crc@construct.emap.co.uk

BR 440
ISBN 1 86081 540 5

© Copyright BRE 2002
First published 2002

BRE is committed to providing impartial and authoritative information on all aspects of the built environment for clients, designers, contractors, engineers, manufacturers, occupants, etc. We make every effort to ensure the accuracy and quality of information and guidance when it is first published. However, we can take no responsibility for the subsequent use of this information, nor for any errors or omissions it may contain.

Published by
Construction Research
Communications Ltd
by permission of
Building Research
Establishment Ltd

Requests to copy any part of this publication should be made to:
CRC Ltd
Building Research
Establishment
Bucknalls Lane
Watford, WD25 9XX

BRE material is also published quarterly on CD

Each CD contains BRE material published in the current year, including reports, specialist reports, and the Professional Development publications: Digests, Good Building Guides, Good Repair Guides and Information Papers.

The CD collection gives you the opportunity to build a comprehensive library of BRE material at a fraction of the cost of printed copies.

As a subscriber you also benefit from a 25% discount on other BRE titles.

For more information contact:
CRC Customer Services on 020 7505 6622

Construction Research Communications

CRC supplies a wide range of building and construction related information products from BRE and other highly respected organisations.

Contact:
post: CRC Ltd
           151 Rosebery Avenue
           London, EC1R 4GB

fax: 020 7505 6606
phone: 020 7505 6622
email: crc@construct.emap.co.uk
website: www.constructionplus.co.uk

Contents

	Preface	v
	Readership	v
	Scope of the book	v
	Design criteria	vi
	Some important definitions	vi
	Acknowledgements	vii
0	**Introduction**	**1**
	Records of failures and faults in buildings	2
	BRE Defects Database records	2
	BRE publications on foundations, basements and external works	4
	Changes in construction practice over the years – a historical note	6
	Summary of main changes in common practice since the 1950s	11
	Building user requirements in the third millenium	14
1	**The basic features of sites**	**15**
1.1	Geology and topography	17
1.2	Fill, contaminated land, methane and radon	43
1.3	Surface water drainage requirements, water tables and groundwater	57
1.4	Site microclimates; windbreaks etc	62
1.5	The effects of vegetation on the ground	70
2	**Foundations**	**77**
2.1	General points on foundations	79
2.2	Old brick and stone footings	109
2.3	Concrete strips, pads etc	112
2.4	Piles	121
3	**Basements, cellars and underground buildings**	**129**
3.1	Structure	132
3.2	Waterproofing	138
3.3	Other aspects of performance	150

4	**Public and other utilities**		**157**
	4.1	Water supply	158
	4.2	Wastewater drainage	162
	4.3	Surface water drainage, soakaways and flood storage	185
	4.4	Other utilities	193
5	**Walls, fencing and security devices**		**195**
	5.1	Embankments and retaining walls	196
	5.2	Freestanding walls	206
	5.3	Fencing	212
	5.4	Exterior lighting and security devices	220
6	**Hard and soft landscaping**		**225**
	6.1	Pavings	227
	6.2	Trees, plants and grass	233
	References and further reading		**239**
	Index		**249**

Preface

This book is the fifth of the BRE Building Elements books and completes the planned series which was begun in 1996. The first four books are *Roofs and roofing* (first published in 1996), *Floors and flooring* (1997), *Walls, windows and doors* (1998) and *Building services* (2000).

Readership
As with the other books, *Foundations, basements and external works* is addressed primarily to building surveyors and other professionals performing similar functions; for example, architects and builders who maintain, repair, extend and renew the national building stock. Surveyors and architects undertaking routine surveys of existing buildings may identify site problems or issues – in particular concerning the behaviour of foundations – which need to be referred to specialists for advice. The larger non-domestic buildings built in the second half of the twentieth century, and perhaps even some in the first half of the century – the standard textbooks of the time were quite explicit – will most likely have had the benefit of adequately designed foundations. In the case, though, of relatively small buildings such as houses, many decisions concerning ground and foundations will have been made by building professionals who had little or no training in geotechnical engineering.

Some topics, such as deterioration in drainage installations, may often be amenable to straightforward rectification, but other aspects, such as the installation of dampproofing in basements, may be outside the experience of individual surveyors. Clients need to be advised on when to call in other consultants to rectify existing problems. The book is certainly not addressed to the geotechnical engineer or the landscape architect, though it will in all probability find application in the education field.

Scope of the book
The descriptions and advice given in this book concentrate on practical details. But there also needs to be sufficient discussion of principles to impart understanding of the reason for certain practices. In previous books in this series, some of the information which applied generally to the subject matter of the book was given in Chapter 1, but here the topics, though all closely related to the site, are rather more disparate, and the principles which govern practice will in consequence be found dispersed in individual chapters.

Included in foundations and basements is all work below DPC level, including strip foundations, piles, retaining walls to basements; but not including ground floor slabs or rafts, which were included in *Floors and flooring*, and building services within the footprint of the building, which were included in *Building services*.

Many points relating to the use of particular materials in close proximity to the ground, such as the durability of brick, block and concrete have been dealt with in *Walls, windows and doors* to which reference can be made.

Large civil engineering structures such as port installations, bridges and tunnels, underground car parks and very large non-building structures such as storage tanks are excluded from the scope of this book.

Included in external works are all items outside the building footprint but inside the site boundary, encompassing wastewater and surface water drains, supply of utilities (eg gas, electricity and cabled services), footpaths, and access for vehicles including car parks and hard standings to be found in the vicinity of buildings. Perimeter and freestanding boundary walls are also dealt with, as is security fencing and, in outline only, lighting; CCTV surveillance systems, though, are normally left in the hands of specialist consultants and contractors, and are therefore not covered in detail.

With such a broad scope, it will be apparent that only brief reference can be made to most topics, and the text therefore concentrates on those aspects with which BRE has been most heavily involved, whether in laboratory research, site investigation or development of legislation.

In principle, all types of buildings are included. However, it is inevitable that the nature of foundations becomes very sophisticated in some building types such as those which are very tall, and these installations rarely lend themselves to simple guidance for use by non-specialists. Indeed, there may well be no professional role whatsoever for them in this respect.

However, even the relatively simple systems used in the majority of domestic construction provide adequate potential for improvement.

Foundations, basements and external works is not a manual of construction practice, nor does it provide the reader with the information necessary to design foundations, basements or external works. Both good and bad features of these elements are described, and sources of further information and advice are offered. The drawings are not working drawings but merely show either those aspects to which the particular attention of readers needs to be drawn, or simply provide typical details to support text. The discussion, for the most part, is deliberately neutral on matters of style and aesthetics and is wary of suggesting that there is ever a unique optimum solution.

In a similar fashion to the other books in this series which deal with other building elements, the present text concentrates on those aspects relating to the subject matter of the book, in this case the site, and which, in the experience of BRE, lead to the greatest number of problems or greatest potential expense if carried out unsatisfactorily. It follows that these problems will be picked up most frequently by maintenance surveyors and others specifying and carrying out remedial work. Occasionally there is information relating to an item, perhaps a fault, which is infrequently encountered, and about which it may be difficult to locate information. Although most of the information relates to older buildings, much material concerning observations by BRE investigators of new buildings under construction in the period from 1985 to 1995 is also included.

The case studies provided in some of the chapters are selected from the files of the BRE Advisory Service, and the former Housing Defects Prevention Unit, and represent the most frequent kinds of problems on which BRE has been consulted.

The standard headings within the chapters are repeated only where there is a need to refer the reader to earlier statements or where there is something relevant to add to what has gone before. An exception to the sequence of standard headings occurs in Chapter 2.1 where the amount of material to be described requires a further breakdown into diagnosis, monitoring and remedial work. Chapter 3 also does not follow the standard headings; it was found to be more appropriate to deal separately with structure and waterproofing of basements which is reflected in the sub-chapter headings.

In the United Kingdom, there are three different sets of building regulations: The Building Regulations 1991 which apply to England and Wales; The Building Standards (Scotland) Regulations 1990; and The Building Regulations (Northern Ireland) 1994. There are many common provisions between the three sets, but there are also major differences. The book has been written against the background of the building regulations for England and Wales, since, although there has been an active Advisory Service for Scotland and Northern Ireland, the highest proportion of site inspections has been carried out in England and Wales. The fact that the majority of references to building regulations are to those for England and Wales should not make the book inapplicable to Scotland and Northern Ireland.

Although practically all topics relating to the construction of buildings are encompassed in the Construction (Design and Management) Regulations 1994, the ramifications for each of the topics covered in this book are quite different. It is therefore not practical to spell them out, beyond noting that there must be a Health and Safety Plan and File for all construction work which should include information on how to manage health and safety issues after the installation is completed and throughout its life until demolition[1].

Design criteria

While this book is mainly about existing buildings and not specifically about the design of new buildings, it has been necessary in several circumstances to give some design criteria so that subsequent performance of the completed building may be assessed against what was required or intended.

Some important definitions

The term 'footprint' has been used to describe the area actually covered by the building fabric. Note that this term is not synonymous with the term curtilage, as used in legislation, which in its normally accepted meaning includes any ground forming a part of the enclosure within which the building stands.

The term 'surcharge' as used in this book has two distinct meanings: a preloading of the ground (eg to induce consolidation), and a condition in which water is held under pressure within a gravity drain, but which does not escape to cause flooding.

So far as water terms are concerned, there has been a significant change in usage since the 1980s. The term 'potable water' to describe water of a quality suitable for drinking is now no longer popular though it is still contained in current Standards, and the words 'drinking' and 'wholesome' to describe water are preferred.

The use of the term 'foul water' in relation to sewage has fallen out of official use, being superseded by the term 'wastewater' which is often more closely defined as 'greywater' or 'blackwater'; 'greywater' is wastewater not containing faecal matter or urine, 'blackwater' is wastewater that contains faecal matter or urine. However, for the purposes of this book, we tend to retain the term foul water when referring to greywater and blackwater; wastewater, though, can include surface water (eg run-off from car parks).

The term 'aerobic' indicates conditions in which free oxygen is present, and 'anaerobic' in which it is not present.

Preface

Since the book is to a considerable extent about the problems that can occur in the below-ground fabric of buildings, two words, 'fault' and 'defect', need precise definition. Fault describes a departure from good practice in design or execution of design; it is used for any departure from requirements specified in building regulations, British Standards and codes of practice, and the published recommendations of authoritative organisations. A defect – a shortfall in performance – is the product of a fault, but while such a consequence cannot always be predicted with certainty, all faults have the potential for leading to defects. The word failure has occasionally been used to signify more serious defects and catastrophes.

Where 'investigator' has been used, it covers a variety of roles including a member of BRE's Advisory Service, a BRE researcher or a consultant working under contract to BRE.

Particular terms relating to topics under discussion will be found in the various chapters which follow.

Acknowledgements

Photographs which do not bear an attribution have been provided from our own collections or from the BRE Photographic Archive, a unique collection dating from the early 1920s.

To the following colleagues, and former colleagues, who have suggested material for this book or commented on drafts, or both, we offer our thanks:

R B Bonshor, Lesley Bonshor, R Cox, Maggie Davidson, R M C Driscoll, Hilary Graves, J Griggs, M S T Lillywhite, Dr R Orsler, C R Scivyer, J Seller and E Suttie, all of the Building Research Establishment.

We are particularly grateful to Richard Driscoll, who contributed much of Chapters 1.1 to 1.3, 1.5, and the majority of Chapter 2; to Ron and Lesley Bonshor who prepared historical and other material on foundations and geotechnics, and on cracking; to Mike Lillywhite who prepared some of the historical studies on foul drainage upon which we have drawn; and to John Ramsay, Member of The Landscape Institute, who commented on Chapters 1.4 and 1.5 from the point of view of the landscape architect, and also contributed much of Chapter 6.2.

HWH
PMT
December 2001

Chapter 0 Introduction

It may be something of a truism to say that all sites are different – perhaps even that every site is unique. The most obvious differences between sites are to be found in the physical composition of the land, and its geology and topography, but they can also be found in its location, climate and microclimate. There can even be major differences across individual sites: even the smallest site intended for a single dwelling. The consequences of this uniqueness are that, to perform satisfactorily, it is arguably the case that the buildings built on those sites need to be different too, even if the user requirements for the fabric above ground are identical. Indeed, in former times, these circumstances were common, and it would have been rare to find two buildings exactly alike in adjacent locations until Victorian times (Figure 0.1), with the possible exception of developments such as the Georgian terraces, and a few almshouses and the cottages built for their employees by the owners of large estates.

These differences are reflected not only in the superstructures but also in the enormous variety of foundations which may need to be adopted to accommodate different ground conditions. They are also reflected in aspects such as hard and soft landscaping, perimeter walls and fencing and shelter belts, and even surrounding construction.

While BRE has concentrated on the fabric of buildings, of which foundations and basements in different ground conditions form an

Figure 0.1 A rich variety of building designs each on a unique site: Bracknell in 1950

integral part, it has also accumulated a wealth of information on other topics selected for inclusion in this book, more particularly on site microclimates, and the performance of perimeter and retaining walls, pavings and fencing. In addition to specific research topics, over the years, many hundreds of site inspections carried out by BRE investigators have yielded relevant information which has been drawn upon in these pages.

These investigations include surveys of housing, of both new construction and rehabilitation, evidence from the United Kingdom house condition surveys (undertaken every five years), and also, particularly for building types other than housing, from the past and current commissions of the BRE Advisory Service and of BRE's Geotechnics Division.

0 Introduction

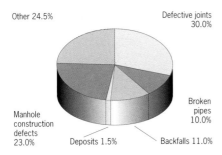

Figure 0.2 Distribution of defects identified in BRE site investigations of drainage installations, 1979

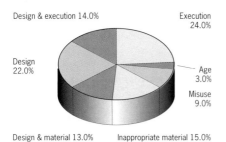

Figure 0.3 Responsibility for defects identified in BRE site investigations of drainage installations, 1979

Records of failures and faults in buildings

BRE Advisory Service records

The most common causes of difficulty with foundations are weak and compressible soils, and exceptionally heavy local loads (especially if accompanied by vibrations); however, the variety of both loads and the structures which carry them is very great, and the design of foundations before the nineteenth century was often a matter for trial and error and local experience. In certain areas, shrinkable clay soils commonly give rise to problems if the original design of the foundations is not suitable for the conditions.

Foundations may cause trouble either because they impose loads which are too heavy or otherwise inappropriate to the ground conditions or because they follow movement occurring in the ground independently of any imposed load. These two classes of problem are distinct and are so treated in this book.

Although a database has been maintained at BRE of faults and defects occurring with certain types of foundations, it has not proved practical to summarise this information for this book.

The BRE Advisory Service has made a special study of basements, in particular the measurement and control of dampness. The experience accumulated from several hundred special investigations has been drawn upon in the book.

There are very few references to external works, including landscaping, in the BRE Advisory Service records, with the exception of items relating to drains and sewers, and slip resistance of pavings.

In the late 1970s a survey was made by BRE investigators of some 70 sites where drains had been subject to recurring blockages (Figure 0.2). When the existing drains were checked against the original drawings, a number of discrepancies were found. Drain cleaning was frequently necessary before a CCTV survey camera could be inserted into the drain, let alone pulled through it. It is notable that only one of the drains examined had no discernible defect, with blockages for the most part being due to excessive fats. One of the worst cases encountered in this study involved discharge from a stack serving 34 sinks in a 17-storey block of flats[2].

In Figure 0.2, 'other' includes manhole location and positioning defects, and invasion by vegetation roots. Most of the jointing defects occurred with traditional joints – only two cases were seen with defective 'O' ring joints – and lack of skill in making the joints appeared to be a main factor. All the broken pipes were in clayware, but it was possible to diagnose the specific cause in only three cases, two of ground subsidence and one of impact loading. Back-falls were not observed to be a direct cause of blockages. Manhole defects occurred mainly where heavily loaded pipe flows deposited detritus in the channels directly opposite to channels carrying lightly loaded pipe flows, where the detritus set. Other defects included rough and broken benching, especially where this was associated with changes in direction (knuckle bends) sharper than 90°.

Figure 0.3 shows the attribution of responsibility for defects occurring in drains.

BRE Defects Database records

Some information on faults which occur in foundations, basements and external works in housing is available in the BRE Defects Prevention Unit (DPU) Quality in Housing database[3] which records non-compliance with requirements whatever their origin, whether Building Regulations, codes of practice, British and industry standards or other authoritative requirements. This database records actual inspections by BRE investigators or by consultants working under BRE supervision, and gathered on nearly 140 sites over the years 1980–90. The criteria used in these inspections, against which faults and defects were identified, included foundations and external works, although very few sites with cellars or basements were included in this programme as they were then out of fashion.

The inspections of the sites were thorough, ranging from just under one man-day for individual dwellings up to six or more for the larger sites, and the investigating team included building surveyors, architects, various kinds of scientists, engineers and quantity surveyors. The inspection reports record types of faults only, with actual numbers of infringements estimated on the scale:
1 universal (that is to say, the fault occurred on all dwellings inspected on that site)
1 frequent (more than half of all dwellings)
1 occasional (a few only)
1 unique (only one on a site)

Each fault type therefore represents a number of actual infringements, though the precise totals of these cannot be deduced with accuracy. Nevertheless, it can be observed that when a fault in foundations was recorded, it tended to be classified as universal and to be found on all dwellings on that site.

The data have been analysed according to date of original construction of the dwelling, broadly into two categories: those dwellings being built new at the time of inspection, and those being refurbished. The analysis for the new-build investigations of 1980 is given in Figure 0.4; and there is very little difference between those figures and those recorded in 1990 (Figure 0.5). Therefore it can be seen that, although the investigations of the former DPU at BRE revealed numbers of faults relating to the subject matter of this book, the fault types recorded, at less than 3% of the total, were far fewer than were observed for other elements of the fabric such as roofs, floors, windows and walls. If it is assumed that the numbers of types represent a reasonable proxy for the total number of faults occurring on those sites, as was done for the other books in this series, it will be obvious that it is not worth drawing detailed pie charts. However, if a comparison is made between this data and that from the house condition surveys to be described in the next section, it can be reasonably concluded that the apparent situation encountered in the BRE studies considerably underestimates the actual.

When faults observed in rehabilitation schemes are considered, although there were more sites having cellars and basements, largely represented in those dwellings older than 1919, the totals are again too small to justify the drawing of pie charts.

In spite of the lack of an acceptably detailed statistical analysis, some general observations can be made with confidence. With respect to foundations and substructures generally, approximately half of all observed types of faults in both new-build and rehabilitated housing related primarily to cracking and settlement, one third to dampness and deficiencies in DPCs, and one sixth to durability of masonry below DPC level. Several sites had waterlogged trenches, leading to difficulties in establishing firm foundations. One dwelling even had a masonry porch pier which could be moved by hand on its base in a sea of mud.

For cellars and basements, approximately half of all observed types of faults related to dampproofing and tanking problems, with most of the remainder relating to access and headroom difficulties. There were a few structural problems – with one particularly serious case occurring in the presence of BRE investigators when a structural wall in a basement partially collapsed as it was being worked on!

Most of the observed types of faults in drains outside the building footprint related to broken or damaged pipes and subsequent washout of subsoil leading to foundation settlement; and a few of these were due to lack of clearance or provision of flexible joints where the drain passed through the external wall.

Half of the observed types of faults in other external works related to deficiencies in freestanding screen or boundary walls, with structural faults in the majority, though there were some potential and actual durability problems too. The other half related mainly to ground, paving level and drainage faults, but a few to durability of paving or other materials; for example, those used inappropriately in retaining and boundary walls.

House condition surveys
The 1996 English, 1996 Scottish, 1991 Welsh and 1996 Northern Ireland House Condition Surveys[4,5,6,7] provide information relating to the subject of this book. The surveys are based on structured sampling of the complete national stocks of dwellings – 20,321,000 dwellings in England, 2,123,000 dwellings in Scotland, and 573,370 dwellings in Northern Ireland. Up to date figures for Wales were not available at the time of writing.

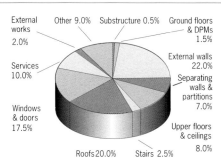

Figure 0.4 Distribution of faults identified in BRE DPU site investigations of new-build housing, according to element, 1980

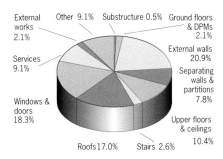

Figure 0.5 Distribution of faults identified in BRE DPU site investigations of new-build housing, according to element, 1990

Although information was gathered on the incidence of structural problems, this has been related specifically to walls rather than ascribed specifically to foundations. Both settlement and differential movement in walls were recorded. In the data for England for 1991, for example, settlement was observed in 5.2% of dwellings (that is to say, representing over 1 million dwellings in the total stock), and differential movement in 2.7% (Table 6.5 in the English House Condition Survey 1991), and it is much the same for 1996. The question of the severity of this damage and whether anything needed to be done about these occurrences is another matter entirely, for the surveyors had only a short time in which to reach a conclusion on a matter which really needed long term observations. Although it would be possible to carry out further analysis of the data on occurrences – for example, to relate it to age of dwelling – this has not been done for this book.

0 Introduction

Of the dwellings in England categorised as unfit, some 41,000 (3%) were unfit by reason of the water supply, and some 134,000 (10%) because of drainage problems, with, as one might expect, about twice as many dating from before 1919 as after 1919. Other external works do not appear in the EHCS figures.

Observations were also carried out on the perceived quality of the site; for example in relation to industrial pollutants, condition of road and footpath surfaces, and condition of street furniture. These data were taken into account in the assessment of the environment. In England, for example, over 2 million dwellings were assessed as having at least one significant environmental problem, although of course not all of these problems relate to the subject of this book.

Non-domestic buildings

The records of the BRE Advisory Service noted above, together with information available through the Construction Quality Forum (CQF), go some way to providing useful information on the occurrence of relevant faults in non-domestic buildings.

CQF data for 1994, to take one year as an example, indicated that foundations, basements and external works figured in 12 of the 111 (say 11%) of items reported on for residential buildings, and 26 of the 301 (say 9%) of items reported on non-residential[8].

Evidence from other organisations

The National House-Building Council provides insurance to home purchasers against defects and major damage for the first 10 years of the building's life, and they have reported that problems associated with the ground predominate. Foundation deficiencies are usually very much more expensive to rectify than are problems with, say, roof timbers.

BRE publications on foundations, basements and external works

Principles of modern building

Although the last revision to Volume 1 of *Principles of modern building*[9] was published as long ago as 1959, since it deals with principles, in some respects it is less out of date than might be expected.

Introductory paragraphs from *Principles of modern building* describing the interaction between superstructure, foundations and soil are quoted below. Much of this material stands up well to the test of time.

'The stresses in any structural system are dependent not only on the dead and imposed loadings, but also on the way the system is supported. The forces acting on a building superstructure include those developed between it and its foundation. These in turn are dependent on the behaviour of the foundation in resisting the forces from the building and the reactions from the soil. The superstructure, foundation and soil form a complex system, and any complete study of the strength and stability of a building must include an assessment of the behaviour of this whole system.

A simple basis of design of the superstructure is to assume that the foundation is rigid. Although deformation of the foundation and settlement due to soil movement are bound to occur, the assumption is equivalent to ignoring any differential settlement of the foundation, and may be reasonable if suitable precautions are taken. The load should be spread over a sufficient area for safe bearing and should be distributed as evenly as possible between different parts of the foundation.

It is impossible to eliminate all differential settlement, however carefully the loads are distributed, particularly as the soil properties themselves may vary considerably over a building site. It may be assumed roughly that the likely differential movements are proportional to the average settlement of a building, which can be estimated fairly accurately according to the established principles of soil mechanics. For a multi-storey block of flats, for example, the differential movements are not likely to be serious if the theoretical average settlement is no greater than 25 mm.

Although differential settlement can lead to appreciable increases in stress in parts of a building structure, eg at the junctions of beams and columns in framed structures, these changes will often have no important influence on the ultimate resistance of the structure to collapse. However, the distortions of the building can result in the cracking of partitions or finishes and it is this effect which must particularly be guarded against. Differential settlement of structures on shallow foundations, such as houses, may also occur as a result of shrinkage or swelling of the soil.

The choice of type of foundation depends upon the disposition of the loadbearing elements of the structure and the bearing capacity of the soil. If the loads from floors are transmitted to the foundation by walls, strip foundations are usually adequate. On poor soil, these strips may need to be much wider than the wall footings, and transverse reinforcement must be used to spread the load over a sufficient area.

If the building is of framed construction, the columns are normally supported on isolated pad foundations, reinforced if necessary. These isolated foundations are usually proportioned so that the bearing pressures on the soil are approximately the same for all of them; the possibility of differential settlement can be further reduced if allowance is made for the fact that the settlement under a foundation pad is dependent, not only on the average pressure on the soil, but also to some extent on the area of the pad.

Raft foundations may be desirable, whatever the type of superstructure, on sites where the soil is weak and the load from the building must be spread over a large area. If the soil is particularly weak, it may be necessary to transfer the load to a lower stratum better able to support it, and piled foundations are used. Fairly short piles (2.5 to 3 m) can also be used to avoid supporting the building on unstable soil such as clay at shallow depths.'

0 Introduction

BRE Digests and other papers
The various BRE publications, such as Digests, Information Papers, Good Building Guides and Good Repair Guides which are listed in the references and further reading lists, together cover foundations, basements and external works fairly comprehensively; though, since they have been published over a time span of many years, they can have received very little cross-referencing between them. All these publications have been drawn upon to a considerable extent in this book. This book therefore provides a key to the main BRE publications relevant to foundations, basements and external works.

BRE Reports
Apart from the series of publications referred to above and the more specialised reports listed in the references, there are a number of other BRE books or reports which examine particular aspects of foundations, basements and external works. They include the *Housing design handbook*[10], *Assessing traditional housing for rehabilitation*[11] and *Cracking in buildings*[12].

Case study

New foundations for York Minster
In the mid-1960s, part of the fabric of York Minster was found to be in danger of collapse. Estimates were made of the rate of settlement of the structure over the years since the late eleventh century. The first nave had settled around 110 mm over the first 50 years of its existence but the replacement nave built in later years appeared not to have moved (Figure 0.6). The Norman tower had collapsed around 1400 when settlement was estimated to have reached nearly 300 mm, and the new tower erected in its place settled a further 75 mm.

The foundations of the cathedral, built by Archbishop Thomas of Bayeux in the late eleventh century, were excavated in part during the works to strengthen the tower footings in 1967–72. Much investigation work was undertaken including the sinking of boreholes to provide information on the subsoil and these studies confirmed that the structure was founded on a stratum of firm brown clay. Firm rock, a sandstone, was encountered at a depth of 24–27 m.

The excavations exposed, typically, a 1.8 m wide trench on which mortared rubble was laid to a depth of 0.8 m; a timber grillage consisting of large cross pieces; then massive baulks 300 mm square and up to 13.7 m long were laid longitudinally along the length of the wall with the end joints scarfed. Retaining walls were erected and infilled with a mortared rubble core. The main walls of the superstructure were then built off these foundations[15] (Figure 0.7).

Examination of the subsoil revealed that adequate bearing capacity existed just below the old footings; remedial work was undertaken and new mass concrete foundations cast around the old Norman footings, with the loads from the central tower being then transferred to the new construction[16].

The solution was, in effect, to double the bearing area of the original foundations in the areas which had undergone differential settlement. The old masonry was cased in concrete with the new concrete stitched to the old foundations by means of stainless steel rods, and the load redistributed by means of flat jacks placed between the mass concrete and a lower compression pad.

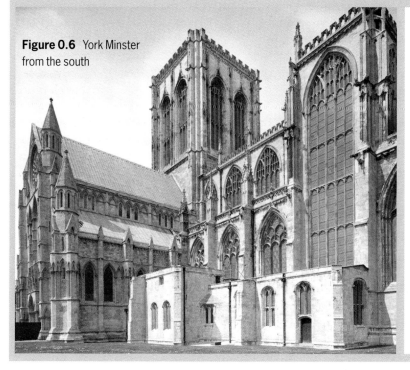

Figure 0.6 York Minster from the south

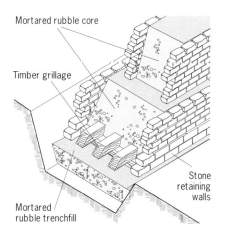

Figure 0.7 The medieval foundations for the walls of York Minster. These foundations needed strengthening in the 1960s

Changes in construction practice over the years – a historical note

Foundations

The period of trial and error

Medieval builders understood well enough the need to carry the loads of the buildings they constructed down to ground level in such a manner that the building would not collapse (Figure 0.8). What they lacked was any precise means of calculating the magnitude of those loads and exactly how they would be transmitted through the fabric. Trial and error, albeit sometimes disastrous error, enabled them to improve their empirical knowledge over the years, and the collective experience of many minds at least ensured that the majority of buildings did not collapse at too early a stage in their lives. Nevertheless, many collapses of Norman and medieval structures did occur due to foundation failure; amongst them can be cited the central tower of the cathedral of Winchester in 1175, one of the west towers at Gloucester around the same time *'owing to a defect in its foundations'*, and the central towers at Wells, York, Salisbury and Canterbury in subsequent years. Lincoln's central tower fell around the year 1240. Ely's central tower fell in 1321, and was replaced by a much lighter construction in timber framing carried on the original foundations – the famous octagon.

Other towers had to be strengthened before collapse occurred, such as those at Hereford in the fourteenth and fifteenth centuries and again in 1841, and others at Wells, Salisbury and Canterbury. Peterborough's Norman central tower proved too heavy for its foundations, and was partially demolished before it too fell. At Worcester, by the 1850s, *'settlements of the piers and arches in the Early English work had attained so alarming a magnitude as to threaten the stability of the structure'*[13].

A review carried out in 1861 noted: *'...the foundations of Norman buildings are rarely consolidated or prepared with proper care, and hence, for the most part, the whole structure will be found to have sunk bodily into the compressible ground, and the heavier tower piers necessarily one or more inches more than the rest'*[14].

There is no doubt that those responsible for the fabric of major buildings had to keep a close eye on the possibility of differential movement of foundations. For example, in 1762 a plumb line survey of a tower and spire revealed that they were out-of-plumb by approximately 23 inches; fortunately, no further movement was detected over the following century.

Of course, not all collapses were due to foundation failure, sometimes the wind proved too much for a dilapidated tower, such as those which collapsed at Ripon in 1660, and Chichester in 1861. There were also many disastrous fires, which may have destroyed the superstructure but which will rarely have damaged the foundations, even though the crypt vaults may have collapsed. The builders of these early periods frequently rebuilt on surviving old foundations, most probably largely irrespective of their condition.

The foundations of domestic buildings developed from posts sunk into the ground, through to what might almost be described as primitive ground beams taking the loads of timber framed structures, and kept out of the mud by plinths of stone (Figure 0.9). But the larger buildings of stone masonry needed much wider footings than the width of the wall above to spread the loads over a sufficient area of the ground, and this exposes the second area of lack of precise knowledge of these early builders: that is to say, knowledge of the exact bearing capacity of all the multitudinous varieties of subsoil they encountered.

From the earliest times until the late Middle Ages, large masonry buildings built on indifferent ground were prone to collapse. Perhaps the main saving grace was that most of these buildings were usually built over a long period of time, so that the ground beneath them consolidated very slowly as the imposed loads gradually increased – the soil would then be able to take the loads of such magnitude as would have caused it to fail in its uncompressed state. Nevertheless, water tables do alter over the years, and it will sometimes be necessary to introduce emergency measures to prevent collapse. See the case study on page 5.

Figure 0.9 The decay of this ancient Carmarthenshire farmhouse has exposed the footings for the clom walls. They were founded on what appears to be a mixture of mortared local stone tied in with timber ground beams forming a semi-basement

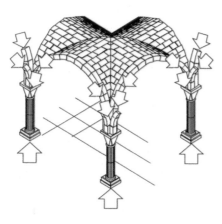

Figure 0.8 A Norman vault. Masons in the eleventh and twelfth centuries knew how to resolve forces though they lacked the means of calculating them

0 Introduction

Figure 0.10 Masonry on the south transept of St Paul's Cathedral. It is said that these walls were founded on brick piers extending down to the gravel bed above the London Clay

Renaissance times

Systematic study of groundbearing conditions seems to have begun in earnest during the later years of the seventeenth century. When Sir Christopher Wren began his preliminary studies for the rebuilding of St Paul's after the Great Fire, he observed that the foundations of the old church had stood upon a layer of very close and hard pot earth, and he concluded *'that the same ground which had borne so weighty a building might reasonably be trusted again'*. Nevertheless he had trial pits dug (he called them 'wells') and discovered that the so-called pot earth declined in thickness from 2–1 m across the site. Underneath was dry sand, then a further 13 m down was the water table with gravel below that, and below that again was the London Clay, *'the natural hard clay, which lies under the City and country and Thames far and wide'*. He also discovered, by digging a chain of these wells across the site, that there was a pit *'which had been robbed by the potters of old times'* and filled with broken shards and other loose rubble, clearly needing to be excavated or bridged.

Although his artificers proposed to him that he should pile the site with timber piles, he refused *'for, though the piles may last for ever when always in water (otherwise London Bridge would fail) yet if they are driven through dry sand, though somewhat moist, they will rot'*.

In the end he built 3 metre-square masonry piers in excavated shafts from the gravel bed above the clay to within 5 m of ground level, and bridged between them with masonry arches to carry the loadbearing walls[17] (Figure 0.10).

Bonded footings for walls continued to be built in stone or brick until Victorian times, when the invention of Portland cement in the 1830s enabled the use of concrete strips beneath the walls. This did not, however, mean the elimination of bonded masonry footings, and the construction textbooks of around the end of the nineteenth century consistently show the old style corbelled footings but carried on concrete strips.

The use of timber piles, already referred to, seems to have been relatively common in certain areas of the UK.

Piles of concrete rather than timber are very much a development of the twentieth century. BRE investigations of the performance of reinforced concrete piles began in 1929 with a study of their durability in seawater[18].

Remedial work by means of underpinning seems to have been exploited during Victorian times; for example, it is understood that the foundations of the north wall of the nave of Rochester Cathedral were underpinned with mass concrete in 1875.

The year 1933 saw the first soil mechanics laboratory in the UK set up at the Building Research Station (BRS). What had hitherto been regarded by many engineers with considerable suspicion now began progressively to be demonstrated as a very serious discipline, subject to the normal scientific processes in which tests are repeatable and reproducible, and in which reliable data are obtained and sound procedures evolved for their evaluation and interpretation. The science of soil mechanics, later to be called geotechnics, enabled BRS, the precursor to BRE, to contribute substantially to the industry's understanding of the subject and its practical applications.

Investigations undertaken in 1935 at BRS, with the collaboration of the industry, into the behaviour of reinforced concrete piles were initiated as a direct result of troubles experienced while driving piles through a hard stratum to a set in firm ground below. It had been found difficult to construct precast piles of sufficient strength to resist damage during driving in severe conditions (Figure 0.11), and the attention of BRS was called to several cases where failure had occurred.

Figure 0.11 A steam piling rig being used for driving concrete piles in the 1930s. Heads of driven piles can be seen, but the remains of broken piles litter the site

During the 1939–45 war, consideration had to be given to the kind of information required to assess the influence of the disturbance of the foundation soil resulting from an underground bomb explosion on the stability of structures standing nearby, and where damage to the structure could be brought about by movements transmitted to the building through the earth. Also of concern was the stability of structures which might subsequently be erected over or near a crater.

Although there had been a general awareness of the effects of movements in clay soils on the foundations of buildings, this knowledge began to be considerably expanded in the immediate post-war period, and work was also carried out on the effect of fast growing trees and shrubs on shallow foundations.

Basements and underground buildings

Some of the earliest basements to survive are the dungeons built into the lower storeys of the keeps of medieval castles which were often cut into the solid bedrock of the site.

In later centuries, crypts were commonly built under churches and cathedrals, some dating from Saxon times. The Normans too were great crypt builders; that at Gloucester Cathedral was originally built with walls 3 m thick and with the floor 2.5 m below the level of the surrounding ground surface.

In the tradition of their stone age ancestors, people, since medieval times, have taken advantage of soft rocks in their locality to carve out underground houses. Perhaps the best known of these in the UK is the small group of houses at Holy Austin Rock at Kinver near Stourbridge which were inhabited until the middle of the nineteenth century, fully serviced with piped gas and water (Figure 0.12). A few of these dwellings are now preserved by the National Trust.

When town growth filled the valley bottoms in the seventeenth and eighteenth centuries, construction moved up the hillsides, often following contour lines. In Hebden Bridge, for example, five-storey back-to-earth houses were built with a road at the lowest level serving the first two stories and a higher road at the rear for the upper three storeys (Figure 0.13). A two-

Figure 0.13 Single aspect maisonettes at Hebden Bridge. The properties on the left hand side of the street are in fact the upper stories of a five-storey block. The right hand side of the street shows the full five-storey height of the terrace

Figure 0.12 The empty shells of underground houses at Holy Austin Rock, Kinver, near Stourbridge

Figure 0.14 Semi-basement rooms on the garden front of Ffynone, a Nash mansion in Pembrokeshire

0 Introduction

External works
Pavings etc

The use of hard stone paving to provide access to major buildings was introduced in Roman times. Traditionally the best pavings were of York or Caithness flags, or of granite in the form of setts. Cobbles of riverine origin were also used to pave medieval streets in the larger towns (Figure 0.15). However, even in urban areas, stone macadam was relatively rare until Victorian times.

Walls and fencing

Perimeter walls on building sites in earlier times were usually primarily for defensive purposes. Unfortunately this particular need still exists (Figure 0.16)!

Before the advent of the railways made long distance haulage an economic possibility, freestanding walls were in the main constructed of the same locally available materials used to construct the adjacent buildings (Figure 0.17).

There is less evidence of fencing surviving from earlier times, being constructed in the main of less durable materials than those used for walls. Occasional examples of local ingenuity can be found, though (Figure 0.18 on page 10).

Figure 0.15 Cobbled paving in the cathedral precinct at Norwich

storey light well of glazed bricks incorporating a grille at street level gave limited light and ventilation to the back rooms. This inadequate provision for lighting and ventilation, coupled with damp ingress, rendered these back rooms increasingly unacceptable as habitable accommodation.

Many of the Georgian terraces in London and other cities appear to have basements, but levels of the mews at the rear indicate that the roads at the front are at much higher levels than the original ground. Most of the walls of these houses are not in fact earth retaining, apart from the access area at the front.

The rural counterpart of the Georgian urban terrace, the country mansion, might also have a similar arrangement of semi-basements. At this time it was the fashion to build these large mansions with the *piano nobile* accessed directly by the main entrance doorway at or just above ground level at the front, while at the rear a grand staircase gave access from the terrace (Figure 0.14). The rooms in the semi-basement under the main entrance were storage cellars, whereas those on the garden front were habitable.

Figure 0.17 A freestanding wall built largely of flint dividing the rear gardens of artisans' cottages in Castle Acre, Norfolk

Figure 0.16 A recent defensive modification: broken glass set in mortar surmounts this freestanding brick perimeter wall

Water supply

The supply of water to towns some distance from river sources by means of aqueducts predates even Roman times, both in the Middle East and in Central and South America; but of course it was the Romans who developed water supply to a fine art, producing sufficient water for daily consumption in quantities not far removed from what is consumed today[19].

Wells with chain and bucket extraction, or leats from river banks, formed the main means of conveying water for domestic consumption in the UK until the invention of the hand pump. A few localities had piped supplies on a small scale from the thirteenth century onwards, with the pipes being made of lead, wood, or leather. Steam pumping engines began to be used around 1800.

Although cast iron pipes had been made earlier, the first ones in London were laid for the conveyance of water in the mid-eighteenth century. Spun iron and steel came later, and these materials were extensively used in such great metropolitan undertakings as those conveying and distributing water 68 miles from Lake Vyrnwy to Liverpool, 80 miles from the Elan Valley (Figure 0.19) to Birmingham, and 100 miles from Thirlmere and Haweswater to Manchester in the late nineteenth century.

It was noted in *Building services*[20] that perhaps the most significant innovation with regard to the means of supplying water to properties has been the introduction of polyethylene mains in substitution for lead; and plastics drainage pipes in substitution for cast iron, asbestos cement and pitch fibre. Fired clay, now with push-fit polypropylene flexible couplings, is still used on a significant scale.

Wastewater drains

Urban sanitation changed little for several centuries before the advent of acceptable systems for waterborne sewage, and the 'night-soil man', with his horse drawn tank, was a common sight in the streets until late Victorian times. Disposal was simply by tipping the sewage into the nearest water course, and little seems to have been done about the ensuing problems until 'The Great Stink' from the River Thames flowing past the Houses of Parliament prompted action by that legislature in 1856.

Victorian public health legislation arising from an increased awareness of the sources and spread of disease in the population, and the accelerated suburban development and widespread speculative house building during the latter part of the nineteenth century, led to the building of extensive networks of sewers and drainage systems, many of which are still in use today, albeit deteriorating.

Drainage systems adjacent to the houses of this era often involved very deep pipework and connections to the sewers, probably the result of cellars and semi-basements, and the Victorian reluctance to accept flat gradients. Many of the houses constructed at the time were drained through 150 or 225 mm private drains or sewers which collected the wastewater from individual 100 mm drains to dwellings in the terrace before joining the, probably deep, sewer. Access to these drains is often very inadequate; normally no manholes were provided at the connection to the public sewer and often the only other access was via the intercepting trap manhole.

The intercepting trap was an invention of the Victorian age and invariably fitted to pre-1919 house drainage systems to isolate the drain from noxious sewer air and hopefully to discourage rats from entering the drain from the sewer. The device subsequently proved to be a constant source of blockage, particularly with low flow rates. It was usually fitted at the end of the drain run on the edge of the building plot or under the pavement, and often close to the sewer connection. For private drains serving a number of houses only one such trap would be used. An illustration of an intercepting trap appears later in Chapter 4.

The well known Maguire's Rule for gradients (eg 4 inch diameter pipe to be laid at a gradient of 1 in 40, 6 inch at 1 in 60 etc) became established by the end of the 1914–18 war. Manholes at this time were not normally provided at connections to public sewers unless the drain diameter was close to that of the public sewer.

During the interwar years the two-pipe above-ground system, with separation of foul and storm water, began to be more widely used. Manholes were generally provided at all junctions and all changes of

Figure 0.18 Caithness flags used to form a fence. The slabs were not sufficiently high to be stock-proof

directions in pipe runs, these bends being entirely within the manholes. Interceptor traps were still used even though their disadvantages were known by this time; a report produced as early as 1912 by a local government board stated that these fittings were unnecessary and caused blockage problems. Drain pipe materials did not change from those used in the pre-war period. Manholes in the house plots were still constructed of brick, usually engineering grade, 9 inches thick, in two separate half brick thicknesses, jointed with cement mortar, hopefully to make them less likely to leak if the flow became blocked and the effluent level rose above the benching.

Important and very influential pieces of legislation of this era which affected drainage practice were the Public Health Act 1936 and the Public Health (Drainage of Trade Premises) Act 1937.

Other utilities

With regard to the other main utilities, gas supplies to buildings, at first for lighting purposes only, predate the supply of electricity by only a few years, though rural areas were considerably behind urban areas in receiving supplies. Small local plants, even serving villages, would operate for a number of years, but became uneconomic, and disappeared when larger plants were built to serve urban areas at the end of the nineteenth century. As one example, The Beauly (Inverness-shire) Gas-Light Company operated its small works between 1869 and 1892, which explains the presence of the few remains of gas carcassing in local buildings, but since that time the village has not had a mains supply. While practically all buildings now are supplied with electricity, there are still large areas of the UK – particularly, but not only, buildings in rural areas – where it is too costly to provide mains gas, and buildings can only use gas provided from bottled or tanked equipment.

Summary of main changes in common practice since the 1950s

Foundations

In the early 1950s, when deep strip footings for house foundations on clay sites had been found to be satisfactory but expensive and not always easy to construct, BRE was at the forefront of the development of short bored piles spanned by lightweight ground beams for this application. Around this time, results of several long term observations on the movements of the ground under different conditions of soil, weather, vegetation and shelter were recorded in order to develop an understanding of ground movements and transpiration of vegetation. These measurements provided data for the rational design of house foundations in Britain[21]. A little later came developments in trench fill; that is to say, concrete poured into mechanically excavated narrow trenches.

BRE's geotechnics researchers had by the 1970s studied the problems associated with deep excavations in densely built-up locations. Two relevant BRE developments were:

- devising suitable measuring techniques and instruments
- developing the 'finite element' method of numerical analysis

These developments played an important part in the design and construction of, for example, an underground car park at New Palace Yard, Westminster, London. The construction needed to be carried out within a short distance of the foundations of both Westminster Hall and the Clock Tower ('Big Ben'), and the stability of both was a matter for great concern. While the work was in progress, the measurements made by BRE, including porewater pressures and deflections in retaining walls, became an integral part of safety considerations of the surrounding buildings.

An appraisal of likely future problems affecting foundations and substructures carried out in the 1980s identified – correctly, as subsequent events were to prove – the increasing scarcity of good building land. Pressures to build on ground previously thought unsuitable were expected to increase; land such as domestic and industrial waste landfill sites, or soft deep alluvial soils, would, it was thought, be used for building in the absence of better

Figure 0.19 The dam at one of the Elan Valley reservoirs from which Birmingham takes part of its water requirements

ground. Research effort in the geotechnics field therefore increasingly concentrated on the development of ground investigation techniques needed for building sites, and on methods of designing and constructing safe but economic foundations for particular conditions. With this objective in mind, deep landfill sites were instrumented to help in the assessment of likely ground behaviour. Deep vibratory ground improvement techniques, known collectively as vibro compaction, began to be used to improve foundation conditions at many housing and light industrial developments.

Geotechnics researchers also investigated methods of protecting buildings which might be subject to ground movements such as could occur over abandoned mine workings. In a full-scale trial an old limestone mine was backfilled by pumping colliery waste into it. A system was also designed for monitoring ground movements during the trials. The system, computer controlled, was able to raise an alarm via a telephone line if ground movement exceeded preset levels.

Shrinkage and swelling of clay soils due to changes in moisture content and resulting from tree growth or removal had long been known as major causes of foundation movements and damage in low-rise buildings, with shrinkage perhaps easier to deal with than expansion. The changes that occur in soils during swelling, to enable better prediction of the forces that operate, can be quantified.

Foundation movements in houses built on clay with trees nearby has been common in parts of the UK for some years. An important test measures the state of desiccation in a clay soil. This simple test, in which a laboratory filter paper is sandwiched between discs of clay and the weight of water transferred to the filter paper is measured, was found to provide a better indicator of desiccation than conventional water content measurements[22].

Analysis was completed during 1987 of concrete exposed for 15 years to a fissured sulfate-bearing clay soil at a site north west of London. The results indicated that concrete buried in the ground could resist sulfate attack if it was well compacted, contained an adequate cement content and had a low water-to-cement ratio. If these conditions were not met, or if sulfates were present in very high concentrations, or if there was a hydrostatic head across the concrete, some attack was likely. The results of the work and of other research on slag and pulverised fuel ash (PFA) cements enabled BRE to improve its recommendations for specification of concrete in sulfate-bearing soils.

BRE's geotechnics investigators devised another test method to provide data for foundation design. The extent of any initial movements that occur in foundations depends on the stiffness of the ground. Ground stiffness is consequently an important parameter, but its measurement normally requires laboratory tests to be carried out on soil samples. A further complication is that the samples may differ in character from place to place, reflecting local variability in stiffness. The test devised provided a quick and relatively cheap technique for assessing ground stiffness in situ, using induced vibration in the ground. The technique has enabled large areas to be assessed comprehensively[23].

More detail of the history of BRE involvement in the development of geotechnical engineering is covered in *Seventy-five years of building research: geotechnical aspects*[24].

So far as the last decades of the twentieth century are concerned, there has been a noticeable tendency for brownfield sites to be used in preference to greenfield sites, and it is inevitable that many of these sites will offer problems. Moreover, less suitable sites are being used in greater numbers, involving poor, filled or steeply sloping ground which would have been rejected in former years.

There have also been significant developments in techniques for the improvement of poor land to provide more economical foundation design, in particular the further development of the vibro technique described earlier. Projects using vibro have ranged in size from a few treatment points beneath strip footings for a pair of semi-detached houses to the treatment of large areas with a uniform pattern of treatment points. The technique is particularly suited to the redevelopment of sites, and can be used relatively close to existing structures (eg much closer than dynamic compaction could be safely used).

Basements

Basements and cellars were built in large numbers in late Victorian and Edwardian times. However, relatively few basements and cellars for individual houses have been built following the 1914–18 war – they went completely out of fashion – although in recent years there has been some revival of interest. The vast majority of basements during the twentieth century have been constructed in other building types, where basements have been seen as a very convenient location for storage areas, boiler rooms and car parking.

Figure 0.20 The War Room, Whitehall

During the early years of the twentieth century the science of waterproofing underground structures made considerable progress, so that by the middle of the century it was possible to build fully serviced, dry, very large underground structures. The 1939–45 war produced a number of examples, of which the War Room under Whitehall is perhaps one of the most famous (Figure 0.20).

It is now possible to provide perfectly acceptable accommodation below ground level, even to the extent of building whole dwellings underground where the need arises.

External works
Mains water

As noted in *Building services*, lead was used commonly for mains water supplies to dwellings right up until the 1920s, and sporadically even later. Some dwellings still have supply pipes of this material in use. In Scotland, for example, the number of dwellings where lead was detectable in the water supply fell from just over 1 in 10 in 1991 to around 1 in 14 in 1996[5]. However, there is still some way to go to achieve complete elimination.

Following the realisation that lead was injurious to health, iron supply pipes gradually began to replace the lead, particularly just before and after the 1939–45 war.

Again, as noted earlier in this book and in *Building services*, perhaps the most significant innovation with regard to the means of supplying water to properties has been the introduction of polyethylene mains in substitution for lead, and plastics drainage and wastes in substitution for cast iron, asbestos cement and copper. One consequence of the introduction of polyethylene pipes has been the removal of a convenient means of earthing the electricity supply, though it is arguable that alternative means of earthing would have happened for other reasons too.

Wastewater drainage

In rural areas of the UK, until waterborne septic tank installations became widespread in the middle years of the twentieth century, it was common to provide earth closets for receiving faecal matter. Provision was made for these in the Model Byelaws, including positioning to avoid contamination of drinking water supplies, a limit of 2 ft^3 on the size of the leak-proof receiving tank, a suitable vessel to contain the dry earth, and a suitable means of applying it to the faecal matter. Earlier installations did not even conform to these basic requirements. One installation was found in the garden of a farm in Worcestershire by a BRE investigator in the late 1970s; it was sited in a brick built, pitch roofed 'three-holer' positioned over a 2 m deep brick lined shaft with access at the foot of the sloping site, with no provision for earth cover, but it was cleaned out annually and the faeces dumped onto the manure heap with those of the animals (Figure 0.21). Only within the last five years had it been taken out of use and a septic tank installed. There has been a nostalgic revival of interest in these so-called 'privies' and a number of books depicting local examples have been produced.

Under the old byelaws there were some fairly loosely expressed requirements for drains and private sewers, using such terms as undue, suitable, proper and adequate, but there were also specific requirements such as the minimum diameter of a drain *'intended for the conveyance of foul water'* of 4 inches. When the byelaws were replaced by the England and Wales Building Regulations in 1965, many of the former criteria were reused; in particular the one requiring inspection chambers to be watertight, which could rarely be achieved in practice until brickwork inspection chambers began to be replaced with plastics in the 1980s[25].

After the 1939–45 war, the extensive house rebuilding and slum clearance required led to numerous studies to simplify housing construction, including a re-examination of drainage design practice. Major changes in former practice under local byelaws based on the Model Building Byelaws made under the Public Health Act 1936 were introduced with the publication of CP 301 in 1950, revised twenty years later and further revised and issued as BS 8301[26] in 1985. The new code indicated that *'means of access to drains and sewers should be reduced to the minimum necessary for each section to be separately tested, cleaned or rodded for the clearance of stoppages'*. Subsequent studies and reports indicated that drain gradients could be flatter without causing increased blockage; fallen rendering from manhole sides and intercepting traps were also common causes of blockages. BRS publications also reported the successful use of 100 mm drains serving 20 dwellings laid at gradients of 1 in 70, and a 150 mm pipe was considered satisfactory for up to 100 dwellings when laid to a gradient of not less than 1 in 150. Drain gradients had generally become flatter and

Figure 0.21 A brick built privy which had been in use for at least 100 years and probably longer

manholes were no longer always positioned at every junction or change in drain direction. Drain depths, because of the choice of gradients, were usually small. Interceptors were generally no longer specified.

PVC pipework with 'O' ring jointing came into use by the mid-1960s. Pitch fibre pipework also enjoyed a brief period of popularity. Manhole construction generally continued to be traditional in form (rendering was sometimes omitted) even when PVC or pitch fibre pipework was used, and normally branch drain invert levels were built above the invert level of the main drain run. Preformed concrete chambers were sometimes used.

Between 1996 and 1998 the various parts of BS EN 752[27], have been published.

Recent tightening of legislation governing work in confined spaces has led to a reappraisal of the need to enter manholes for maintenance purposes, and much of the inspection and remedial work formerly carried out manually can now be carried out using machinery. Design procedures for urban drainage systems are currently undergoing revision which in turn may lead to changes in regulations[28,29].

See also sustainable urban drainage systems in Chapter 4.3.

Building user requirements in the third millennium

The droughts of the late 1970s, the floods of the late 1990s, and increased demand for more homes in already overcrowded areas brought about by an ageing population and growth in the numbers of one-person households, have focused attention on the part of government, local planning authorities, individual householders and the media on factors in the environment which can affect the provision of housing, sometimes detrimentally. These factors include:
- proximity to hazardous industrial plants
- building on landfill which may be hazardous
- building on a flood plain and potential risk of flooding
- building on ground liable to subsidence or landslip
- risk of exposure to radon
- exposure to wind damage

It was noted earlier that the EHCS survey recorded that there were over 2 million dwellings in England alone which had at least one environmental problem.

All of the above factors relate to the site and its location, which is the main focus of this book. Proximity to hazardous industrial plants is not within the present terms of reference. However, the next four in the above list are. Radon has been dealt with in *Floors and flooring*, though further information specific to basements will be found in this book, and the last item, wind damage, has been dealt with in *Roofs and roofing*.

The state of the nation's housing stock has become a subject of more or less permanent media interest. At the turn of the twentieth century, it has been estimated[30] that:
- 166,500 homes are within 5 km of hazardous industrial plants
- 1,850,000 business and domestic properties are at risk of flooding
- 1 in 50 buildings, including 440,000 homes, are on ground liable to subsidence
- 100,000 homes are on sites liable to landslips
- 429,000 homes are exposed to radon*

While there is some room for debate on the overall numbers of dwellings at real rather than imagined risk within the post code areas detailed, there is no doubt at all that there is increasing public concern at these potential risks and the effect they have on the value of properties. Television pictures have shown widespread flooding in Kent, Sussex, North Yorkshire and the Severn valley; and public concern has been expressed at continuing construction of dwellings in river flood plains. The Environment Agency was reported as concluding that in addition to 1.9 million houses, there were also 35,000 commercial properties situated in river flood plains[31].

Without doubt, in the third millennium new challenges such as these, and new opportunities stemming from the invention of new materials and techniques, will shape the form and content of newly built buildings. However, although many existing buildings will increasingly no longer measure up to changing user requirements and rising expectations, the nation will simply not be able to afford, year by year, to demolish and rebuild anew more than a very small fraction of the existing building stock, currently worth at least £1,500 billion. Maintenance, repair and refurbishment will be required for the foreseeable future[32].

---

* The National Radiological Protection Board (NRPB) have estimated that only 100,000 dwellings in the UK have indoor radon levels in excess of the UK recommended action level of 200 Bq/m³.

Chapter 1
The basic features of sites

Figure 1.1 Bracknell town centre in 1975. In the 25 years since the photograph in Figure 0.1 was taken, the whole of the town centre has changed, and virtually all prime building land has been developed

This first chapter deals with features and characteristics of existing sites having the most significant effects on the types of structure which have been, and can be, built and serviced on them. It does not deal with the criteria which govern the selection of greenfield sites for new construction. This book therefore is not the right place to provide comprehensive advice to building professionals on how to assess the condition of virgin ground before any construction on it begins. However, if something subsequently goes awry with that construction due to inadequate previous consideration of ground conditions, there will be a need to discover the reasons why defects have occurred, and this need provides the main theme of the chapter.

In Britain, particularly in urban areas, most of the readily available prime building land had already been used for that purpose by the last quarter of the twentieth century (Figure 1.1). Since that time greater use both has been and is being made of recycled land (brownfield land) and land hitherto considered too unsuitable to use for building; these sites can contain a wide variety of geotechnical and other hazards.

There is no substitute for thorough investigation of the ground before construction begins, and this policy is becoming increasingly important. Problems encountered both during and after construction because of inadequate ground information can be both extensive and costly to rectify. However, an understanding of the main factors relating to the ground should enable surveyors and architects at least to realise when geotechnical and other engineers should be called in to advise.

Before any construction work begins, the site should have been thoroughly investigated. Site investigation consists essentially of two stages. The first stage is the collection of available information on the conditions at the site; this is normally handled by a desk study and a walk-over survey. The second stage is the collection of further information required for detailed engineering design and reassessment of the information obtained during the desk study and the walk-over survey. This can be done by the so-called direct methods of ground investigation: boring, trial-pitting, and soil sampling and testing, although these are occasionally supplemented by indirect or geophysical measurements.

A typical ground investigation may be broken down into the following activities:
- preliminary desk studies
- examination and interpretation of aerial photographs
- the walk-over survey
- design of a ground exploration programme
- exploration by trial pits and boreholes
- soil and rock classification by sample description and index testing
- in situ and laboratory testing of soils or rocks for their mechanical and chemical properties
- preparation of a report

In the past, expenditure on ground investigations for small projects such as housing developments has amounted to between 0.1 and 0.2 % of the construction costs, and most of this money has been spent on trial pits or boreholes. In the present climate of increasing building litigation, with increasing risks to all parties involved in the construction process, expenditure on ground investigation should be a minimum of 0.2 % of the cost of the project.

For most buildings constructed within the 1970s, 1980s and 1990s, a report on the original site investigations should be available, either from the building owner or from the original consultants, and this will be a starting point for diagnostic work. For other buildings of earlier construction dates, for which such information is not available, it may mean that some or all of the investigative steps detailed above will need to be repeated, perhaps involving the appointment of specialist geotechnical engineers.

It is possible to alter the characteristics of poor soils before construction begins, and some buildings may have been built on land which has been improved.

Chapter 1.1   Geology and topography

This chapter includes a summary of the main categories of subsoils and bearing conditions, ground movements etc. There are also brief mentions of earthquake risk, archaeological remains and the importance of obtaining a knowledge of the past history of any site where further building or excavation work is to be undertaken.

In low lying land, difficulties which may be encountered include soft soils, high flood levels, and water pressure on basements due to high water tables. In clay soils, shrinkage due to water demand from vegetation, or due to heat from boilers or kilns may be encountered. In addition, in hilly areas, some clay soils can be unstable. Silts, sands and gravels may suffer loss of fines, or they may be susceptible to frost heave. Reclaimed or filled land may suffer from natural consolidation of the fill, rot of vegetable matter in fill and subsequent generation of methane, and even fire in some circumstances. Swallow holes or other cavities are a risk in chalk and limestone areas, and in former mining areas there may still be some subsidence where old underground workings collapse (Figure 1.2).

Low-rise structures have often been founded on relatively simple foundations at relatively shallow depths where soil tends to be both more variable and more compressible than it is at greater depths. Even for these buildings, design work should be preceded by site investigations. Competent foundation design for any structure, however small, requires an initial desk study. Valuable information can be obtained at low cost and desk research is a particularly good investment where the costs of site investigation must be kept low [12].

Summary of sources of information and methods of investigation relating to the topography, vegetation and drainage of a site

Slopes
Current Ordnance Survey maps (at scales of 1:10,000, 1:10,560 and 1:25,000).

Water
Ordnance Survey maps. Water is shown in blue on the 1:25,000 scale maps. This allows easy recognition of springs, ponds, rivers and other drainage features. Water is also coloured blue on some old 6-inch-to-1-mile (1:10,560) maps; old maps may show changes in the position of watercourses when compared with more recent maps. On large scale maps – 1:10,560, 1:10,000, 1:2500 and (in urban areas only) 1:1250 – high groundwater or disrupted drainage may be indicated by symbols for boggy land or heathland.

Vegetation
Aerial photographs provide a permanent record of site vegetation. For a given site, air photography may be available about every 10 years from 1946. Aerial photography specialists can estimate the height of trees or bushes from stereo aerial photo cover. Large scale topographical maps will show the positions of hedgelines, woodland and, occasionally, of isolated trees.

Made ground
Comparison of large scale topographical maps made at different dates will allow identification of changes on the site such as infilling of hollows of old pits, removal of vegetation or demolition of old buildings. Maps of 1:10,560 scale from about 1840 are available; they provide a record of site development over the years. British Geological Survey maps at 1:10,000 scale compiled after about 1965 delimit made ground.

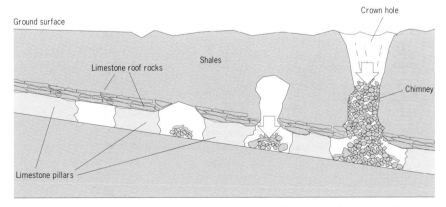

Figure 1.2 Void migration above old limestone workings followed by formation of chimney and crown holes

> **Summary of sources of information and methods of investigation relating to ground conditions**
>
> **Geology**
> The geology of a site can be determined from geological maps published by the British Geological Survey (BGS), normally at 1:10,560 or 1:10,000 basic mapping scales and summarised at smaller scales of 1:63,360 or 1:50,000. Each 1:50,000 geological map has a descriptive memoir giving details of the local rocks. Where there are extensive Drift deposits, then the Solid geology map is supplemented by a Drift edition. Where the superficial deposits are sparse, a single edition of Solid and Drift will show both forms of geology. Site investigation is generally concerned with superficial deposits. Full collections of maps are available at the London and Keyworth, Notts, offices of the BGS; Scottish maps are available at the BGS Edinburgh office. Requests for information relating to specific investigations should be made to the BGS.
>
> **Head**
> Head is poorly structured rock debris with shears as well developed planes of weakness. Geological Drift maps, especially at large scale, may identify significant thicknesses (ie greater than 1–2 m) of Head, but will not necessarily give any warning of the presence of thin layers of materials; Head must be checked by trial pitting. Geotechnical properties of these types of subsoil will be very variable. The only reliable sources of information available at desk study stage are likely to be reports of subsoil investigations carried out in the area (possibly available from the local authority) or technical papers in civil engineering or geological journals.
>
> **Landslips**
> Landslipping is outlined on some 1-inch-to-1-mile, 1:50,000 and 1:25,000 scale geological maps. It is also shown, where recognised, by notes (eg 'foundered strata') on the manuscript 6 inch County Series maps. Unfortunately, not all landslips have been recorded on these maps. Many can be recognised as hummocky terrain on the ground, but additional help can be obtained from aerial photographs of the area. The preferred scale for aerial photograph interpretation of this kind is 1:2000, although material up to a scale of 1:10,000 may provide useful information. Black and white photographs are normally the most readily available and should be viewed in stereo for the best effect. To have the best chance of detecting landslipped ground, photographs taken at as many different times as possible should be examined. Landslips are often associated with clays at the base of steep scarp slopes such as the Gault Clay at the base of the chalk in the North Downs. Geological and geotechnical literature is also useful in identifying landslipping.
>
> **Chemical attack**
> Chemical attack can result from the natural composition of soils, rock or groundwater in which foundations are to be placed, or from chemical waste in fill (BRE Special Digest SD 1[33]). Much the commonest form of attack is solution by acidic groundwater of carbonate rocks such as chalk or limestones, or the solution of rocks with carbonate cements. Evaporites, such as rock salt or gypsum anhydrite, are dissolved even more rapidly by groundwater. Many British clays contain sulfates. The likelihood of attack from contaminated ground can be determined by examining topographical maps of different ages to try to detect infilling on the site, and by looking for evidence of old factories or works and trying to assess their potential for pollution (eg former gas works have often left the ground contaminated).
>
> **Mining**
> Many different types of mineral have been mined or quarried in the United Kingdom. Geological maps at 1:10,000 or 1:50,000 scales will identify layers which contain minerals (eg coal). Unfortunately there are no abandonment plans of many disused mines. Guidance on the position of old mine shafts can sometimes be obtained from old topographical maps. Other records may be found in local archives such as the County Records Offices, or may be held by central government (eg the Mining Records Office of the Health and Safety Executive). Other sources include the Catalogue of Abandoned Mines (1928 to 1939) and the Directory of Quarries and Pits (1973). Records of abandoned coal mines are kept at British Coal Archives.

The assessment and classification of visible damage resulting from structural distortion is dealt with in BRE Digest 251[34]. The assessment is based on a description of work considered necessary to repair the building fabric; classification into six categories is recommended, taking into account the nature, location and type of damage. The various causes of ground movement giving rise to damage are described later in Chapter 2 where the most common causes of damage are discussed. Where the resulting damage is slight, cracking may result from a combination of causes which are difficult to identify, and the cost and effort involved in carrying out an identification would be disproportionate to the scale of damage, except for circumstances where the movement is likely to be progressive. It is rare for damage to be significant, and when it is, ground movement is usually the cause.

On the other hand, some large buildings can give rise to particular problems, especially where they are built on difficult ground conditions and in proximity to other existing structures with deep basements or foundations. These can present considerable challenges, even to experienced geotechnical engineers. To give but one example, although it was initially estimated that excavations in the London Clay for the large basements of the Shell Centre on the south bank of the River Thames would create upward displacement of the underlying Bakerloo line tunnels of between 20 and 30 mm, the maximum heave measured by BRE investigators up to 1986 was 50 mm in the Southbound Bakerloo tunnel immediately under the basement and 41 mm in the lower Northbound tunnel. By the mid-1980s the rate of heave showed little sign of decreasing although the excavations had taken place over 27 years earlier[35].

1.1 Geology and topography

Characteristic details

The geology of a particular site

Although the geology of a site can be usually determined from published geological maps, manuscript maps for some parts of the country can be viewed and copied at the BGS library in London, and other records may also be obtained from local BGS offices. The nature of the strata on the various maps can be identified from descriptions in the British Regional Geology handbooks available from the Stationery Office or from the published and out-of-print BGS Sheet Memoirs, which relate to specific 1-inch-to-1-mile sheets.

One problem with the use of geological maps and records is that geologists divide the soils and rocks according to their age, and not according to their nature (ie whether they are sands, clays or rocks). Therefore a single geological unit may in fact vary in composition and hence, in its behaviour under foundations, from one location to another. This type of variation can usually be discovered by reading written geological records, or memoirs, pertaining to the area. For example, while the Reading Beds are normally associated with highly shrinkable clays, in some parts of the country they are predominantly sandy.

The most important information that can be found from geological records includes:
- the nature of the subsoil (ie whether it is sand, clay or rock)
- the thickness of the different soil or rock strata in the vicinity of the site
- adverse ground conditions associated either with particular soil or rock types, or with a combination of the subsoil and the characteristics of the site

Landslips

Many natural and man-made slopes exist in a state of only marginal stability and down-slope movements of ground can be triggered by a relatively minor change of circumstances. Instability of this kind, which is normally referred to as landslide or landslip, may result in a sudden and irresistible movement of a large mass of soil or rock, or alternatively a series of large ground strains. In the latter case, any building in the path of the movement is likely to act as a retaining structure, albeit inadequately, unless it has been specifically designed for this purpose. In addition to these obvious movements, the surface layer of some clay slopes may be prone to a continual, insidious movement, a process known as slope creep.

Movement occurs in natural, cut or filled slopes when the total disturbing force acting on a block of soil or rock exceeds the total resistance in the ground (Figure 1.3). The disturbing forces can include the weight of the block (including the weight of any water), seepage forces (which normally act downhill), seismic forces and superimposed loads while the resisting force is the shear strength of the soil or rock acting over the total area of the potential failure surface.

Where the slope is very long compared with its height, the failure surface is often more or less parallel to the ground surface, occurring at a depth where the strength of the surface soil or rock has been weakened by weathering. However, where the slope is of limited length, the failure surface will tend to be approximately circular, unless there are existing planes of weakness or less competent strata in the ground which will influence both the location and the shape of the failure. Failure surfaces in rocks, in particular, tend to follow pre-existing defects such as bedding planes, joints and faults; and where soil overlies a sloping rock surface there is a tendency for the interface between the two materials to form the slip surface.

Failure most commonly occurs as a result of a reduction in the resisting force, which can be brought about in a variety of ways:
- an increase in the porewater pressures in the ground, reducing the effective stresses acting on the slip surface

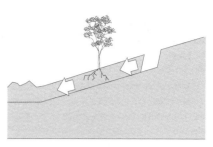

Slab translational slide

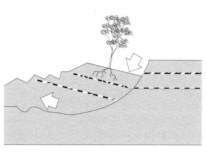

Circular rotational slide

Figure 1.3 Examples of slip surfaces

- excavation at the toe or crest of the slope reducing the area of the potential failure surface; natural surface erosion by wind and water may have a similar effect
- subsurface erosion by groundwater flow
- weathering of the surface layer of soil or rock, leading to a reduction in strength
- recurrent small strains leading to a long term reduction in strength; this process, which is known as progressive failure, is particularly relevant to over-consolidated clays where the reduction in strength from peak to residual values can be large
- removal of natural cementing material in the ground by seepage water; a process known as leaching which is particularly relevant to some soft clays
- removal of vegetation and killing of roots, which may have had a binding action on the surface soil
- high frequency vibration which can cause loose granular soils to reduce in strength due to a buildup of porewater pressure

The weight imposed on a slope by buildings is normally insignificant when compared with the weight of the soil mass, and is therefore unlikely to cause more than a localised failure.

However, building activities may have serious implications for the stability of the slope; for example:
- the excavation of foundations and basements either near the toe or the crest of a slope may reduce the area of the potential sliding surface
- cut-and-fill construction techniques, which are commonly used for low-rise construction on sloping ground, can change the stresses in the soil and result in a local steepening of the slope
- the construction of retaining walls may interrupt existing drainage
- the addition of fill material to the top of a slope can lead to instability of the slope, as well as of the fill itself. Conversely, adding fill material to the toe of a slope or removing it from the top of a slope will usually reduce the likelihood of instability

In many cases, the modifications may be insufficient to cause an instantaneous failure, and substantial movements may occur only several years later as a result of some triggering event.

Slope creep
It is rare in the UK for low-rise properties to be affected by large irresistible landslides; a far more common cause of damage is the insidious process of slope creep.

Slope creep is a phenomenon that affects the top metre or two of certain natural clay slopes, migrating slowly down the hill over long time scales. In London Clay, for example, slope creep typically produces average movements of no more than 1 cm per year. However, it is possible that, rather than being a continuous slow activity, the movements may be triggered by certain climatic conditions and therefore occur only intermittently. One possibility is that slope creep is related to exceptional desiccation during prolonged periods of dry weather. In particular, the down-slope movements may be produced by the incomplete closure of desiccation cracks, which would explain why only the surface layer of soil is affected.

The cost of stabilising a slope that is creeping is likely to be prohibitive and, if such areas cannot be avoided, the best policy is usually to adopt an appropriate foundation design for the buildings to take account of the movements. This can be done in one of two ways using:
- deep foundations to anchor the structure in the underlying stable soil
- a raft to allow the whole structure to move as a single unit with provision for correcting its level if necessary

Old foundations and wells
Where sites that have been already built on are being redeveloped, old foundations may need to be removed and deeper features (eg basements and wells) to be filled. For smaller features, it is likely to prove difficult to carry out the backfilling under controlled conditions. As a result, the ground conditions across the site can be variable and it may be necessary to reinforce strip footings to allow them to bridge soft spots. In some instances it may be more practical to use a suitably reinforced raft foundation. Alternatively, piles or piers may be used to ensure that the new building is founded well below the influence of the previous foundations.

Mines, quarries and solution features
For the purposes of this book, solution features are cavities in the ground caused by the previous action of running water below ground level.

As already noted, many different types of material have been mined or quarried in the UK, from flint in Neolithic times to coal at present. It is believed that there are about 100,000 old mine workings in the UK, about 30 % of which are uncharted. The majority of these are the legacy of coal extraction during the seventeenth, eighteenth and nineteenth centuries, although a variety of other resources such as silver, lead, tin, iron ore, flints, limestone, shale, salt and refractory clays have also been mined at various times in the UK. Many of the coal seams that were worked during the Industrial Revolution now underlie urban areas.

Information on the extent of old workings is not always available, although sometimes the records will have been preserved. As one example, Figure 1.4 shows the extent and dates of working of several anthracite collieries in the Dulais Valley in South Wales, obtained from a variety of sources. Since there is no single repository for this information, a certain amount of detective work will normally be involved.

Areas thought to contain solution features such as swallow holes or old mine workings should be properly surveyed to locate any cavities. The positions of the proposed buildings should then be adjusted to avoid any cavities that are identified or, where this is not practical, the cavities should filled. Alternatively, where filling is impractical, buildings can be founded on piers taken down to a stable stratum or on a heavy raft foundation which is capable of bridging over a local collapse. It will usually not be possible to found the building below the level of the mining activity. If the former solution is adopted, the piers must be protected from the potential downdrag forces that would be generated by a subsidence of the overburden supported by the mine workings. This can be achieved by providing a space between the shaft of the pier and the overburden filled with a suitable low friction material (eg bitumen).

A number of general principles can be adopted to help reduce the level of damage caused. These are listed in *Foundation design and construction*[36] and may be summarised as follows:
- structures should be either completely rigid or completely flexible. Simply supported spans and flexible superstructures should be used whenever possible
- the shallow raft foundation is the best method of protection against tension or compression strains in the ground surface

1.1 Geology and topography

- large structures should be divided into independent units. The width of the gaps between the units can be calculated from a knowledge of the tensile ground strains derived from the predicted ground subsidence
- small buildings should be kept separate from one another, avoiding linkage by connecting wing walls, outbuildings or concrete drives

Raft foundations should be designed so that the horizontal stresses produced by lateral ground movements acting on the edge of the raft are limited. The raft should also be cast onto a membrane to reduce the friction on the underside of the raft. Similarly, where strip footings are considered adequate, they should be cast onto a layer of sand providing for longitudinal movement at the ends of the foundation trenches.

Piled foundations are not automatically suitable for areas liable to mining subsidence because the lateral ground movements can shear the piles or cause tensile failure of the ground beams connecting the piles.

Sloping sites

With increasing problems of land availability comes greater likelihood that buildings are built on land previously thought to be expensive or having awkward features (eg steeply sloping sites which may be unstable). Although building on unstable sites is to be avoided wherever possible, in some cases it will be inevitable. In general, it is more practical to improve the stability of slopes rather than to design foundations to withstand or accommodate down-slope movements.

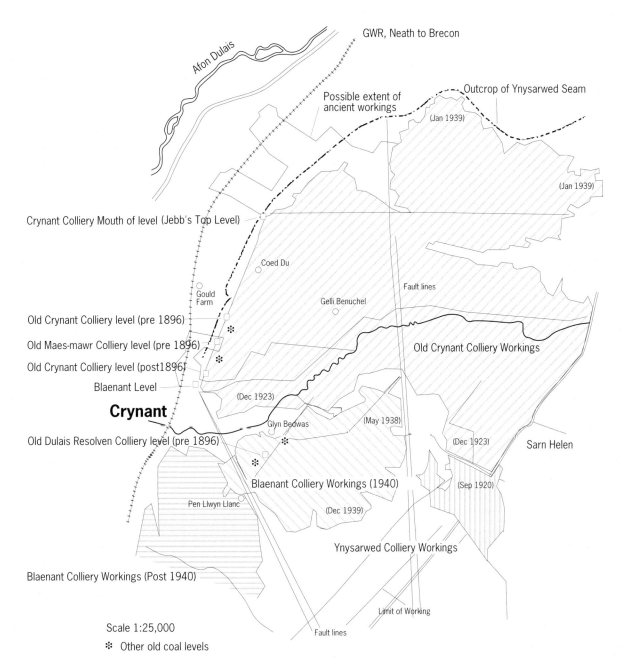

Figure 1.4 Old colliery workings in the Dulais Valley. (Information obtained from Ordnance Survey 6 in Glam IX NW (first edition), Coal Board plans, Glamorgan Record Office D/D DJ 58/23, and Blaenant Colliery No 2 Rhondda Seam, Plan No 6, Cefn Coed Colliery Museum)

Where slope stabilisation is impractical, there are two options available to the designer:
- the building can be founded on a rigid raft so that the entire structure moves as a single unit, assuming that service connections can still be maintained
- the building can be founded on piles designed to withstand the lateral forces generated by the sliding soil

The variability of the ground

While a great deal is known about most of the materials from which buildings are formed, the same cannot be said about the ground. Not only are soils and rocks very much more difficult to access and test than structural materials, their properties can vary enormously from one location to another and, even over a particular site, from one side to the other. Furthermore, these properties may vary over time. Taking for example the behaviour of a soft clay: when loaded, the clay will undergo a process of consolidation whereby the water is squeezed out and it reduces in volume, stiffening in the process. On the other hand, a structural member such as a steel lintel is much more consistent over time. It can, at worst, be expected to undergo very small creep strains over decades. The lintel can readily be tested in the laboratory in circumstances very similar to those experienced in its final location; the same cannot be said for the soft clay site from which samples will be taken and tested in ways substantially different from what occurs at the site during and after construction. A further difficulty is that the samples themselves will be affected by the process of excavation and removal from the site.

For these purposes a simple classification of soils etc can be adopted:
- rock
- granular soil
- cohesive soil
- fill

The first three types refer to naturally occurring material and the fourth to man-made ground which could be composed of natural, granular or cohesive soil, rock or waste products. Fill is dealt with in Chapter 1.2.

There is a basic difference in behaviour between coarse, granular soil and fine, cohesive soil. The percentage of silt and clay size particles (or fines smaller than 0.06 mm) in a soil is important since, when this percentage is high, the soil will cease to behave as a granular soil. The critical magnitude of the fines content is likely to be in the range of 10–25%, and typically about 15%. In the British Soil Classification System for Engineering Purposes in BS 5930[37], coarse soil is defined as having less than 35% of the material finer than 0.06 mm and fine soil as having more than 35% of the material finer than 0.06 mm. There are other differences in addition to particle size.

The *Review of foundation conditions in Great Britain*[38] prepared by the DoE contains:
- a nationwide database which consists of site-specific information on ground conditions abstracted and condensed from site investigation records
- maps showing the general distribution of selected foundation conditions on a topographic base at scales of 1:250,000 and 1:625,000
- a report describing the background to the review, suggesting how the results may be used and providing an introduction to the subject of ground conditions and associated activity

The Building Regulations Approved Document A[39], Section 4.9, mentions conditions of ground instability from such features as geological faults, landslides and disused mines.

Rock

Rock is divided into three categories: igneous, sedimentary and metamorphic.
- Igneous rock is composed predominantly of three minerals: quartz (SiO_2), feldspars and biotite. Depending on its mineral composition, the rock may be acid, intermediate or basic. Acid rock tends to be lighter in colour than basic rock
- Sedimentary rock is formed from broken down rock (ie soil) under high pressure and temperature, and possibly chemical action. The constituent particles may be primary minerals (eg silica which forms sandstone or calcium carbonate which forms chalk and limestone); alternatively, they may be secondary minerals which form mudstone and shales. Sedimentary rock may also be organic (eg lignite or coal)
- Metamorphic rock is derived from either sedimentary or igneous rock through the action of extreme high pressures and temperatures that can be generated deep in the ground by geological events

So far as foundation design is concerned, The Model Building Byelaws, the Building Regulations 1965, 1972 and 1976, and the Building Regulations Approved Document A which followed, all cited rock as being *'not inferior to sandstone, limestone or firm chalk, and requiring at least a pneumatic or other mechanically operated pick for excavation'* as a deemed-to-satisfy provision not needing a foundation wider than that of the wall to be carried, irrespective of the imposed load.

Although most types of rock will deform slightly when loaded, for the most part these deformations are small, and there is little need for elaborate precautions to be taken unless major faults intervene. To give one example, BRE investigators measured the structural movements in a sandstone stratum adjacent to an atomic power station that had experienced applied stresses of 100 kN/m²; that is to say, well in

1.1 Geology and topography

excess of the values quoted in the Building Regulations. The measurements were made by precise levelling to an accuracy better than 0.025 mm. The greatest settlement was less than 2 mm. There was no evidence of settlement at depth nor of creep due to these stresses, thus suggesting the use of simple foundations[40].

The nature of soils
Soil is made up of solid particles, ranging vastly in size from a few micrometres for clay particles up to metres for boulders, which are surrounded by spaces or voids filled with air, water, or other material. When a soil is loaded it deforms primarily by a rearrangement of the particles; it is therefore the interactions between the particles, and especially the friction that is generated between particles as they slide over one another, that governs the deformation and strength properties of the soil.

The solid particles may be rock or they may be mineral. A mineral is a naturally occurring chemical compound whose composition can be expressed by a simple formula, such as quartz (SiO_2), whereas a rock is a combination of minerals. Many small particles, such as are to be found in clay, exist as tiny, flat plates. Larger particles tend to be more nearly spherical, although, depending on their composition and their origin, particles may be rounded or may contain sharp edges; in fact, five classifications are used to describe particle shape – angular, sub-angular, sub-rounded, rounded, and well rounded.

Where the pores are entirely filled with water, the soil is described as saturated and where the pores contain a mixture of air and water, it is described as partially saturated. The pore fluid exerts a critical influence on the behaviour of the soil and accounts for most of the differences between the very fine grained soils (eg clay and silt) and the coarser grained soils (eg sand and gravel).

It is useful to draw a distinction between the very fine grained soils (eg clay) that have significant quantities of water chemically bound to the soil particles and the coarser grained soils (eg sand and gravel) that do not. One of the consequences of the low permeability of fine grained soils is that, when a sample is removed from the ground, it is held together for a considerable length of time by cohesion; this effect is created principally by suctions in the porewater but may also stem from a chemical bonding between solid particles. Soils that possess cohesive properties are known as cohesive soils and those that do not are known either as cohesionless or granular.

The density or compactness of a soil depends on its geological origin, on the nature of its formation and on its subsequent stress history. The compactness may be described by four closely related parameters:
- voids ratio
- porosity
- dry density
- dry unit weight

An additional parameter which can be used to describe the degree of packing in coarser grained soils (eg sand and gravel, or fill) is relative density, sometimes called the density index.

Particle size
Granular soils are normally classified according to the size of the constituent particles and whether the soil is well graded (ie it contains a wide range of particle sizes) or whether it is poorly graded (ie it has a large proportion of particles of a similar size) (Table 1.1 on page 24).

The distribution of particle sizes is expressed as the percentage by weight greater than a given particle diameter. Consequently a soil with a predominance of particles in the range 0.06–2 mm would be described as a sand.

For the particles with a diameter greater than about 0.06 mm, the distribution is determined by passing the soil sample through a series of sieves with diminishing apertures and measuring the weights of material retained on each sieve. Dry sieving is only suitable for sands and gravels that do not contain any clay, and wet sieving is normally required to separate the finer particles and prevent them from sticking to the coarse fraction.

The distribution of the finer particles with a diameter of less than about 0.1 mm is determined by sedimentation. This test relies on the fact that, when dispersed in water, particles of different sizes will settle at different rates. After shaking to evenly distribute the soil, the density of the upper part of the suspension is measured at given intervals, either by sampling with a pipette and weighing the dried residue or by using a hydrometer. Applying Stokes's Law, the maximum particle size remaining in the upper part of the suspension at any given time can be calculated. So, by adjusting the time intervals according to the specific gravity of the soil particles, the measurements can be used to determine the percentage of material falling within a particular size range. Normally the time intervals would be chosen to measure the proportion of medium silt (0.02–0.006 mm), fine silt (0.006–0.002 mm) and clay (less than 0.002 mm).

Cohesive soils (clay and shale)
Residual clay (eg the china clay deposits in Cornwall) is very rarely encountered elsewhere in the UK. It is the in situ product of weathering of rock and minerals, and may retain some of the characteristics of the parent rock and quite often contain pieces of the parent rock in the form of 'floaters'. These may give considerable trouble during site investigations and in sinking piles or in excavations for foundations.

Boulder clay is extremely heterogeneous material deposited by glaciers. As they traverse the land surface, glaciers scour and erode rock and soil they come in contact with, and pick up various types of material. These are then deposited together as the glaciers melt and the resulting material often consists of pieces of rock, shell, chalk, flint etc in a clay matrix.

Table 1.1 Particle size classification (after BS 5930[37])

Soil type	Particle size (mm)	Typical properties
Boulder fragments	> 200	Rounded or angular, bulky, hard rock. Particles very stable; used for fill, ballast and to stabilise slopes. Because of size and weight their occurrence in natural deposits tends to improve the stability of foundations. Angularity of particles increases stability
Cobble	60–200	
Gravel:		
Coarse	20–60	Rounded or angular, bulky, hard, rock fragments. Easy to compact, insensitive to moisture and frost action. Gravel is more resistant to erosion than sand. Well graded sand and gravel are generally less pervious and more stable than those with uniform particle size. Irregularity of particles increases stability slightly.
Medium	6–20	
Fine	2–6	
Sand:		
Coarse	0.6–2	Finer uniform sand approaches characteristics of silt (ie decreases in permeability and reduces in stability with increases in moisture content)
Medium	0.2–0.6	
Fine	0.06–0.2	
Silt:		
Coarse	0.02–0.06	Regardless of moisture content, soil exhibits little or no plastic behaviour, and strength when air-dried is low. Inherently unstable, particularly when moisture is increased with a tendency to become quick (as in quicksand) when saturated. Relatively impervious, difficult to compact, highly susceptible to frost heave, easily erodible. Bulky grains reduce compressibility, flaky grains (eg mica) increase compressibility producing an elastic silt
Medium	0.006–0.02	
Fine	0.002–0.006	
Clay	< 0.002	Soil exhibits plastic behaviour within a certain range of moisture contents and considerable strength when air dried. Soil has cohesion or cohesive strength which increases with decreasing moisture content. Low permeability making it difficult to compact when wet and impossible to drain by ordinary means. When compacted, is resistant to erosion and piping, and is not susceptible to frost heave, but is subject to expansion and shrinkage with changes in moisture. Properties are influenced not only by size and shape of particles (ie flat plate-like), but also by their mineral composition (ie type of clay mineral) and chemical environment

Alluvial clay and silty clay are deposited from still or slow moving water found typically in lakes, swamps, river flood plains or under marine conditions where water velocities are generally no more than about 1 mm/s and often much less than this.

Silt consists of material greater than 0.002 mm in diameter, but less than 0.06 mm. It usually consists of bulky shaped particles of primary minerals (ie minerals that have not been affected by weathering, such as quartz). Clay is composed primarily of flat or elongated particles of secondary minerals, especially kaolinite, illite or montmorillonite, less than 0.002 mm in diameter.

Deposition under marine conditions, except near river mouths, usually leads to a very homogenous deposit because the clay particles flocculate together to form larger particles which settle out uniformly with the silt and, in some cases, fine sand. In addition to this, marine conditions remain relatively stable over long periods of time. Much of the clay that is prevalent throughout south east England, such as London Clay, Oxford Clay, Gault Clay and the Woolwich and Reading Beds, is marine clay.

Varved clay consists of alternating layers of silt and clay of comparable thickness and are deposited in lakes fed by ice. The layers of clay and silt are typically 100–500 mm thick. Where still, swampy conditions have existed for long periods of time with occasional sudden flows of water, the resulting deposit is usually a uniform clay or silty clay with very thin laminations of coarse silt or sand. These may not be more than a grain or two thick and are usually not thicker than 10 mm. Often they are extremely difficult to detect by eye in the soil at its natural moisture content. On partial drying, however, the coarser seams dry first to a light colour and can then be detected.

Thin organic seams may also be present, marking periods of accelerated swamp growth. Again these seams have an influence on the permeability of the clay and may influence the strength of the mass where an organic seam corresponds to a plane of high shear stress. Organic matter can also occur as pockets of material or as fine material dispersed through the clay. Where the organic material becomes predominant, the soil is a peat, which is discussed later.

The Model Building Byelaws, the Building Regulations 1965, 1972 and 1976, and the Building Regulations Approved Document A which followed, all cited three categories of clay and sandy clay: stiff, firm or loose:

- Stiff – *'Cannot be moulded with the fingers and requires a pick or other mechanically operated spade for its removal'*
- Firm – *'Can be moulded by substantial pressure with the fingers and can be excavated with graft or spade'*
- Loose – *'Can be excavated with a spade. Wooden peg 50 mm square in cross-section can be easily driven'*

For the last case, foundations did not fall within the provisions of the deemed-to-satisfy regulation if the total load exceeded 30 kN/m.

A rather more precise method of describing the characteristics of clay in a field test is available in Table 1.2 (BS 5930).

In practice, soils composed of pure clay minerals are never encountered. Deposits typically referred to as clay contain 30–70% clay minerals. In some cases, soils containing no more than 15% clay minerals can exhibit the classic characteristics of clay described in Table 1.3.

Heavy foundation loads can cause plastic flow, particularly in clay soils, and this may lead to troublesome settling. Where large structures are built on soft estuarine clays, they can provide too high a loading for that type of soil and cause plastic flow.

Consolidation properties in cohesive soils are usually measured by sampling. The soil sample, which is typically 76 mm diameter and 19 mm high, is contained in a steel ring to prevent any lateral movement, and is loaded vertically by means of a lever acting on a platen. The vertical compression of the sample is measured using a dial gauge. Porous discs placed in contact with the top and bottom faces of the specimen ensure that the specimen can drain freely from both ends. The results of individual loadings are used to derive the coefficient of consolidation which is used in the calculation of the timescales associated with settlement.

The Rowe cell is a variant of this test which allows a larger specimen to be tested under a controlled back pressure. Vertical pressure is applied to the specimen, which is confined in a cast bronze ring by means of hydraulic pressure applied to a rubber bellows. The apparatus can be modified to measure horizontal permeability by either excavating a sand drain in the centre of the specimen to provide inward drainage, or by lining the inside of the cell wall with a 1.5 mm thick porous plastics material to provide outward drainage. For some soils their horizontal permeability may be significantly different from their vertical permeability; this may have important implications for predicting the behaviour of a field situation where the drainage is predominantly lateral, such as the consolidation of a thin layer of alluvium under an earth embankment.

Both the tests referred to above measure one-dimensional consolidation parameters (ie no lateral strain is allowed to occur during the consolidation). This models what happens to a soil layer in nature as it consolidates as the result of subsequent deposition and is also applicable to certain field conditions such as the consolidation that occurs under the middle of a large building.

Shrinkable (expansive) clay
Some clays are expansive, and these stiff, over-consolidated clay deposits cover large areas of England (Figure 1.5 on page 26). The relatively extreme drought conditions experienced in 1975–76, 1983–85, 1989–91 and 1995–96 had widespread continuing repercussions through the UK because of the greater incidence of damaging foundation movements. The level of insurance claims in the UK from these causes is now approximately £350 million annually[41,42]. The problems in many cases are associated with high water demand from vegetation, in particular from broad-leaved trees. The effects of vegetation on ground conditions are dealt with in Chapter 1.5.

Clay soils contain a significant proportion of extremely small particles with diameters of less than 0.002 mm. As previously noted, many of these particles are derived from three minerals: kaolinite, illite and montmorillonite whose molecular structure tends to form small plates. Unlike coarser grained soils where the water simply fills the available voids, these small plates can hold the water within their molecular structure, much as a jelly does. An increase in moisture content forces the plates apart causing the soil to expand. Conversely, a reduction in moisture content allows the particles to move closer together causing the soil to shrink.

The volume of a clay soil can be reduced by:
- increasing the imposed loading produced; for example, by raising ground levels or applying foundation loads
- reducing the porewater pressure

Table 1.2 Field test for consistency of clays

Consistency	Simple field test	Undrained shear strength range (kPa)
Very soft	Finger easily pushed in up to 25 mm	< 20
Soft	Finger pushed in up to 10 mm	20–40
Firm	Thumb makes impression easily	40–75
Stiff	Can be indented slightly by thumb	75–150
Very stiff	Can be indented by thumb nail	150–300
Hard (or very weak mudstone)	Can be scratched by thumbnail	> 300

Table 1.3 Liquid and plastic limits for some UK clay soils

Deposit	Location	Liquid limit (%)	Plastic limit (%)	Plasticity index (%)
London Clay	Middlesex	63–82	28–30	35–52
London Clay	North Kent	88	25	63
Gault Clay	Cambridge	79–86	29–31	50–55
Glacial till	Humberside	36–42	16–22	20
Glacial till	Garston (Herts)	45–47	18–20	25–28
Alluvium	Grangemouth	62–78	26–32	34–50

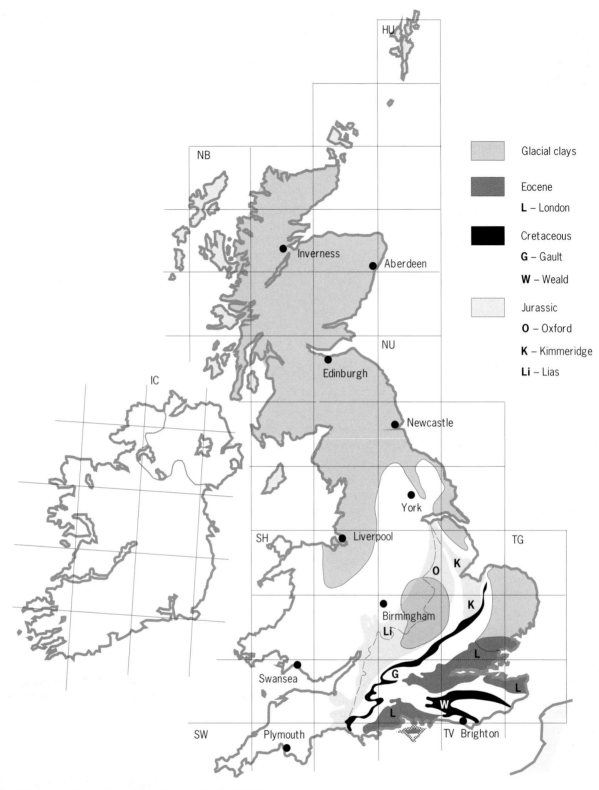

Figure 1.5 The location of firm shrinkable clays in Great Britain

Once the porewater pressure reduces below atmospheric, the water goes into a state of tension and imposes a suction on the soil particles. High suction is produced by evaporation and extraction through the roots of transpiring vegetation. Substantial suction has also been detected beneath the sites of furnaces, where the high temperatures have driven out the soil water.

If, though, the applied load is reduced, or 'free' water is made available to soil with high suction, the available moisture will move back into the soil. Whether the clay is swelling or shrinking, because of its limited permeability, the resulting volume changes can only occur slowly, often over a period of years.

For a particular soil, the volume change potential depends on both the proportion of clay particles relative to any silt or other coarser grained material and the mineral composition of the clay particles since each clay mineral has its own characteristics. Shrinkable clays usually contain a relatively large proportion of montmorillonite a mineral which has considerable potential for absorbing water and hence for changing volume.

Most of the firm, over-consolidated clays that are prevalent throughout south east England can be classed as shrinkable, examples being London, Gault, Weald, Kimmeridge, Oxford, Woolwich and Reading, Lias, Barton Clays, and the Drift Clays that are derived from them by glaciation (eg the chalky boulder clays of East Anglia). Their moisture contents, due to the effects of over-consolidation, are close to the plastic limit. Close to the ground surface, the moisture contents are influenced by evaporation, transpiration and rainfall and, as a consequence, may fluctuate from as little as 15% in dry summer weather to 40% in wet winters.

Some shrinkable clays occur further north; for example, those derived from the weathering and glaciation of Carboniferous shales around Sunderland and north of Shrewsbury. However, in the north the surface clays are generally sandy and their potential shrinkage is therefore smaller.

In addition to the firm clays, there are soft, alluvial clays found in and around estuaries, lakes and river courses. Examples of these locations are the Fens, the Somerset levels, the Kent and Essex marshes alongside the Thames, and the clays of Firths of Forth and Clyde. These clay soils have a crust of firm clay that has been strengthened by over-consolidation brought about by successive seasons of wetting and drying over previous millennia. The soil of the crust has a lower water content than the softer clay beneath. Clay shrinkage is not the only foundation problem in these areas: excessive settlement due to loading the underlying softer clay and peat can also occur[43].

For the action of vegetation including trees on shrinkable clay soils, see Chapter 1.5.

Pyritic shales
Many shales contain iron pyrites (FeS_2) which may be oxidised to sulfuric acid by *Thiobacillus ferro-oxidans*. If a source of acid-soluble calcium (eg calcite) exists within the shale, gypsum ($CaSO_4.2H_2O$) can form, forcing the layers apart and causing heaving problems. The process normally starts by allowing access of air to the shale: either by using the shale as a fill material or, in bedrock, by digging foundations, services etc[44].

Chalk
It has already been noted that 'firm chalk, requiring at least a pneumatic or other mechanically operated pick for excavation' was a deemed-to-satisfy provision not needing a foundation wider than that of the wall to be carried, irrespective of the imposed load[45].

The bearing capacities measured on loose chalk can vary considerably. BRE investigators have measured values of 400 kPa which were less than one third of the ultimate values, though settlements could be fairly high due to closing of partially open discontinuities[46].

Cohesionless (non-cohesive or granular) soils
The shape of sand and gravel grains reflects the mode of transportation prior to deposition. For example, river gravel is usually rounded or sub-angular as a result of abrasion during rapid transportation. The pieces of gravel are usually quartz or resistant pieces of the parent rock. In the UK, flint, which is formed naturally as a hard siliceous inclusion in chalk, is common in gravel deposits. Glacial gravel may be angular and may show striated faces. Residual gravel or scree slope gravel may be sharp and angular or may be flat fragments of native rock.

Most sand consists mainly of rounded or sub-angular, strong quartz grains as other softer minerals seldom survive the abrasion suffered by the quartz grains. Occasionally there may be small quantities of feldspars, hornblendes, other minerals or calcareous material consisting of shell fragments. In some areas, sand deposits are made up entirely of calcareous grains which are weaker than quartz grains.

The engineering behaviour of sand and gravel is determined largely by the nature and shape of the grains, particle size distribution, the density of packing of the grains and the effective stress level in the soil. The two most important engineering features are the high permeability and the low strength at low confining stresses.

The Model Building Byelaws, the Building Regulations 1965, 1972 and 1976, and the Building Regulations Approved Document A which followed, said that a compact gravel and sand mixture: *'Requires a pick for excavation. Wooden peg 50 mm square in cross-section hard to drive beyond 150 mm'* (Table 1.4).

Where the soil is granular, an excessive load can cause complete failure in shear, the soil being displaced laterally from beneath the foundations.

Table 1.4 Compactness of coarse granular soils			
Compaction state	Description of field test	Standard Penetration Test 'N' value	Relative density (%)
Very loose	Very easy to excavate with a spade	< 4	< 20
Loose	Fairly easy to excavate with a spade or penetrate with a crowbar	4–10	20–33
Medium dense	Difficult to excavate with a spade or penetrate with a crowbar	10–30	33–66
Dense	Very difficult to penetrate with a crowbar; requires a pick for excavation	30–50	66–90
Very dense	Difficult to dig with a pick	> 50	90–100

Peat

Peat is a decomposed organic material, usually fibrous. Most peat is extremely compressible and structures built over peat may settle by very large amounts over long periods of time.

However, although peat beds are compressible, it is still possible to build successfully on them, though movements can be considerable. As one example, BRE investigators carried out precise levelling surveys of the settlements during construction and after occupancy of a two storey school building. A major water course passed through the site. The accumulation of organic matter resulted in a layer of peat being formed within alluvial deposits. This peat layer existed beneath the entire building and had a thickness of about 1.50 m. The two-storey steel framed building was supported by shallow reinforced concrete spread foundations founded on the granular alluvial material[47].

In addition to problems of bearing capacity, it is known that the chemical constituents in peat can attack concrete.

Alluvium

The composition of alluvium can vary considerably between samples; each case needs to be examined individually. In one example, at a soft alluvial site near the Thames in East London, ground loading and water drawdown tests were conducted by BRE investigators in conjunction with a programme of careful laboratory and in situ testing. The measured settlements were predicted quite accurately from measured soil properties using a simple finite element analysis. The water drawdown test showed the ground to be fairly impermeable. The predicted long term settlements were sufficiently small to show that shallow foundations could be considered for lightweight buildings[48].

Swallow holes

Erosion of material below the ground surface can occur by natural flow of water, and caverns (more specifically known as swallow holes) can form in layers of chalk and limestone by this means. A surface layer of soil, perhaps a metre or so thick, can suddenly collapse into one of these holes. Stable conditions over an existing swallow hole may be upset by changed drainage conditions.

A similar source of subsidence, which has in the past caused considerable damage in one or two areas of the country, is the winning of salt by dissolving it in water and pumping it out of the ground, causing collapse of ground at the surface. Restrictions on brine pumping introduced in the 1950s considerably reduced the numbers of these incidents.

Archaeological and other remains

Every so often attention is drawn to archaeological investigations carried out in a hurry while building work is held up. There can be few long established urban sites in the UK which do not have some past artefacts of human habitation buried beneath them (Figure 1.6).

Remains of old buildings have been a source of difficulty over the ages. When, for example, Sir Christopher Wren first became involved in the rebuilding of St Paul's, it was said that one of his first tasks was to investigate the old foundations which criss-crossed the site[17]. Lincoln Cathedral was said to have been built over Roman remains, and even parish churches in relatively rural areas requiring domestic scale extensions to their facilities, such as that at Mancetter, Warwickshire, in 1995, will require to be archaeologically rescued either before or during the progress of building works. At this particular site extensive Roman remains were uncovered of the hitherto undiscovered original Legionary fort of Manduessedum. While it may be obvious that careful watch needs to be kept on sites in known archaeological areas, it is equally necessary whenever and wherever any excavations are being carried out in the UK[49].

Main performance requirements and defects

Strength and stability

The strength of most building materials can usually be expressed in terms of a tensile strength (eg for metals or plastics) or an unconfined compressive strength (eg for concrete or ceramics). In the case of soils, however, where the individual particles are either unconnected or only lightly bonded together, these parameters are of little relevance. Nevertheless, a soil can exhibit considerable strength; supporting the weight of a high-rise building for example. This strength is developed by virtue of the confining stresses; in other words, a mass of sand or clay is able to support large vertical loads because of the horizontal stresses exerted by the properties of the materials which surround and confine it.

A useful concept in expressing the strength of a material such as soil is shear strength, which is the maximum shear stress that can be sustained by the soil under given confining stresses. A soil may have at least three strengths:
- peak
- constant volume
- residual

These strengths depend on how much deformation has occurred to the soil. The peak value is normally described as 'failure', although the residual condition may be more appropriate for some circumstances

Figure 1.6 A medieval burial exposed during an archaeological rescue excavation

where large movements are acceptable (eg in earth slopes). Depending on the particle size distribution, the shape of particles, the density of packing of the particles and the state of confining stress in the ground, a soil undergoing shearing may try to expand or contract. Whether the volume change is up or down will alter how its resistance develops as shearing progresses. For example, after reaching a peak value a soil may lose resistance as shearing progresses. Eventually, at high strain levels, the soil reaches a constant volume, though straining is still occurring; at this point the soil is said to have reached a critical state. At extremely high strain levels, for example on the failure surface in a landslide, the strength may have fallen to its lowest possible level, the residual strength.

Properties of rock
Rock is relatively strong compared with soil. Its strength is normally expressed as an unconfined compressive strength rather than a shear strength. However, the strength and deformation characteristics of a rock mass are strongly influenced by the flaws or discontinuities within the rock mass, and often bear little resemblance to the properties of the intact rock.

Where there has been little separation or shear movement along a discontinuity, it is referred to as a **joint**. The orientation of the joint is defined by two angles, the **dip** and the **strike**: the strike is the compass direction of a horizontal line drawn on the plane of the joint and the dip is the angle between the plane and the horizontal measured in a direction perpendicular to the strike. Most rock contains sets of joints that have similar dips and strikes, so the joint sets form a regular pattern in the rock.

Joints can be produced by a variety of processes, including cooling of igneous rock, of which some of the most spectacular examples occur in the UK in the Giants' Causeway and Fingal's Cave, and in folding due to tectonic forces and in stress relief near the sides of valleys. Joints may be unfilled, filled with soft soil or filled with a cementitious material such as quartz, limonite or calcite. The strength characteristics of a joint may therefore be determined either by the properties of the rock or the infill material.

The mechanical properties of near-surface rock are also altered by weathering which results from a variety of factors; these include abrasion by wind, water and ice, differential thermal expansion of the constituent minerals, frost expansion of water in crevices and chemical action. Reaction with carbon dioxide is particularly important as it converts the primary rock minerals to secondary clay minerals, such as illite, kaolinite and montmorillonite.

Properties of granular soils
The behaviour of granular soil is governed by:
- the mineralogy of the particles
- the shape and size distribution of the particles
- the effective stress level
- the voids ratio

The mineralogy can be determined by simple mineralogical tests and the shape and size of the particles by inspection. It is usually sufficient to describe the particles as rounded, sub-angular or angular, although other shapes such as flaky (eg mica) or rod shaped may be described.

The most important properties of granular soil to be determined from a site investigation are the relative density and the angle of shearing resistance. There are four approaches to quantifying these parameters:
- by measuring voids ratio and other parameters in undisturbed tube samples of the soil. This is possible over only a limited range of fine medium sand and even in these cases may be difficult. Even if this operation is successful, however, further difficulties arise through disturbance during transportation and again when attempting to trim or extrude the soil in the laboratory to measure its volume and dry weight
- by excavating a certain amount of soil, and measuring its dry weight and the volume of the hole from which it came. This method is difficult in dry or completely saturated sand because it is difficult to form a hole without disturbing adjacent material. It is also expensive to measure the density at depths of more than 1 or 2 m because a shaft has to be sunk within which a man can work. The method is not suitable in gravel
- by using gamma radiation. A gamma radiation probe can be lowered into a casing put down into the soil, or a surface unit can be used for near-surface measurements
- by using the results of in situ tests, such as the dynamic standard penetration test (SPT) or static cone penetration test (CPT)

Properties of cohesive soils
The consolidation of an argillaceous sediment may be simply described as the process by which it is compressed from mud into a clay and finally, if the properties of the overburden allow, into a mudstone. This process is important in the interpretation of geological structure, and, on a smaller scale, it has an application in the design of foundations for engineering structures since the weight of a structure founded on clay will cause settlements due to consolidation and the magnitude of these settlements must be known[50].

Clay consolidation is accompanied by the expulsion of water from the soil pores. This will occur even under quite moderate loads and may take many years to complete. Clays can give very satisfactory foundations, but unequal loadings or irregular soil conditions can make settlement vary in amount from place to place under a building. It is this differential settlement that can cause damage.

The engineering properties of residual clay are partly governed by the high bonding stresses which existed in the parent rock so that the soil will usually be fairly stiff and

1 The basic features of sites

Table 1.5 Characteristics of cohesive soils	
Soft-to-firm clay $c_u < 80$ kPa	**Stiff-to-hard clay** $c_u > 80$ kPa
Usually lightly over-consolidated near the surface as a result of secondary or delayed consolidation, groundwater movements, influence of vegetation etc	Usually heavily over-consolidated as a result of erosion of overburden
Foundation design will almost invariably be governed by settlement where foundation width is more than 1.5 m or where excavation penetrates the surface crust of stiffer soil	Foundation design may be governed by settlement or bearing capacity
For structural foundations on deep beds of clay, immediate settlements will usually be 20–33% of total, and consolidation settlements will continue over a long period of time	Up to 75% of settlement will usually occur during loading or within a few weeks after loading
Heave in excavation will occur immediately	Heave in excavation may continue over a long period
Excavations exposing a steep face in soft ground will usually collapse very quickly. If they do not collapse quickly, they will probably remain stable over a long period of time	Excavations exposing a steep face will usually remain stable for a few weeks or months, even to great depths. Over a long period of time they may collapse, sometimes quite suddenly (eg railway cuttings in London Clay)
Immediate stability is likely to govern slope design and either undrained or drained analysis may be appropriate	Long term stability is likely to govern slope design and drained analysis should be used. Undrained analysis may be appropriate for short term behaviour if the effects of fissuring etc are taken into account
Soft cohesive soil usually gains strength over a long period of time. Stage construction can often be adopted to make use of this	Hard soil may lose strength over a long period of time
Fissuring is usually not strongly developed in soft clay, but laminations, varves or organic inclusions may govern soil behaviour to a large degree, particularly through their influence on permeability	Fissuring is often strongly developed in hard clay and may form surfaces of weakness which will govern behaviour to a large extent. Laminations may also influence behaviour
Shrinking problems may sometimes be important in soft soil but swelling will not	Shrinking and swelling problems may be particularly important in hard clay

The single factor having the greatest influence on the behaviour of cohesive alluvial soil is the degree of over-consolidation in the soil. Soft clay (ie with shear strength less than about 25 kPa) will be normally consolidated or, more probably, lightly over-consolidated near the surface as a result of groundwater movements, influence of climate or vegetation, and the effects of delayed or secondary consolidation. Most soft deposits will not have suffered very much erosion, whereas most stiff-to-hard clay (ie with shear strength greater than about 80 kPa) will have been pre-consolidated under considerable depth of overburden, perhaps 200 m or more, which will have been subsequently partly or wholly eroded away. Some common characteristics of these two types of soil are given in Table 1.5.

The shear strength of peat is variable and sometimes surprisingly high, although very large shear strains may be necessary before the full strength is developed. Usually the shear strength of peat is in the range 7–50 kPa. The permeability of peat can often be quite high.

Allowable bearing pressures

The allowable bearing pressure is the maximum intensity of foundation loading that can be safely applied to the ground without causing settlements that will lead, in turn, to distress in the building. The allowable bearing pressure is therefore a function both of the material properties of the ground and the amount of distortion that can be tolerated by the structure. Approximate values (kN/m²) for a range of ground types are given in BS 8004 and these are summarised in Table 1.6.

To determine the allowable bearing pressure more accurately it is necessary to consider both the load required to cause failure in the soil (the ultimate bearing capacity) and the deformation characteristics (settlement). The broad principles of the analysis, which is somewhat different for deep and shallow foundations, are described in the following sections.

behave as a strongly over-consolidated clay. Occasionally, however, the soil may become soft in joints subject to persistent groundwater seepage or where residual ground occupies low lying waterlogged areas. This soil may vary, therefore, from a soft to a hard material intermediate between soil and rock.

Most boulder clay is fairly stiff and over-consolidated. In general, therefore, it is a fairly consistent material of high strength. In its natural state the main problem encountered with boulder clay is in driving piles or making excavations as some layers in the material may be very hard.

A feature of varved clay is that it may be highly permeable in a lateral direction and, where exposed, be subjected to rapid erosion because of the presence of the silt layers.

Thin laminations within cohesive soils do not significantly affect the strength and deformation characteristics of the soil, but can lead to much higher horizontal permeability of the mass than expected. They can also transmit porewater pressures and, in this way, influence the behaviour of the soil. In stiff, over-consolidated clay, these laminations create planes of weakness.

1.1 Geology and topography

Qualitative descriptions of soils are not very satisfactory; for example a clay soil may be described as stiff by one person and as firm by another. With non-cohesive soils, the state of packing and other characteristics are important in relation to bearing capacity.

The behaviour under load of non-cohesive soils (eg sands and gravels) depends partly on the degree of lateral confinement of the soil. The safe bearing capacity increases if the foundation width or depth below the free ground surface is increased. It should be noted, however, that settlement may increase with greater foundation width but decrease with greater foundation depth.

Foundations for many low-rise buildings are typically 1–2 m deep and require a minimum undrained shear strength of soil of about 50 kN/m² to support a foundation pressure of 100 kN/m² with a good margin for safety. The loads imposed by a typical low-rise building on the underlying soil are likely to be in the range 50–100 kN/m². Where this is the case, only soft cohesive or peaty soils pose a problem of excessive settlement or bearing capacity failure. These soils are typically found as water-deposited alluvium in river valleys, or as Head deposits. In alluvium the groundwater level is often close to the surface. Both alluvium and Head tend to vary considerably in thickness and composition over short distances. Sands, gravels and rocks, on the other hand, are likely to offer a good layer for founding low-rise buildings.

The low strength of non-cohesive soils (eg sands) under low confining stresses leads to low bearing capacity for surface footings of narrow width. The bearing capacity is further reduced if the water table is within the 'stress bulb' under the foundations as buoyancy forces reduce the effective stresses in the ground.

The ultimate bearing capacity of a soil can be determined from measurements of its shear strength obtained from tests on suitable samples or on the site. The maximum safe bearing capacities which normally have been used in the past have been about 50% of the ultimate bearing capacity and therefore give a factor of safety of about 2. This factor of safety resulted from experience and was not determined by calculations based on assumed allowable settlement.

Table 1.6 Allowable bearing pressures for different soil conditions

Description	kN/m²
Rocks*:	
Strong igneous and gneissic rocks	10 000
Strong limestones and sandstones	4 000
Schists and slates	3 000
Strong shales, mudstones and strong siltstones	2 000
Non-cohesive soils†:	
Dense gravel or dense sand and gravel	> 600
Medium dense gravel or sand and gravel	<200–600
Loose gravel or sand and gravel	<200
Compact sand	>300
Medium dense sand	100–300
Loose sand	<100
Cohesive soils:	
Very stiff boulder clay and hard clay	300–600
Stiff clay	150–300
Firm clay	75–150
Soft clay and silts	< 75
Very soft clay and silts	not applicable

Notes:
* Assuming that foundations are taken down to unweathered rock.
† Provided the width of foundations is not less than 1 m and groundwater level is at least one foundation width below the base of foundations.

Alterations in load conditions

Any increase of load on a soil will cause some settlement. With coarse grained soils it may occur almost immediately, but with clays it may extend over many years. Greater loads may cause localised overstressing and plastic flow of the soil, with consequent large settlements; and extremely large movements can be expected if the load exceeds the ultimate bearing capacity of the soil so that shear failure occurs.

Compaction

There is a fundamental difference between the deformation behaviour of cohesive soils such as clay, and granular soils such as sand and gravel. Because the water is 'bound' to the very fine clay particles in cohesive soils, an increase in applied loading tends to force the soil grains closer together, thereby squeezing out a small amount of water. However, this change in moisture content can occur only slowly due to the limited permeability of the soil. Consequently, the initial effect of applying a load to a saturated clay is simply to increase the pore pressures without changing the effective stresses. The flow of water out of the affected area is then controlled by the hydraulic gradient generated by these pore pressures and the drainage conditions at the soil boundaries. As the water flows away from the loaded area, the pore pressures dissipate and the soil compresses; this process is known as consolidation (Figure 1.7 on page 32).

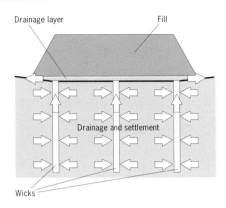

Figure 1.7 Ground consolidation by loading with fill in conjunction with drainage. The drainage water may need to be pumped out

Applying a load to a saturated granular soil surrounded by an impermeable barrier produces a similar effect: the applied load is carried entirely by an increase in pore pressure without affecting the effective stresses. Allowing the water to slowly drain through the barrier results in a gradual dissipation of pore pressure and a corresponding increase in effective stress. Only a minute flow of water is required to allow this transfer to take place because the soil grains are already in contact with one another. Moreover, while these conditions can be simulated in the laboratory, the drainage conditions and dissolved gases that exist in the porewater in the ground, allow the water to flow or to compress instantaneously. Granular soils do not therefore consolidate in the way that clay soils do. The improvement of ground by the application of static loads is dealt with in more detail in Chapter 2.1.

In addition to compression under static loads, soils can be deformed by vibration or by repeated applications of shock loading (Figure 1.8). This process, which is especially important for fill materials that often have relatively high void ratios and are not fully saturated, is known as compaction. The improvement of ground by the application of dynamic loads is dealt with in more detail in Chapter 2.1.

Different soils have different compressibility values. Some more common examples are:

- heavily over-consolidated boulder clays (eg many Scottish boulder clays) and stiff weathered rocks (eg weathered siltstone), hard London Clay, Gault Clay and Oxford Clay at depth. These have very low compressibility, with a coefficient of volume compressibility, m_v (m^2/MN) < 0.05
- boulder clays (eg Tees-side and Cheshire) and very stiff blue London Clay, Oxford Clay and Keuper Marl. These have very low compressibility, m_v (m^2/MN) 0.05–0.10
- upper blue London Clay, weathered brown London Clay, fluvio-glacial clays, lake clays, weathered Oxford Clay, weathered boulder clay, weathered Keuper Marl and normally consolidated clay at depth. These have medium compressibility, m_v (m^2/MN) 0.10–0.30
- normally consolidated alluvial clays (eg estuarine clays of Thames, Firth of Forth and Bristol Channel) at depth. These have high compressibility, m_v (m^2/MN) 0.30–1.50
- very organic alluvial clays and peats. These have very high compressibility, m_v (m^2/MN) > 1.50

Under natural conditions on a geological timescale, a clay soil undergoes consolidation as the result of subsequent deposition above it. As most soils in the UK were deposited under marine conditions, the applied vertical effective stress at any stage in the consolidation was simply the effective overburden pressure.

Many UK clay soils from the early Tertiary period have been extensively eroded, especially during glacial periods, so that what are now near-surface deposits once existed at depths of several hundred metres. These deposits therefore have pre-consolidation pressures that are far greater than their current overburden pressure and are described as heavily over-consolidated.

Soils which have not been significantly eroded are described as normally consolidated. However, the surface crust of these soils is often slightly over-consolidated as a result of fluctuations in groundwater level or desiccation. Some increase in density may also occur as a result of delayed or secondary consolidation.

Collapse of underground mines and cavities

Cases of sudden collapse do occur, often without much warning, and they are not confined to sites of old buildings. One failure under the corner of a school built in the 1950s took place as the result of a concentrated discharge of water from a soakaway into the overlying soil. In another case, one end of a swimming pool was wrecked by subsidence attributed to the collapse of the overburden into a swallow hole which, in turn, was caused by the flow of water from a soakaway taking the discharge from the pool. It is unwise to discharge any water into the ground near buildings in areas liable to feature swallow holes.

Where construction in areas known to contain underground cavities is unavoidable, appropriate foundation design should have been adopted to minimise the susceptibility of structures to movement. In certain circumstances

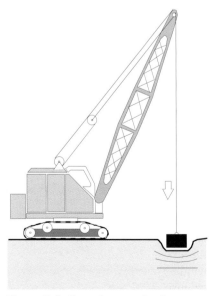

Figure 1.8 Ground compaction by repeated shock loading

– for example, where a major development had to be protected – it might also be cost effective to fill the cavities in order to reduce the potential for future movements to acceptable levels.

The principal cause of natural cavities in the UK is gradual dissolution by percolating groundwater, a process which tends to affect carbonates such as chalk and limestone. The cavities produced by this process are referred to as solution features and are typically caves or vertical pipes with a diameter of about 1 m which may subsequently have been adventitiously filled with loose soil. Collapse of the infill material leading to sudden movement of the ground surface can be triggered by the application of foundation loads or, more commonly, by a sudden influx of water; for example, following a period of heavy rainfall. Badly sited soakaways, broken drains or fractured water supplies may aggravate existing solution features and are a common cause of subsidence damage for buildings founded on chalk or limestone. The risks are greatest where the chalk or limestone is overlain by a thin layer of relatively permeable sediment such as sand. In these circumstances, the enlargement of the solution feature is likely to go unnoticed until it reaches a size where the material above it can no longer arch across the hole and it collapses creating a crater several metres in diameter.

The collapse of abandoned mine shafts or bell pits can cause similar problems to those created by the collapse of solution features. Bell pits, which date back to the thirteenth century, consist typically of an unsupported shaft 1.0–1.2 m in diameter widening to a diameter of up to 20 m at its base. Pits were sunk as often as required to exploit a seam. The depth of bell pits was generally limited to about 12 m, although some later pits were as deep as 30 m. With the advent of mechanical pumps and winding gear in the eighteenth century, deeper working became practical, and mines tended to consist of one or two vertical shafts providing access and ventilation to a number of horizontal or near-horizontal galleries.

It has been estimated that there are some 80,000 old shafts in Britain, many of which are uncharted. Many of these were loosely backfilled and covered making their location and identification difficult. The modern practice is to seal redundant shafts at the surface of the bedrock using a reinforced concrete slab whose thickness is at least twice the diameter of the shaft. In older shafts, however, the seal usually consisted of a timber platform covered with uncompacted fill materials to restore the general ground level; in some cases, the shaft was sealed using whatever materials were to hand, such as mine cars or trees. Even where the platform remains intact, there is still a possibility that failure may occur in the supporting shaft lining or surrounding rock. The collapse of the shaft or the gradual decay of a timber platform can cause a sudden failure, resulting in the appearance of a crater at ground level. Where the rockhead is near the surface and the collapse arises from the failure of the platform supporting the fill material, the resulting subsidence is likely to be severe but localised. Where the collapse has been caused by a breakdown of the shaft lining, or where the shaft is covered by a thick deposit of unconsolidated material, the subsidence is likely to be less severe but to affect a larger area.

Many old mines were left with pillars still supporting the roof of the overlying strata; these propping techniques were referred to by various terms including room-and-pillar, pillar-and-stall or post-and-bank mining. At depths greater than about 60 m, the pillars tended to collapse soon after abandonment; at shallower depths, however, the overlying strata can be supported for several hundred years, until gradual degradation, and the erosion of infill material by groundwater, eventually leads to failure of the pillars, producing a general subsidence of the ground surface. At the same time, roof strata between pillars can collapse producing a larger dome, which in turn collapses allowing the void to migrate upwards. This process continues until the void reaches the ground surface creating a plump hole, unless the collapsed material, which bulks as it breaks up, completely fills the void or the dome reaches a stable and competent stratum. The likelihood of the void reaching the surface clearly reduces with increasing amounts of cover, but is also dependent on the nature of the overlying soil. Loose, waterbearing granular deposits, such as gravels and sands, may be expected to accentuate the ground disturbance caused by mine workings, while stiff cohesive deposits, such as glacial till, are likely to provide an arching action.

Old room-and-pillar workings, particularly the deeper ones, are prone to collapse at any time, producing large and sudden deformations at the surface which may have severe effects on local buildings. A collapse may occur as a single event or sporadically over a period of time and is not necessarily associated with the application of foundation loads, although construction activity, particularly piling, can promote the conditions that lead to instability. A number of rules of thumb have been developed by engineers to assess the likelihood of a mine collapse causing surface subsidence. These include the safe depth approach, where it is assumed that no damage will result if old workings are covered by at least 15 m of rock and the 100 year rule where it is assumed that if no subsidence has occurred within 100 years, then no subsidence will occur thereafter. However, current thinking is that rules of this kind are of limited applicability and their use is deprecated.

Current and recent deep mining of coal in the UK consists almost entirely of the longwall method where the coal seam is removed completely and the gallery then collapsed deliberately. In certain parts of the UK, such as Shropshire, this technique has been in use since the seventeenth century.

Case study

Measuring foundation movements in a terrace of four houses on a clay site

A terrace block of four houses constructed on a clay site has had a history of damage dating back to shortly after construction in 1961. BRE investigators monitored the movements of this block by installing precise levelling and tape extensometer stations on the building. An electrolevel inclinometer was installed on one of the internal walls of the unoccupied building and a borehole inclinometer in its garden. The history of the site was compiled covering the period 1949 to 1973.

It was concluded that the properties had been affected by at least two periods of shrinking and swelling of the clay underlying the foundations, and more specifically related to growth and felling of adjacent trees. No direct evidence of creep of the clay slope was found though it was likely that the substantial swelling of the clay after construction had a lateral component, pointing to some sideways movement.

Figure 1.9 The only solution to saving buildings built near eroding coastlines may be to move them lock, stock and barrel to a site further inland. Here, the cliff edge has reached the building at the centre left of the photograph. A number of other buildings further up the coastline are also at risk

Longwall mining produces a wave of subsidence whose characteristics are well established and, to an extent, predictable. Damage is often associated with the lateral strains generated in the surface soil, rather than a loss of support; any building in the path of this wave experiences first a tensile force at foundation level followed, as the mine advances, by a compressive force. The surface movements associated with longwall mining are generally complete twelve to eighteen months after the advancing face has passed and it is therefore possible to carry out permanent repairs to any damaged properties at this stage. Further information is given later in this chapter under the heading 'Mining subsidence'.

Subsidence and heave

In 1971 insurance companies responded to building societies anxious to safeguard the security of their loans by adding subsidence, and later heave, to most domestic building policies. It was assumed that claims would be relatively few – indeed it was claimed at the time that there was no additional premium charged for this cover. However, although relatively few buildings are so severely damaged by subsidence or heave of the ground that major rebuilding is required, insurance policies do not specify how much movement is required for a valid subsidence or heave claim. As a result, many owners of low and medium-rise buildings have made legitimate claims for relatively minor instability and consequent damage. Typical wording in a policy might be; *'damage to buildings caused by subsidence and/or heave of the site on which the buildings stands and/or landslip.'* Cover is limited to subsidence and heave *'of the site'*, to exclude damage caused by movements within the building itself. The emergence of subsidence as a major insurance risk in the late 1970s heightened general awareness of the risks that ground activity posed to buildings of all types and sizes.

However, cracks in a building, from whatever cause, are often categorised without adequate investigation, as structural cracking caused by foundation movements. The consequences of this superficial approach can be made worse by fears of a significant reduction in the value of properties when the time comes to sell them, and the need for surveyors to avoid the risk of litigation. To quote from *Cracking in buildings*[12]: *'Court judgements have, no doubt unintentionally, reinforced defensive attitudes: the professional who judges the foundations of a cracked building to be sound may appear to be taking an avoidable risk compared with one who recommends the adoption of remedial measures, whether needed or not. The consequence of overreaction has been wastage of time and money – the number of occasions when work has been done which need not have been done far outnumber those when work has not been done which should have been done.*

An objective approach should first recognise that:
- *only a minority of cracks have any structural significance*
- *only a proportion of cracks of structural significance is attributable to foundation movements*
- *only a proportion of foundation movements produce appreciable cracking*
- *only a proportion of cracking due to foundation movements represents a worsening state of affairs'*

It is well known that later extensions to existing buildings, even if founded at the same depth, are likely to move independently of the existing buildings and produce cracks at their junctions. Less obvious is that extensions that are unheated, including garages and screen wall enclosures, may be subject to frost heave. For both reasons, therefore, it is usually considered safer not to tooth in or otherwise rigidly connect the two buildings but to provide a suitable shear joint. Any crack that subsequently forms at the junction can be flexibly sealed if necessary.

1.1 Geology and topography

Landslips
In addition to instability from movements of clay, buildings may be affected by being built on ground which is liable to progressive movement (Figure 1.9). A sinking coastline has been, and still is, the cause of a number of buildings, and even whole communities, disappearing. Of course, little can be done about coastal erosion short of temporary expedients such as massive sea defences, but where the movements are due, for example, to slippage on clay slopes, it may be possible to provide remedial measures.

The commonest cause of failure from landslips on sloping ground is an influx of water which tends to increase the total weight of the soil mass, reduce effective stresses and increase the seepage force. It follows that special care is needed in the design and maintenance of drainage systems for buildings sited on potentially unstable slopes, particularly with regard to stormwater drainage. As a general rule, soakaways should not be used on sloping ground if there is even a small likelihood of instability. Special care is also needed where sewers, water supplies and drains cross areas where there is evidence of past movement or a likelihood of even small movements in the future. In such circumstances, flexible joints are essential to eliminate the possibility of a cracked pipe feeding leaking water directly into the ground at a point on the potential failure surface.

Porewater pressure
Where questions of stability are involved, the measurement of the distribution and magnitude of the porewater pressure gives a closer insight into what is happening in the middle of a soil deposit than can be obtained by any other means. The application of pore pressure measurements can be a considerable aid to the control of construction of earthworks, and in certain special cases for the loading of foundations.

Seismic forces
Although it is not normally necessary in the UK to specifically design to resist seismic forces, there is still a slight risk of their occurrence. For example, in 1185 the Norman cathedral of Lincoln was reported by Benedict of Peterborough to have been *'cleft from top to bottom by an earthquake'*[51].

In the UK, Scotland, particularly the Highlands, is one of the areas most at risk from earthquakes, and tremors regularly occur. Fortunately they are not usually severe, though they have been known to disturb loose items on shelves. To give a few examples at random, one tremor took place in Leadhills on 14 February 1749 – *'the people left all their houses'* and the tremor *'lasted about a minute and a half'*. There was no report of damage[52]. Another tremor took place in August 1816, which was felt over large areas of Scotland and which actually *'rattled the slates on roofs in Inverness'*[53]. An earthquake lasting for several minutes was felt throughout the Highlands on 7 June 1931. Another was felt in the Western Highlands on Christmas Day 1946, and yet another in April 1952[54].

Earthquakes in the UK are not just confined to the past, nor indeed to Scotland. On 23 September 2000 there were several media reports of an earthquake measuring 4.2 on the Richter scale waking Warwickshire citizens from their sleep. There were no reports of structural damage.

Seismic loads generate similar forces (ie lateral and shear forces on walls) to wind loads but potentially over a much greater dynamic range. They are not explicitly designed for by current UK practice. Although they do occur, experience suggests that their intensity in the UK is always less than wind forces. Eurocode 8[55] provides rules for seismic design in Europe where this is necessary.

Case study

Landslips on Eocene Clay sites in London and on the Isle of Wight
BRE investigators carried out studies on a site on the London Clay slopes and concluded that the then existing 11° south slope was liable to further shallow successive rotational sliding. It was recommended that surface drainage measures be installed to slow down the rate of encroachment of the landslips, and that the behaviour of the structures and adjacent slopes be monitored to give warning of local instability.

In another case, BRE investigators were involved in the assessment of a site in the Isle of Wight where massive landslips had occurred following earlier building work. The building owner was advised that further construction on that site should not be undertaken. It will often be a matter for engineering judgement when it is safe to proceed, and the possibility must be faced, therefore, that sometimes earlier decisions on the suitability of sites prove to have been faulty.

Unwanted side effects
Cracking
A major manifestation of problems with foundations is that of differential movement, frequently accompanied by cracking in the fabric. Since cracking is a direct result of decisions on foundation design, it is dealt with in Chapter 2.

Groundborne vibration
A characteristic of loose, granular soil and, in particular, loose sand is its sensitivity to vibration. Loose sand can settle markedly under vibratory loading, such as under machine foundations. It may also suffer liquefaction, which is a state where the porewater pressures in the sand equal the total external and body stresses so that the effective stress is zero. This state can develop, for example during earthquakes, in saturated sand layers.

The effect of groundborne vibration in buildings will be a direct result of decisions on foundation design, and so is also dealt with in Chapter 2.

Principal techniques used in the UK for measuring soil strength

The laboratory vane

This is a scaled down version of the instrument used to measure strengths in situ; a typical laboratory vane has a height and width of 12.7 mm. Because of the indeterminate effects of soil fabric and variable boundary conditions, the accuracy of the laboratory vane is questionable and, as a consequence, the results are seldom used for obtaining parameters for use in design.

The direct shear test

A soil sample is placed in the box of the apparatus, which is split at its mid-height to allow the top half to slide horizontally with respect to the base and a normal pressure applied to the upper surface by means of a loading platen. The sample is then sheared by displacing the bottom half of the box and the shear load registered by a load cell or proving ring restraining the upper half of the box. The test results are normally presented as a function of shear displacement and usually the change in thickness of the sample is also recorded.

The shear box may be either square or circular in cross section and is typically 25 mm high with an area of 20–25 cm^2. While both granular and cohesive soils can be tested in the direct shear box, the sample is not sealed and porous stones must be provided to allow free drainage of the top and bottom surfaces of cohesive samples.

The interpretation of the direct shear test suffers from a number of difficulties. Nevertheless, modified versions of the direct shear test, such as the reversing shear box or ring shear apparatus, can measure residual strength parameters that may be of relevance – for example, to calculating the stability of a slope containing existing slip surfaces.

Triaxial test

The most common and versatile method of measuring the stress–strain properties of soils, and in particular the undrained shear strength of clays, is the triaxial test which has now largely replaced the shear box as the preferred method of measuring strength.

The triaxial apparatus allows the vertical stress and radial confining stress to be varied independently. Because the cylindrical soil sample is contained within an impermeable membrane, it is possible to control the drainage from the sample. With the drainage blocked off the sample can be tested under undrained conditions while measuring the pore pressure inside the sample. Hence the vertical and radial effective stresses can be determined.

Samples can be consolidated to different stress conditions and then loaded to failure under either undrained or drained conditions.

For a given sample of clay, a drained triaxial test will produce a far higher failure strength than an undrained test because the sample consolidates as a result of the pore pressures generated by the shearing. Undrained behaviour is of practical importance for saturated clay soils when applied loads change faster than pore pressures dissipate; for example, when construction is rapid.

For tests on cohesive soils, the sample can be allowed to consolidate under the confining pressure before it is tested. The test can then be performed either slowly, allowing the specimen to drain, or quickly, allowing no drainage. For undrained tests, the pore pressure inside the specimen can be monitored using a pressure cell and this enables the effective stresses acting on the specimen to be calculated. The three common forms of test are:

- unconsolidated, undrained triaxial compression test without porewater pressure measurement
- consolidated, undrained triaxial compression test with porewater pressure measurement
- consolidated, drained triaxial compression test with volume change measurement

The undrained shear strength of the soil is normally based on the results of unconsolidated, undrained compression tests made on undisturbed samples. The confining pressure is selected to ensure that any gas in the sample is dissolved, and would normally be approximately twice the estimated overburden pressure acting on the sample when it was in the ground.

Alternatively, the undrained strength can be based on consolidated undrained triaxial compression tests where the sample has been consolidated back to the estimated in-situ over-burden pressure. These tests are normally performed at a slower strain rate than the unconsolidated tests to allow the pore pressure to equalise so that meaningful measurements of the effective stresses acting on the specimen can be made.

Because of the influence of soil fabric on strength, the size of the sample can also have a significant effect on the shear strength determined in the triaxial test; as a general rule, larger specimens will contain more discontinuities and will therefore yield lower strengths.

Mining subsidence

Subsidence occurs when mine workings collapse, and any structure connected to the ground surface has to undergo the same deformation as the ground itself. Very irregular settlements can occur. If a geological fault exists in the area it is likely to be set in motion by mine working, and any building that spans across the fault may show spectacular distortions.

Although the incidence of subsidence due to mining operations is expected to reduce in proportion to the massive reduction in coal mining operations in the UK in the 1970s and 1980s, there is evidence that old workings were not always backfilled when collieries closed, and that a risk of subsidence will continue for a considerable number of years.

The design of foundations in mining areas is a matter for the specialist using details of past and future workings.

Satisfactory buildings have been designed for areas of mining subsidence by making them either strong enough to withstand the strains in the ground or flexible enough to accommodate them. A significant development of the latter approach was in the development of foundations for the Consortium of Local Authorities Special Programme (CLASP) system of school buildings initiated by the then County Architect of Nottingham, D E E Gibson in 1955. *Following extensive discussion with the Building Research Station and other bodies, he demonstrated the suitability for such sites of a pin jointed, single bolted, light steel frame erected on a slab cast in separate sections but continuously reinforced with a light mesh in the middle of its thickness, and with no deep foundations. A primary school was built in 1958 near Nottingham in a period of six months. It was built on a 5 in thick slab, lightly reinforced, and separated into sections 10 ft × 10 ft*'[56]. Since that time, a considerable number of buildings have been built on the same principle.

Guidance on the design of foundations for buildings in areas subject to subsidence was consolidated in a report, *Mining subsidence*[57], published in 1959 by the Institution of Civil Engineers, where guidance was given on the siting, design and construction of buildings, and on precautions which needed be taken. Following the publication of this last report, *'it became possible in some areas, by the incorporation of modern methods, to erect buildings several storeys high without undue risk'*[56].

Testing

BS 1377 specifies standard procedures for carrying out most commonly used soil tests. Part 2 of the Standard[58] describes classification tests for soils.

Laboratory tests

The principal techniques which have been and are being used in the UK to measure the strength of soils and rocks are:
- the laboratory vane
- various forms of direct shear and triaxial apparatus

See the feature panel opposite.
Less frequently used in relation to buildings, and therefore not described here, are:
- the California Bearing Ratio (CBR) apparatus
- the Franklin Point Load Test

In situ soil testing

The soil can be tested in situ, thereby removing the effects of sampling disturbance (Figure 1.10). There is a wide range of in situ tests available; those that are commonly adopted for site investigation in the UK are described below.

The Standard Penetration Test or SPT is the most common form of in situ test. The equipment for the SPT consists of a 63.5 kg hammer which drops 760 mm onto an anvil and drives a thick walled, split barrel sampler via the drive rods into the base of the borehole. The test is performed by counting the number of blows required to drive the sampler 300 mm after a seating drive of 150 mm. Considerable care must be taken when assessing the results of an SPT since the values obtained are not free from operator influence. Furthermore the equipment made by different manufacturers can give slightly different results.

Cone penetration tests and soundings
A sounding device is essentially a sharpened rod which is hammered into the ground; the number of blows that are required to drive the point a given distance is then a measure of the strength of cohesive deposits or the density of granular materials. A typical lightweight sounding device is the Mackintosh Probe which consists of a sliding hammer of 4 kg falling 0.3 m which drives a 27 mm diameter cone into the soil. The number of blows required to penetrate each 0.3 m interval are recorded, thereby producing a profile of number of blows with depth. This tool is particularly useful for identifying the depth of soil overlying rock, locating a competent stratum at shallow depth, assessing the extent of a soft layer on a site, or extending the cover of more expensive tests by calibration and correlation.

The latest cone to evolve is the electric cone which can continuously measure friction and cone resistance as it is pushed into the soil at a constant rate of penetration. Some electric cones have been fitted with transducers to measure porewater pressures generated in cohesive soils (these devices are normally referred to as a piezocones) thereby improving the profiling capability.

Vane shear test

The vane shear test is an in situ test for measuring the undrained shear strength of soft or firm cohesive soils. A deep cruciform shaped blade is pushed into the soil to a depth equal to at least three times the blade height, and the torque required to rotate it at a slow constant angular velocity is measured by the torque head. The dimensions of the blade and the maximum torque recorded are used to calculate the undrained shear strength of the soil.

Figure 1.10 Soil conditions under test

Long term monitoring of unstable slopes

Slope stabilisation is expensive and, in cases where there is no risk of a sudden irresistible movement, it is often preferable to monitor the rate of down-slope movement to determine the likelihood of the movements causing unacceptable damage to buildings before deciding on the scope of remedial work.

Slip indicators

The simplest way of detecting movement is the slip indicator which consists of a small diameter tube grouted into a pre-drilled hole. Any movement causes the tube to kink and the depth of the slip surface can then be determined by lowering a weight down the tube until it meets the obstruction; a second weight attached to a thin cord can be left down the hole to determine the thickness of the shear zone. The disadvantages of the slip indicator are that it may fail to detect small movements and that it gives no measure of the amount of movement that has occurred.

Inclinometers

A more satisfactory, but expensive, device for monitoring down-slope movements is an inclinometer which consists of a 1 m long torpedo instrumented with an extremely sensitive strain-gauged pendulum. The torpedo is fitted with small wheels at each end, which locate in specially slotted aluminium or plastics tube (or tubes) that is permanently grouted into a pre-drilled hole. Readings are taken at 1 m intervals as the torpedo is lowered down the tube, which enables the deviation of the tube from the vertical to be computed. The tube normally has four equally spaced slots which allow the inclinometer to profile the verticality of the tube in two orthogonal planes. The torpedo can also be reversed to average out instrument errors. The main disadvantages of the inclinometer are the cost of both the boreholes and the demountable instrumentation, and the risk of the tubes being distorted and rendered impassable by even small ground movements.

Piezometers

In addition to movement, information is sometimes required on the porewater pressures in the fill; for example, to enable a full stability analysis to be performed or to determine the rate of consolidation under a surcharge. Measurements of this kind can be made with piezometers. The simplest piezometer is the Casagrande standpipe type which consists of a plastics or metal tube with a porous tip. These can be driven into soft soil. Alternatively, a standpipe can be installed into a borehole and the filter formed by pouring a layer of sand around the tip; the remainder of the borehole can then be filled with a cement and bentonite grout placed via a tremie pipe. The water rises in the standpipe until it reaches its equilibrium level. The piezometer is read simply by lowering an electrical dipper down the standpipe; the dipper buzzes when it reaches the level of the water.

While the standpipe piezometer is adequate for measuring extremely slow changes in groundwater level, it cannot be used to measure rapid changes in porewater pressure and, in general, is not suited to remote or automated reading. Therefore there are a number of other piezometer types that have faster response times and which can be read remotely. These include electrical pressure transducers fitted with porous filters, and hydraulic and pneumatic devices. The method of operation for the hydraulic and pneumatic devices is essentially similar, and both consist of a porous tip connected to a diaphragm; an increasing amount of de-aired water or dry nitrogen gas is then supplied to the opposite side of the diaphragm until it is forced off its seating, letting the fluid return to the readout unit via a second connection. While the response time of all these devices is far quicker than a standpipe type, they are more susceptible to false readings as a result of desaturation of the porous tip. To combat this, a ceramic filter with a high air-entry value is normally used.

Considerable specialist monitoring has been undertaken in geotechnical and foundation engineering which can be applied to monitoring superstructures. Further information is available[59]; for example, the circumstances and stages that monitoring is appropriate, the practical issues of how to proceed in terms of where instrumentation should be situated, and what instrumentation and techniques should be used.

Two instrumentation techniques, both unique to BRE, have been in continuous and long term use. The first uses vibrating wire strain gauges to monitor the buildup of loads in bored, cast-in-place pile foundations, or the pressures acting beneath piled rafts. The second technique is based on electrolevels. Both instrument types can be installed with little disruption to normal working routines. Moreover, they are relatively cheap and can be readily removed and replaced if they malfunction[60].

Work on site

When a large area of ground is to be covered by proposed building work, it is particularly important to know in advance the variations that exist in the site conditions. The foundation design may be varied to suit different parts of the site, and sometimes it will be desirable to plan the location and shape of the building to take account of the variations in the site conditions. These investigations must also take account of any signs or records of flooding, and of the existence of swamp or marsh conditions.

Examining the site

A close examination of the condition of existing buildings on a site, together with a knowledge of the details of their construction, particularly of the foundations, should give a good indication of the feasibility of constructing extensions. However, it should not be assumed that ground conditions are consistent over the whole site.

The surest way of determining the types of building and foundations that can be erected on a site is by using boreholes and trial pits; undisturbed samples are taken of cohesive soils such as clays and silts, and penetration or load tests can be made on non-cohesive soils. The extent of the investigation will depend largely on what is already known about the site from local sources and the examination of nearby buildings. The British Geological Survey can usually indicate the nature of the underlying strata; if these are strong enough, it will only be necessary to carry boreholes into a recognisable stratum.

Inadequate site investigation increases the risk of unforeseen ground conditions being discovered during the construction phase of a development. These discoveries inevitably increase both the cost of the development and the overall construction time.

The most important part of any site investigation is a careful, accurate description of the ground

1.1 Geology and topography

on which the building is to be constructed. Useful information on the likely ground behaviour can be gained from a good description of the soil or rock encountered. To obtain a proper description of the ground, the soil must be examined in its natural state; trial pits and, where appropriate, boreholes should therefore be excavated to a layout plan proposed in advance. The investigation should be based on knowledge gained from the desk study and walk-over survey. It must take into account the variations in the topography and local geology of the site, together with the anticipated layout of the building and its foundations. Further information is contained in BRE Digest 383[64], which also describes how to make an accurate description of soil. It does not consider the description of rocks in any detail, though many aspects of soil description are also applicable to them.

Boreholes and, in particular, shallow trial pits can provide economic and versatile ways of examining and assessing in situ soil conditions; they are further described in BRE Digest 381. The technical requirements of the investigations, rather than the cost, should be the controlling factor in the selection of the investigatory methods.

In addition to the relatively simple and direct investigation techniques such as digging pits, there are other far more sophisticated techniques now coming into the repertoire of the geotechnical engineer. An example of one of these techniques is surface wave velocity which is finding increasing use as a method for determining moduli of the ground in situ at low strain. Measurements are made of surface waves generated in the ground by a vibrator. In this technique a correlator is used to measure the time delay on the passage of the waves between a pair of geophones. The signal from one of the geophones is filtered to select the frequency required. An accuracy similar to that of a spectrum analyser

---

Essential information to be obtained from a site investigation

Slope angles
These can be obtained more accurately during the site visit than from Ordnance Survey maps. An Abney level or a Doctor Dollar clinometer can be used. Slope angles may be interpreted in terms of the types of materials underlying the site; for example, rock stands steeper than clay. Very flat ground near streams or rivers is probably associated with soft or loose alluvium.

Instability
Hummocky, broken or terraced ground, or boggy, poorly drained conditions on hill slopes may be associated with landslips. Except in very windy conditions on exposed sites, trees normally grow vertically; ground movements may cause them to change direction. Kinks in hedge lines may also be a sign of past movements.

Vegetation
If the site is on shrinkable clay, a complete record of the positions of all trees and shrubs is needed, together with their approximate sizes, heights, and girths and, if possible, their species. The position and size of hedges should be recorded, together with the absence of any trees, shrubs or hedges indicated on previous records or aerial photographs.

Made ground
Areas that may have been previously filled should be noted; this can be done by comparing the available Ordnance Survey maps and aerial photographs with what can be seen on the site.

Structures
Structures in the area and on the site should be examined and any damage recorded. Where possible, information on the types of foundations commonly used in the area and of any problems encountered by other builders should be obtained. The location of any structures that are marked on maps or aerial photographs, but which no longer exist should be recorded.

Soil and rock
Exposures of soil or rock (eg in railway cuttings) should be recorded and sampled. Alternatively, shallow holes can be excavated using an auger to obtain some idea of the characteristics of the soil.

Groundwater
The positions of springs, ponds and other water should be noted. The absence of these features, which are shown on Ordnance Survey maps, geological maps and aerial photographs, may indicate that ground levels have been raised by placing fill on the site. If holes are made to examine soil conditions, the presence or absence of groundwater should be also noted.

Mining and quarrying
Signs of mineral extraction in the area should be recorded. These may include old mine buildings, derelict or hummocky land, surface depressions, evidence of infilling and spoil heaps.

Solution features
Land underlain by chalk or limestone may contain naturally occurring voids or pipes filled with soft soils which can collapse or settle under a structure. This type of ground is often associated with dry valleys and surface hollows or with areas where streams disappear into the ground.

Access
Ease of access for drilling rigs or hydraulic excavators which might be needed for detailed ground investigation or construction work. Photographs are useful of the condition of gates and tracks which this plant might be required to negotiate, so that any subsequent damage caused can be properly quantified.

Local enquiries
Valuable information can be obtained by talking to local people and consulting sources of reference material; these include libraries, muniment rooms, county archives, local history societies, natural history societies (who may have information on local geology), planning authorities, universities and polytechnics.

Trial pits

For making an accurate assessment of the soil in situ, a trial pit is often the most suitable method. Trial pits can be excavated to depths of about 5 m, either by hand or by using an excavator with a hydraulically operated back-hoe, always remembering that the excavation must be properly supported to comply with BS 5573[61] and to allow personnel to safely inspect the undisturbed soil strata. It is important to record all information and to identify the features in the soil profile that will have the greatest effect on possible building foundations. These include such features as reduced level, stratum depths, visual log, sketch of pit, soil strength or density, full soil description, samples taken, ease of excavation, ingress of water, date of excavation, pre-existing slip surfaces and stability of the sides (BRE Digest 318[62]).

It is important to assess the variability of each stratum and the major features (eg fissures and fissure spacing, cementation of particles, particle size etc). These will have a considerable effect on the interpretation of laboratory and in situ soil test results. For example, a clay with wide fissure spacing relative to the size of the test device will give variable results, whereas the same test in a highly fissured clay will give more consistent data. Disturbed samples of soil may be obtained from different depths and sent to a laboratory for index testing. Each sample should be put in a suitable plastics bag or container and fully labelled with site, trial pit number and depth. A trial pit also allows small scale in situ tests to be carried out in the sides and base of the pit. In cohesive deposits, hand vane and hand penetrometer tests can be used to obtain a provisional assessment of strength.

Case study

Typical trial pit log

A trial pit was excavated by BRE investigators in clay soil[63] (Figure 1.11). The following facts were recorded: job location, date, investigators, weather, excavator type, bucket width, pit support system, pit stability and groundwater observations.

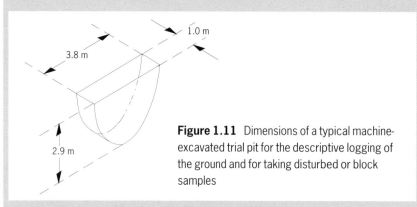

Figure 1.11 Dimensions of a typical machine-excavated trial pit for the descriptive logging of the ground and for taking disturbed or block samples

The inspections revealed the following information relating to soil conditions and the depth below ground level at which they were observed to occur:

Depth (m)	Soil description and sample depths
0–0.30	Grey–brown friable silty clay with fine rootlets and frequent rounded medium gravel (top soil)
0.30–1.35	Soft-to-firm brown, closely-fissured silty clay with occasional rounded flint gravel fragments 0.30–0.50 m. Near vertical cracks to 1.10 m, and occasional fine rootlets to 1.3 m (Head). Disturbed samples at 0.30–0.50 m and 1.00–1.20 m
1.30–1.50	Polished and striated shear surface, dipping at about 0.8°. Striations at N009°E
1.35–2.90	Firm brown closely-fissured silty clay. Some blue–grey staining on fissure surfaces. Fissure spacing increasing to 20–100 mm at 2.20 m and below, with occasional selenite crystals 2.1–2.4 m (London Clay). Disturbed samples at 1.50–2.00 m and 2.50–2.80 m

is obtained with a fair resolution at low frequencies. Damping is measured by using the instrument in the auto correlation mode. The measurements have been used by BRE investigators to calculate settlements for a site on London Clay. On another site on London Clay, the values obtained in situ using surface wave techniques were compared with the values obtained from laboratory samples using the bender elements technique. Provided the in situ stress state and history are taken into account, the values of very small strain stiffness obtained from laboratory tests are well within the experimental scatter of measurements from the surface wave surveys[65].

Questions to be posed in a desk study of the topography, vegetation and drainage site conditions are:
- does the site lie on sloping ground and, if so, what is the maximum slope angle?
- are there springs, ponds, or watercourses on or near the site?
- are, or were, there trees or hedges growing in the area of proposed construction?
- is there evidence of changes in ground level (eg by placement of fill) or of the demolition of old structures?

1.1 Geology and topography

Exploratory boreholes

The depth of boreholes needs to be related to the type of foundations being proposed. Two of the most common cases are illustrated in Figure 1.12.

Using a hand auger is the simplest and cheapest of the boring methods. Hand augers are generally available in diameters up to 200 mm (Figure 1.13). Provided the soil is not too stiff, does not contain boulders or large gravel and does not collapse, then depths in excess of 5 m are easily achievable and an assessment of the soil profile with depth may be obtained. In cohesive soils, small-diameter (38 mm) sample tubes may be driven into the base of an auger hole to obtain a relatively undisturbed soil sample. In free-draining soils, an indication of the water table level may be obtained.

Deeper exploratory boreholes are generally constructed using a rig with one or a combination of three methods:
- lightweight percussion boring (Figure 1.14)
- power auger boring
- rotary core drilling

The first two methods are used for boring through soft-to-stiff soils; the third method is normally reserved for stiff soils and rocks. Of these three methods, the equipment for lightweight percussion boring is the most readily available and cheapest to hire. The rig makes it easy to work in remote and inaccessible areas and can be used to operate equipment for obtaining soil samples and in situ tests.

Disturbed soil samples for soil index tests and identification can be obtained from the boring spoil and put into suitable plastics bags or containers, and each fully labelled with site, borehole number and depth. So-called undisturbed samples suitable for soil strength tests require the use of special samplers, the most common of which is the U100 (undisturbed 100 mm diameter) open drive sampler. This sampler is lowered to the bottom of a suitable cleaned borehole and driven into the soil with a sliding hammer; it has a detachable cutting shoe.

For higher quality samples in medium-to-stiff soils, a thin-walled, open tube sampler would be used. With this type, the wall of the sample tube is much thinner and has a ground cutting edge, thereby reducing soil disturbance. The soil disturbance is further reduced by pushing, rather than hammering, the sampler into the base of the borehole. This is done by using a ground frame and a series of pulleys or an hydraulic ram.

For higher quality samples in soft-to-medium stiff soils a piston sampler would be used. This sampler also uses a thin walled sampling tube, but incorporates a piston to aid recovery and to stop the ingress of debris while the sampler is lowered to the bottom of the borehole.

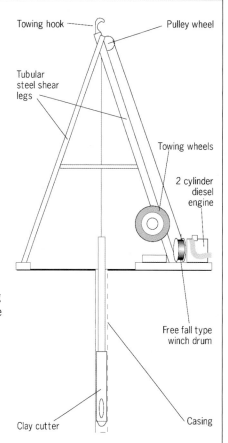

Figure 1.14 Lightweight percussion boring rig

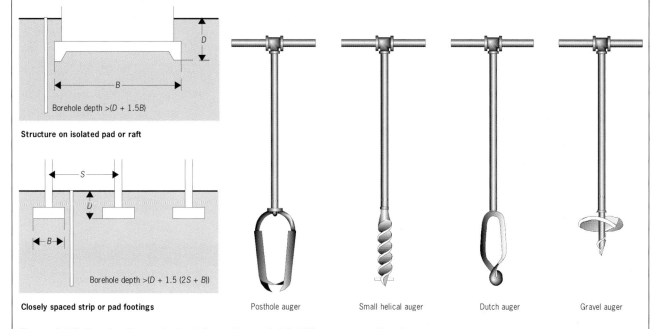

Figure 1.12 Required borehole depth for two common types of foundations

Figure 1.13 Different types of hand auger

Case study

Investigation of the site for a factory

A site investigation was carried out for a proposed factory development on a site in north east England; one part of the site was intended for immediate development, and the remainder was to be developed later. The geological survey indicated that the soil was soft alluvial material associated with a nearby river, overlying heavily consolidated Keuper Marl. The site was reasonably flat, and the whole of it was covered with the same rank grass and weeds which indicates that there were no marked changes in the types of underlying soil.

The borehole investigation was begun along two lines across the area to the north of a road, and soil profiles were obtained. The first seven boreholes revealed a thick lens of peat, with soft clay above and below it, but there was no sign of the red Keuper Marl, even in the deepest borehole which went down over 10 m. The first borehole along the second line was deeper, and finally reached Keuper Marl at a depth of about 15 m. It was found that the surface of the Marl rose gradually until on the south side of the road it was within 1 m of the surface, but still covered with the soft alluvial clay which overlaid the whole area. Numerous shallow holes were then bored to find the boundaries of the area where the Marl was close to the surface.

Many small undisturbed samples (35 mm in diameter) were taken while the borings were being made, and their strength was measured in a field test kit. The maximum safe bearing capacity of the alluvial material was found to be only one tenth of the value given by the Keuper Marl.

This was not the whole picture, however. A soft alluvial clay contains a high proportion of water, and, while the material might not fail (in the sense of sudden movement) if its safe bearing capacity is not exceeded, the water will be slowly squeezed out under load, the clay will consolidate, and the building will settle. A few larger undisturbed samples were therefore taken from the various layers of alluvial material, and laboratory tests were made to assess the amount and rates of settlement to be expected.

It was concluded that the part of the site to the north of the road lay over an old buried channel of the river and was not suitable for the proposed development, whereas the area with Marl near the surface was capable of carrying the heaviest of the proposed buildings without appreciable settlements.

Questions to be posed in a desk study of the ground conditions are:
- what geological strata lie below the site and how thick are they?
- what problems are known to be associated with this geological context?
- is the site covered by alluvium, glacial till (boulder clay) or any other possibly soft deposits?
- what is the strength and compressibility of the ground?
- is the subsoil a shrinkable clay?
- does experience suggest that groundwater in these soil conditions may attack concrete?
- is there evidence of landslipping either on or adjacent to this site or on similar ground nearby?
- is there, or has there ever been, mining or quarrying activity in this area?
- are there coal seams under the site?

Inspection

The problems to look for in diagnostics work are:
◊ inadequacy of original site survey
◊ signs of previous flooding
◊ failure to identify shrinkable clay soils
◊ trees close to buildings

Chapter 1.2

Fill, contaminated land, methane and radon

As has already been noted, there has been a gradually increasing pressure in the 1990s to use brownfield sites rather than greenfield for building purposes. Brownfields sites, though, commonly involve residues from previous occupation which can cause problems for future development. This chapter provides a basic introduction to some of the problems which may be encountered.

The term fill is used to describe ground that has been formed from material generated as a result of human activity. Consequently fill, or made ground as it is sometimes called, differs from natural soil which has its origin in geological processes. Fill may be composed of natural soil and rock or may be formed from industrial, chemical, mining, dredging, building, commercial and domestic wastes. Fill commonly found in the UK includes opencast mining backfill, colliery spoil, pulverised fuel ash (PFA), industrial and chemical wastes including iron and steel slags, building and demolition wastes and domestic refuse. Hydraulic fill is also seen where slurry is pumped into cavities and the water allowed to drain away, leaving solid matter as a residue. Also covered in this chapter is the use of material to fill old underground mine workings.

Fill used beneath ground floor slabs is also described in *Floors and flooring*[66].

The sites of old buildings, even if the buildings were demolished many years ago, may also contain material that will affect new construction; and old foundations may cause additional cost and delay to construction owing to difficulties of demolition or excavation. Old basements may hide voids or uncompacted demolition debris. A factory or other works may have left the ground contaminated by toxic waste, presenting a health hazard to construction workers and a source of chemical attack on new foundations.

In the case of contaminated land and the disposal in land of polluting substances, constraints are to be found in legislation; for example, the various building regulations, regulations on waste disposal and pollution control, and in the Health and Safety at Work etc Act, among others. A comprehensive description of the UK legislative framework for contaminated land may be found in *Contaminated land: problems and solutions*[67].

In October 1993, the Confederation of British Industry warned of the very high cost of cleaning up all industry-contaminated land in the UK (Figure 1.15). Although difficult to put an accurate figure on the extent and price of rehabilitation, it was suggested that as much as 10% of all UK land could be defined as polluted to some extent, that it would cost £600 million even to assess the scale of contamination, and that merely containing the problem might cost £20 billion. If all the main polluted sites were to be completely rehabilitated – made fit for any purpose – the total bill could double.

Figure 1.15 A badly contaminated former industrial site

The development of filled sites can present a wide variety of problems, mostly associated with one or the other of the following two aspects of behaviour:
- the ability of the fill to support the building or other form of construction without excessive settlement
- the presence in the fill of materials which could be hazardous to health or harmful to the environment or the building

A survey of buildings on fill in England and Wales has indicated that 17% of low rise construction in the 1980s and 1990s has been on filled ground. Problems have occurred due to differential settlement. A number of causes of differential movement have been identified and these were usually unconnected with the foundation loading. The various vibrated stone column techniques have been the most popular form of ground treatment, but dynamic compaction and preloading have also been used. Reinforced strip footings and semi-raft foundations are typical shallow foundation solutions. Deep foundations, including piles, also have been adopted.

Table 1.7 Sources of ground pollution
Waste disposal
Gas works
Oil refineries and petrol stations
Power stations
Iron and steel works
Petroleum refineries
Metal products fabrication and metal finishing
Chemical works
Textile plants
Leather tanning works
Timber treatment works
Non-ferrous metals processing
Integrated circuits and semi-conductors manufacture
Sewage works
Asbestos works
Docks and railways land
Paper and printing works
Heavy engineering installations
Installations processing radioactive material

Information from Housing Association Property Mutual (HAPM) indicates that of the 2000-plus sites they inspected during the early 1990s, more than half were on clay, more than half had a significant sulfate content in the soil, just under a third were on landfill, a quarter were on shrinkable clay under the influence of trees, 1 in 10 were on contaminated land, 1 in 14 were on land with a high water table, and 1 in 25 were at risk from swallow holes. So far as sites on contaminated land were concerned, around a fifth of those schemes were considered not to have adequate precautions in place.

The development of derelict and marginal land depends on the properties of the ground and its ability to support foundations. Often the only way to ensure that the ground can perform satisfactorily is to carry out ground treatment. However, systematic testing of the ground is rarely carried out to assess its performance prior to treatment. This tends to overlook the existing properties of the ground which in many cases would have been adequate without ground treatment.

Character of fills etc

Strength and stability
No special consideration is needed where fill has been placed under carefully controlled conditions since the loadbearing characteristics of well compacted fill are likely to be superior to many naturally occurring soils. However, where the fill is poorly compacted, the allowable bearing pressure of the surface soil is likely to be limited by the potential for long term settlement in the fill. In these circumstances there are essentially three options available to the foundation designer:
- to use piers or piles to transfer the foundation loads to a deeper, more competent stratum
- to use wide strip footings or a raft foundation to reduce the bearing pressure of the foundations and to stiffen the structure against differential compression of the fill
- to carry out ground improvement techniques to increase the strength and decrease the compressibility of the fill

Which of these options is the most effective and economical will depend on the thickness of the fill and its compressibility and variability. Where the fill is variable in thickness or composition, footings may need reinforcement to help bridge over soft spots. Wide strip footings and rafts may be ineffective at reducing foundation movement on fill with a significant organic content, which are likely to settle appreciably as a result of biodegradation and self-weight compression. Pile and pier design in these soils will have to take negative skin friction into account. The practicalities of ground improvement techniques are dependent on the properties and composition of the fill material and are discussed in more detail in Chapter 2.1.

Types of sites at risk
Foundations of any kind on made ground or other hazardous ground, should be engineer-designed, and include:
- ground on clay sites with slopes greater than 1 in 10 and therefore be prone to slip or creep
- areas liable to long term consolidation of the ground, particularly where this may be made-up or reclaimed ground, or where layers of peat are encountered
- areas, such as old refuse tips, containing material that is subject to spontaneous combustion, chemical change or bacteriological decay, or which include toxic wastes
- ground near ponds, or over or near underground watercourses or culverts
- ground over or adjacent to existing services such as sewers, gas and water mains, and electricity and telecommunications cables

1.2 Fill, contaminated land, methane and radon

- areas around holes in the ground, both natural (eg swallow holes) or due to mining or quarrying; also old bomb craters and soft spots where trees have been removed
- areas liable to subsidence caused by mining or mineral extraction
- areas around wells, mine shafts etc
- old foundations or other concealed constructions
- on clay soils, close proximity to, or recent removal of, trees or heavy vegetation
- areas liable to flooding or where the water table level is above the expected level of the foundations
- areas where past experience has shown the presence of high sulfate concentrations or other naturally occurring, potentially deleterious substances – mainly in clay soils – in sufficient concentration or in circumstances that would cause damage

BS 8004[68] states that **all made ground should be treated as suspect because of the likelihood of extreme variability**. Where foundations are to be built on new or existing fill, BRE Digests 274[69] and 275[70], and *Cracking in buildings*[12] provide guidance.

Construction on existing filled areas should be preceded by careful investigation of the site. The investigation of a filled site to assess its suitability for development should include two complementary approaches: a historical review and an investigation of the ground. BRE Digest 274 describes techniques of improving the load carrying characteristics of fills and foundation design. BRE Digest 275 discusses these two approaches and describes how ground improvement techniques can be used to increase the density of a fill. The design of foundations on filled ground is also discussed.

Measurements of settlement at a number of opencast mining backfills have demonstrated that these fills, placed without systematic compaction, are usually susceptible to collapse compression on inundation. The inundation could be caused by a rising groundwater table or by surface water penetrating downwards via trenches or other excavation[71].

Contaminated land

Pressure is increasing to use land contaminated by previous industrial activities. This raises a number of issues related to the past use of building sites, including:
- hazards to the health and safety of people working or living on the site
- hazards to the environment
- uncertainties about the engineering characteristics of the site
- vulnerability of building materials to contaminants in the soil and groundwater

The NATO Committee on Challenges to Modern Society has defined contaminated land as *'land that contains substances which when present in sufficient quantities or concentrations are likely to cause harm directly or indirectly to man, to the environment or, on occasion, to other targets'*. The House of Commons Environment Select Committee has endorsed this definition.

A comprehensive list of sources of pollution is given in Table 1.7[72].

A wide range of potentially harmful substances may be associated with activities or sites listed in Table 1.7; the more common contaminants are given in Table 1.8[73].

Further guidance and information on the hazards associated with contaminated land can be obtained from British Standard DD 175[74], from *Building on derelict land*[75], and from *Notes on the fire hazards of contaminated land*[76].

Table 1.8 Types of contaminants and hazards		
Types of contaminants	**Location of contaminants**	**Principal hazards**
Toxic metals (eg cadmium, lead, arsenic, mercury). Other metals (eg copper, nickel, zinc)	Metal mines, iron and steel works, foundries, smelters. Electroplating, anodising and galvanising works. Engineering works (eg shipbuilding). Scrap yards and ship breakers	Harmful to health of humans or animals if digested directly or indirectly. May restrict or prevent the growth of plants
Combustible substances (eg coal dust and coke dust)	Gasworks, power stations, railway land	Underground fires
Flammable gases (eg methane)	Landfill sites, filled dock basins	Explosions within or beneath buildings
Aggressive substances (eg sulfates, chlorides, acids)	Made ground, including slags from blast furnaces	Chemical attack on building materials (eg concrete foundations)
Oily and tarry substances, phenols	Chemical works, refineries, by-products plants, tar distilleries	Contamination of water supplies by deterioration of service mains
Asbestos	Industrial buildings. Waste disposal sites	Dangerous if inhaled

Table 1.9 Composition of typical UK household waste	
Material	Percentage by weight
Paper	22–36
Plastics	5–12
Textiles	1–5
Miscellaneous combustibles	3–6
Miscellaneous non–combustibles	1–7
Glass	6–11
Putrescent matter	16–24
Metals	1–8
Fines (< 10 mm)	6–33
Moisture content	27–37

Landfill sites

Landfilling is the most widely used refuse disposal technique in the UK. The typical composition of refuse is shown in Table 1.9.

The hazards presented by these constituents include:
- substances harmful to health and plant growth
- combustibility of material
- chemical attack on service pipes, cables etc
- emission of flammable, toxic, asphyxiant or corrosive gases
- ground stability problems
- odour, drainage problems, leachate production and surface run-off

Probably the most significant hazard is gas emission. The age of the site has a bearing on the rate of emission. It is not unusual, though, on an old site for emissions to be reactivated by disturbance from, say, construction processes. Further information on landfill gases is given later in this chapter under the heading 'Methane etc (landfill gases)'.

Former industrial sites

Hazards most likely to be met on former industrial sites are:
- physical obstructions
- chemical or biochemical attack involving health and environmental problems, combustibility, corrosion, gas emission and leachate production
- physico-chemical expansive reactions

Of these, chemical contamination is thought to be potentially responsible for the most serious problems for materials on old industrial sites.

Sites formerly used for the manufacture of town gas from coal are among the most numerous and widely distributed examples of contaminated land. They are likely to be contaminated with a wide range of substances derived from the production and purification of gas. Similar contaminants can occur on other sites where coal carbonisation processes were operated; for example, coke ovens, tar distillation plants and by-products works. The presence of contaminants may limit the range of further uses of these sites.

A wide range of organic compounds can be present on gas works sites, the most important being coal tar which is an extremely complex mixture of many substances including benzene, toluene, xylene and phenol. Another important by-product of gas production has been ammoniacal liquor, the principal constituents of interest being ammonium salts, cyanides, sulfides and carbonates.

Sites used for breaking up redundant or obsolete manufactured items such as vehicles, electrical equipment, machinery etc are common. Contamination of these sites is due to various substances, particularly metals and oils, whose presence can significantly affect human health, plant growth, or cause deleterious effects to building materials and essential services. These hazards should have been taken into account when housing, agricultural or amenity redevelopment schemes have been built on these sites. Particular materials and substances which may create hazards include:
- potential combustibility of, for example, oil soaked soil, paper and plastics
- potential chemical attack on building materials, including service pipes and cables
- emission of flammable, asphyxiant or corrosive gases
- possible presence of radioactive materials
- problems associated with odour, site drainage and surface water run-off
- illicit disposal of hazardous wastes

Many redundant small sewage treatment facilities have become available for redevelopment, especially facilities formerly serving small communities in rural areas which become linked to larger installations. Possible hazards to human health and toxicity to plants may be encountered on these sites as well as on agricultural or horticultural land which has been treated by the repeated application of sewage sludge or effluent.

A variety of solid, semi-solid and liquid wastes may be present at sewage works sites: the range of possible contaminants is wide. Deposits of materials on land should be easily identifiable; however, the presence of wastes in lagoons, drying beds, humus tanks etc may be less obvious because, with time, their surfaces may have dried out and become vegetated, so giving the appearance of solid ground. The wastes in sludge form represent the greatest hazard, containing a wide range of metals and other elements in varying and sometimes very high concentrations. However, some of the trace elements found in sewage sludge are essential to the health of crops and animals, as evidenced by the use of sludge fertilisers.

Soil will be normally found to contain certain metals (Table 1.10).

Table 1.10 Metals normally found in soil	
Element	Normal concentration range in soil (mg/kg of dry soil)
Molybdenum	2.0
Arsenic	0.1–40
Selenium	0.2–0.5
Boron	2–100
Manganese	5–500

Concentrations in excess of the figures given in Table 1.10 may indicate contamination due to sewage or other causes. Though the processes of treatment greatly reduce their numbers, some disease-producing organisms can be present; these may include bacteria, viruses and the eggs or cysts of parasites. Pathogenic bacteria such as salmonellae may be present in sewage sludge, as also the eggs of parasites such as beef tapeworm and potato cyst eelworm. Although these hazards are likely to be small on old sites, caution is still required. Sewage sludge contains a significant proportion of organic matter which may subsequently decay to produce methane and other gases.

To identify the contamination at a former industrial site, it is necessary to undertake a thorough and properly phased site investigation. A comprehensive analytical programme should include:
- common elements and determinants such as lead, zinc, mineral oils
- those contaminants usually associated with that particular land use
- adventitious or not easily predicted contaminants

When assessing the possible risks on any site, it is appropriate to concentrate on substances known to be harmful to identified potential targets (including building materials) associated with the present or proposed use. For attack on building materials and services, a minimum analysis should include soil pH, sulfate, sulfide, chloride, ammonium ion, oily and tarry substances, phenols, mineral oils, and solvents and petroleum hydrocarbons such as benzene, toluene, ethylbenzene and xylene.

For further information, see *Performance of building materials in contaminated land*[77].

Methane etc (landfill gases)

An explosion at a bungalow in Loscoe, Derbyshire, in 1986, was believed to have resulted from the seepage of methane from a nearby refuse landfill site, and unacceptably high levels of methane required the evacuation of neighbouring houses. Following this incident, there has been an increase in the application of special geotechnical, and other, measures to protect potentially affected sites.

Fill materials containing organic matter are likely to produce methane gas as they degrade; consequently precautions in the form of natural or forced ventilation below ground floor slabs may be needed. An introduction to the problem of methane in buildings was given in *Floors and flooring*[66] (a brief summary follows). While most of the methane hazard comes from the biodegradation of vegetable material in landfill depositories, the gas may be found in excavations in, for example, coal measures in the Midlands.

The principal components of landfill gas are methane (which is flammable) and carbon dioxide (which is toxic), and so if it enters a building it can pose a risk to both health and safety.

The organic matter in landfills is decomposed by the action of micro-organisms. Decomposition usually begins before the organic matter is deposited. Initially, it is due to the activity of aerobic (oxygen-requiring) micro-organisms. As decomposition progresses, the release of CO_2 and other gases prevents the ingress of air and decreases the supply of oxygen available to the micro-organisms. Eventually, aerobic activity ceases and organisms capable of surviving under anaerobic (oxygen-deficient) conditions become dominant; other decomposition processes then commence. During the early anaerobic phase, CO_2 production increases and some hydrogen (H_2) is produced. Later the gases generated under anaerobic conditions are usually dominated by methane (CH_4) which is flammable, though the presence of other constituents such as CO_2 and residual nitrogen (N_2) may modify the properties of the gas mixtures.

Landfill gas can enter buildings through gaps around service pipes, cracks in walls below ground and floor slabs, construction joints, and wall cavities. It can also accumulate in voids created by settlement beneath floor slabs, in drains and soakaways, and in confined spaces within buildings such as cupboards and subfloor voids.

Preventive measures include very low permeability gas membranes and high permeability layers from which gas can be extracted in a controlled manner.

In order to design counter-methane measures, it is necessary to know concentrations of gases and to have an estimate of the rate of emission from a site. This is usually done by monitoring the atmosphere in specially constructed boreholes over a prolonged period. Counter-methane measures may be required to prevent migration out of a site, migration within a site, and migration into buildings. The general principles to be applied in the last case are:
- to keep methane away from the building
- to prevent it entering
- to monitor to see whether these measures have been successful

Prevention of entry usually requires a combination of a gastight seal, under-building ventilation and attention to detailed design of services, foundations etc. All counter-methane measures for landfill sites should be designed to withstand the inevitable settlement of the site.

Radon

An introduction to the problem of radon in buildings was given in *Floors and flooring*.

Radon is a colourless, odourless gas which comes from the radioactive decay of radium and uranium. Uranium is to be found in small quantities in all soils and rocks, although the amount varies from place to place. It is particularly prevalent in areas of the country where granite or limestone predominate. Radon in the soil and rocks mixes with the air and rises to the surface, where it is quickly diluted by the atmosphere. However, radon which enters enclosed spaces within or underneath buildings can reach relatively high concentrations in some circumstances.

When radon decays it forms tiny radioactive particles which may be breathed into the lungs. Radiations from these particles can cause lung cancer which may take many years to develop. In addition, smoking and exposure to radon are known to work together to greatly increase the risk of lung cancer.

Most buildings in the UK do not have significant radon levels. The problem of radon lies mainly in the south west of England, Derbyshire and Northamptonshire, parts of the Pennines, the Grampians and Highlands of Scotland, and parts of Northern Ireland.

For further guidance on remedial measures in cellars and basements, see Chapter 3 of this book and Chapter 1.5 of *Floors and flooring*.

Underground fires

Wherever combustible material is present below ground there is a possibility that fires may start and continue burning beneath the surface. Occasionally, they may break through to the surface and flames may appear, but most frequently, because the supply of oxygen at depth is limited, fires proceed slowly by smouldering or charring and not by flaming. Slow combustion may proceed for long periods with little evidence at the surface. There may be occasional emissions of steam or smoke, or vegetation may die or become blackened; however, these signs are not always evident or seen. At some sites the vegetation may be particularly lush above the fire due to the increased soil temperatures.

Figure 1.16 Installation of an impermeable membrane into a contamination containment system at a brownfield site

The main hazards from underground fires are:
- production and release of toxic, asphyxiant or noxious gases which can travel considerable distances through the ground
- subsidence of the burnt ground, causing physical damage to any buildings or other structures on or near the site, and creating hidden cavities which may make the extinguishing process or land reclamation more hazardous
- heat damage to buried structures and services (eg power supply cables)

Underground fires may be extremely difficult to extinguish. In the early stages when the area affected is limited, digging out, accompanied by dowsing with water, may extinguish the fire. Once a fire is well established, this process may increase the air supply and cause the fire to advance faster than the digging. The construction of barriers to control the spread is expensive and frequently difficult, especially if the fire is large or the combustible material is deep. In some cases grouting techniques can be used to limit the spread of combustion.

Cut-off walls

Modern disposal sites should have been constructed to contain or remove in a controlled manner any harmful gases or leachate, the liquid that accumulates in the refuse.

However, many older landfill sites will rely on the natural ground conditions to retain the gases and leachate. Unfortunately, especially for the smaller, less well controlled sites that often border built-up areas or development zones, ground conditions cannot always be relied upon to achieve the required containment. In this event, increasing use is being made of deep cut-off walls which are taken down into impermeable strata, where these exist, to prevent lateral migration of leachate and gases.

The walls are constructed by excavating a trench, usually 0.6 m wide, using hydraulic, back-hoe type excavators; these can dig to depths of about 12 m although using modern diaphragm-walling rigs, specialist contractors can install walls to depths exceeding 15 m. The trenches are excavated through a self-hardening slurry of cement, bentonite (a clay mineral that holds very large amounts of water in its molecular structure and is extremely impermeable) and a powdered slag, the residue of blastfurnace steel making. The slurry acts to support the sides of the excavation during its construction and then solidifies within 24 hours to form the permanent wall. Care is required during construction to ensure that the minimum of ground material falls into the mixture, and to avoid areas of weakness and increased permeability, particularly at the junctions of contiguous excavation panels. Additional chemicals may be added to the slurry to help prevent slurry loss and to accelerate or retard the setting time[78].

Most cut-off walls of this kind have been in use only for a relatively short period of time and it remains to be seen how reliable they prove to be in the long term. Increasingly, designers of these walls are specifying a centrally placed plastics membrane as an additional security against the wall being breached through construction joints or cracks that may form in the cement-bentonite-slag material (Figure 1.16).

Geo-membranes

Geo-membranes are flexible membranes made from a variety of polymer resins. They have a relatively low hydraulic conductivity and are highly resistant to a wide range of chemicals.

The incorporation of a geo-membrane into a slurry trench wall is often specified when high concentrations of contaminants, very aggressive chemicals and gases are likely to be encountered. The construction is similar to that described above, although the permeability of the cement and bentonite slurry is often specified at a different value since the slurry can be regarded as a means of placing and protecting the membrane during installation and in service. However, in a composite wall, the slurry still makes a significant contribution to reducing the flow of contaminants across the wall.

Following excavation of the trench under a cement-bentonite slurry, the membrane is installed by mounting the membrane panel, typically between 2 m and 6 m wide and up to 30 m deep, on a metal frame. The frame is mechanically lowered into the trench. The base seal is achieved by locating the whole assembly a predetermined distance into the base layer, ideally by over-excavating into the aquiclude to make an extended flow path past the toe of the membrane. Damage to the membrane and the inter-panel joints must be avoided.

The panel widths are joined using mechanical jointing systems; the integrity of the inter-panel joints is critical to the satisfactory performance of the geo-membrane. The inter-panel joints should be extruded from the same material as the geo-membrane and securely welded to the membrane sheets.

For shallow slurry trench walls of 4–6 m, the membrane can be installed without mechanical joints. The membrane is contained in a roll unwound progressively with excavation, and is lowered to the bottom of the trench with the use of a drag box or drag lines.

Geo-membranes are made from a wide range of synthetic polymer materials:
- butyl rubber
- chlorinated polyethylene (CPE)
- chlorosultonated polyethylene (CSPE)
- ethylene copolymer bitumen (ECB)
- ethylene propylene rubber (EPDM)
- high density polyethylene (HDPE)
- low density polyethylene (LDPE)
- neoprene (chloroprene or polychloroprene)
- polypropylene
- polyvinyl chloride (PVC)

High density polyethylene (HDPE) is the most commonly used material in the UK. It has good chemical resistance, high strength, good durability and low permeability to water vapour and gases. It is a semi-crystalline polymer, based principally on ethylene. Sheets are made by the extrusion processes and can be over 10 m wide and as long as required, subject to handling restrictions.

There is currently no specification on minimum geo-membrane thickness for slurry trench walls. HDPE is available in thicknesses 1–3 mm; the 2 mm thickness is generally used[78].

Main performance requirements and defects

Strength and stability

Problems with fill are predominantly associated with any loose or uncompacted material. A particular hazard for partially saturated fill which has been poorly compacted is the possibility of a reduction in volume on inundation, known as collapse compression.

Many partially saturated soils undergo a reduction in volume when inundated with water. This collapse compression is common in some fill materials, occurring without a change in external loading, if the fill has been placed in a sufficiently loose or dry condition. The process is caused by a weakening of interparticle bonds, weakening of particles in coarse grained fill, and weakening or softening of aggregations of particles in fine grained fill materials such as shale or mudstone.

Materials for fills

Fill that has been placed to an appropriate specification under controlled conditions for subsequent building development can be described as engineered fill. The terms controlled fill and structural fill are also sometimes used. Where an engineered fill has been used as a foundation material, the adoption of a suitable specification for material, placement and compaction plant, layer depth etc, together with adequate supervision of placement and compaction, should have ensured reasonably uniform properties throughout the fill deposit and that the required foundation performance has been achieved.

Non-engineered fill arises as the by-product of various human activities associated with the disposal of waste materials. Little control may have been exercised during placement and consequently there is the possibility, and in some cases the probability, of extreme variability. This makes it very difficult to characterise the engineering properties and predict behaviour. Where non-engineered fill is used as foundation material, problems may be experienced and considerable caution is essential. Ground treatment may be necessary prior to building development. There has been extensive research into the problems of building on fill and the viability of ground treatment methods.

There are two aspects for consideration under this heading:
- materials and techniques used to fill old quarry or opencast workings
- materials used over site reduced levels to raise ground to the underside of the planned oversite

Many clay and gravel pits, quarries and disused docks have been infilled, and these and landfill waste sites are frequently used for building purposes.

Old quarry workings

It has long been recognised that filled ground may present hazards for buildings constructed upon it due to settlement or heave (Figure 1.17). Even when the fill has been engineered with the specific purpose that it should form a foundation, movements have sometimes been unacceptably large and have damaged the structures. This may have been due to an inappropriate specification or to inadequate control of fill placement. Specifications used for engineered fill are usually those developed for highway works and are not necessarily appropriate for fills on which buildings will be constructed. A model specification has been developed for selecting and controlling fills which support structures safely without the occurrence of damaging movements[79].

The design of foundations on filled ground, even for the smallest of buildings, is particularly a task for the geotechnical engineer.

Grouting of old coal mine galleries and roadways

The usual way of stabilising or consolidating old workings is by injecting cementitious grout mixed with fly ash. Where larger cavities are filled, sand or pea gravel may also be added. These operations, which are described briefly in the following paragraphs, are normally undertaken by geotechnical or mining engineers with appropriate knowledge and experience.

The position of the grouting holes is based on the results of exploratory drilling. Typically, for collapsed or partially collapsed workings, primary grouting is carried out through 50–75 mm diameter holes at 6 m centres, followed by secondary grouting at 3 m centres to tighten up. In some circumstances, a closer pattern of grouting at, say, 1.5 m centres may be needed to form a perimeter wall.

Grouting is normally continued until the grout returns to the surface or a limiting pressure is reached. If neither of these conditions has been reached after injecting a large amount of grout, operations are suspended for 12–24 hours to allow the mix to set. Further smaller quantities of grout are then injected, allowing time for each to set, until some back pressure is recorded. If this condition cannot be achieved, further intermediate holes are drilled and injected.

Where open workings are filled, a perimeter wall is essential to minimise grout wastage. This wall is normally formed by using a larger diameter hole (75–100 mm) and adding pea gravel to the grout as it is fed into the hole under low pressure. In general, the work must commence from the deepest section of a coal seam and proceed in opposite directions around the perimeter of the site, leaving the last few holes at the upper end of the seam ungrouted until the infilling has almost been completed. Where there are workings at different levels, the shallowest seam is normally treated first, followed by the deeper seams in descending order, to minimise the amount of casing that is needed.

Figure 1.17 Backfilling of a deep opencast quarry. The 'high-wall' will present potential problems of differential settlement once the site has been developed

1.2 Fill, contaminated land, methane and radon

Filling old mine shafts
Old shafts requiring consolidation fall broadly into three categories:
- open or merely capped at the surface
- completely backfilled
- partially backfilled with the fill resting on a platform or obstruction

Empty shafts are normally consolidated by filling from the base of the shaft using suitable granular material. Sands are satisfactory for dry shafts, but coarser materials such as gravel or hardcore are preferred where there is freestanding water. Colliery waste can also be used provided it is relatively incombustible and chemically inert. It may be necessary to grout any uncollapsed underground roadways or insets to prevent the ingress of fill which could ultimately lead to a loss of support.

Shafts containing loose backfill can be injected with cement grout to fill any major voids and thereby prevent collapse. The standard method of treatment is to drill-and-case one or more holes to the bottom of the shaft and then to inject a cement based grout under pressure as the casing is gradually withdrawn up the shaft. Alternatively a rubber sleeved perforated pipe or tube-manchette can be installed. Grout is then injected via a double ended packer placed inside the pipe which allows individual horizons within the shaft to be treated repeatedly if necessary. While shafts of up to 2.5 m diameter can be treated from a single central hole, larger diameter shafts are likely to need an array of holes to ensure that perimeter areas are properly consolidated.

Where the backfill is resting on a platform or obstruction, the safest way of providing a permanent solution is to drill through the barrier down to the base of the shaft and then backfill with sand or gravel to within a few shaft diameters of the barrier. The remaining void can then be filled with a cement grout, and the fill above the barrier treated if necessary.

Every care should be taken to ensure that the grouting operations do not cause uplift of any surrounding structures. This is normally achieved by using optical levelling or dial gauges suspended from simply supported beams, although more sophisticated systems based on the use of electrolevels are now available which allow the movements to be monitored continuously.

Fill and hardcore below solid ground floors
The choice of materials for fill beneath solid ground floors and hardcore, methods of placing fills and their likely settlement, are all matters which should receive considerable attention. Problems may arise if material used for fill or hardcore is contaminated with substances which can be hazardous to the structure or to health or the environment (BRE Digest 222[80]).

Materials used for hardcore and fill were extensively referred to in *Floors and flooring*. Ideally, they should be granular, and drain and consolidate readily. They should be chemically inert and not affected by water. Sites where a considerable depth of hardcore has been used are particularly vulnerable to settlement. Where the depth of hardcore varies greatly, it is common for this settlement to be uneven. The amount of settlement can vary from a few millimetres up to 100 mm in the worst cases. As the hardcore progressively consolidates with time, it is usual for the rate of settlement to decrease.

With new-build housing it is usual now to specify a suspended floor where the depth of fill is likely to exceed 600 mm at any point.

> **Case study**
> **Movements in buildings founded on fill**
> Cracking of the brickwork, distortion of window frames and extensive superficial internal damage have been investigated in two blocks of two-storey houses founded on fill over sandstone. Voids were found beneath parts of the footings due to scour from a damaged sewer. Small earth-retaining walls collapsed and external steps parted from the wall of one house. Consulting engineers acting for the local authority had recommended that the damaged houses of each block be underpinned by piling to the sandstone. BRE was consulted on the scale of the piling necessary and it suggested that further observations be undertaken on both blocks of houses, including levelling surveys at DPC level and that a more detailed examination of the form of the cracking which had occurred be carried out. Two additional trial pits near the rear corners of the undamaged houses were also recommended. As a result of these further observations, BRE was able to suggest a more economical solution than that at first envisaged.

> **Case study**
> **Settlement of fill in an industrial building**
> A single-storey shed about 250 m long was built on fill which overlaid London Clay which varied in thickness from almost nothing to 10 m. The cavity brick walls of the building were carried on a reinforced concrete ring beam supported by short piles taken into the clay, and the concrete floor was supported by the fill. Settlement of the fill occurred in relation to the walls, and a hump caused by an underground culvert became visible. Cracks were noticed in the floors and walls soon after the structure was completed, and the structure beneath some boiler plant had to be rebuilt and underpinned within only a few years. Progressive cracking of the building and distortion of the floor structure became so serious that consulting engineers were asked to suggest remedial measures. At that time the floor had settled by amounts varying from 75–375 mm; the irregularities were caused by underfloor culverts and variations in the floor loading, as well as by variations in the thickness and composition of the fill. Even the walls resting on the piles had settled from about 100–300 mm. These large movements suggested that the bearing capacity of the piles was exceeded, and the similarity between the settlements of the floor and those of the walls indicated that the pile movements were caused by drag from the fill as it settled.
> Furthermore, the main unit of production plant had settled by amounts varying from 10–75 mm causing the bearings to run hot, power to be wasted in friction, and various components to be broken. The floor was underpinned by new bored piles connected by beams, and showed no further signs of settlement during the next few years.

Difficulties can also be met at edges adjacent to walls where deep trenches have been dug to construct the foundations, even though the average depth of hardcore is small.

In former mining areas the use of colliery shale, or in other areas of pyritic shale, as hardcore under many solid concrete floors led at one time to widespread incidence of swelling in oversites due to sulfate attack, and, consequently, a rash of major failures in slabs. Sulfates may also be present in the ground. The mechanisms were described in *Floor heave in buildings due to the use of pyritic shales as fill material*[81]. Further advice on remedial work is given in *Floors and flooring*.

Figure 1.18 Undersailing of brickwork below the DPC caused by expansion in the floor slab. The amount of movement is shown as about half a finger's length

Settlement

Fill materials usually settle after they are placed, particularly where they contain organic materials or household refuse, and can cause trouble both for foundations and for ground floor slabs.

The amount of settlement will depend on how well the fill is compacted while it is being placed, on the nature of the fill, and on the underlying ground. Fill material containing mainly household refuse, end-tipped on to mud flats, could be expected to produce very large settlements, whereas crushed stone used over rock would have very small settlements. Care needs to be taken in the choice of a suitable material and it must be compacted by appropriate equipment.

As already indicated many times, it is a characteristic of fill materials which have been inadequately compacted or placed excessively dry that, when they are inundated with water for the first time, they suffer a reduction in volume called collapse compression. The associated ground movements can have a serious effect on structures which have been built on the fill prior to inundation. Where building on a non-engineered waste fill is contemplated, the assessment of collapse potential should be one of the most important elements of the ground investigation. Where engineered fill is to be placed as a foundation for buildings, or where non-engineered fill is to be treated prior to building development, the primary objective of the specification and control of fill placing should be to eliminate collapse potential completely. Identifying and measuring collapse potential is rarely easy as it is not feasible to obtain undisturbed samples of many waste fills and the commonly available in-situ testing techniques do not correlate well with collapse potential. Following an investigation of collapse compression by BRE investigators, a methodology for identifying and measuring collapse potential in fills was proposed which included a newly developed procedure for a borehole infiltration test (*Building on fill: geotechnical aspects*[82]).

Non-engineered fills may have poor load carrying characteristics and be vulnerable to large volume changes. Appropriate ground treatment prior to building on these fills can control and limit the differential movements which might damage the buildings founded on the fills. Selection of an appropriate treatment technique requires understanding of the physical, and in some cases chemical and biological, processes which cause volume changes. The effectiveness of ground improvement techniques and the extent to which the load carrying properties of fills can be improved have been evaluated using selected case records[83,84].

Subsidence

Clays change in volume when their water content changes. The amount of this change at undisturbed sites depends on the type of vegetation, climatic conditions, and the topography. The soil dries near the surface during the summer and becomes wet again during the winter, with a corresponding shrinkage and swelling of clays. The effect of this change extends to a depth of around 1 m below grass, and to much greater depths where there are trees. Any part of a structure founded in moving clay soil above this depth will be subjected to some vertical movements and horizontal strains. Extensive damage has been caused to buildings with shallow strip footings or pads in clay soils affected by nearby trees.

Shrinkage may also be induced artificially in clay soils when heat from a building passes into the ground and dries the clay. In past years this has had important consequences in some industries.

1.2 Fill, contaminated land, methane and radon

Expansion (heave)

Some materials, such as over-consolidated clay or unburnt colliery shale, may absorb water and expand after being placed. Burnt colliery shale has also caused trouble, when used as hardcore under house floors, by expansion resulting in heaving of the floors and displacement of enclosing walls; this problem has been described in *Floors and flooring*.

Sulfate attack on the underside of a concrete groundbearing slab can occur when fill below the slab contains sulfate salts and the slab is not isolated from the fill by a dampproof membrane.

It has long been known that water soluble sulfates from hardcore materials could cause disruption to concrete floors laid directly over them. In damp or wet conditions, sulfates can migrate to the underside of the slab and react with the tricalcium aluminate found in Portland cement to form ettringite. The reaction produces expansion in that part of the slab in contact with the fill. This reaction is expansive within the concrete. The first visible signs of sulfate attack are usually some unevenness in the floor, followed by cracking and possible heave. The upper part of the slab is put in tension giving rise to a map pattern of cracking. Containment of the slab at the abutments forces the slab to distort into a domed shape which, with time, can become quite prominent.

In the worst cases the walls bounding the slab can be pushed out. This is visible as walls under-sailing the DPC or as disruption to the masonry (Figure 1.18). Where the wall is of cavity construction, the outward movement of the concrete slab can push the inner leaf towards the outer leaf without necessarily moving the latter. Alternatively, where the inner leaf is built off the slab, it can be pushed upwards.

See also Chapter 2.1 of this book and Chapter 3.1 of *Floors and flooring*.

Very low temperatures can penetrate to ground below buildings, just as does heat, and can lead to frost heave. This occurs in certain fine grained soils, by the formation of ice lenses. The reduced vapour pressure of frozen water in the soil causes water vapour, or even liquid water, to migrate from unfrozen soil to the ice, where it builds up to form lenses which may become 100 mm or more thick. These may cause the ground to swell considerably. In the UK, natural frosts do not usually cause any serious frost heaving, but artificial extraction of heat from the ground by cold stores can result in considerable damage to structures (Figure 1.19).

Reinforcement

The properties of a fill may be improved by structural reinforcement to stiffen and strengthen the fill at appropriate locations. The reinforcement may be linear, planar or in the form of a grid; they may be extensible or non-extensible. The main application of these reinforcing techniques is associated with slope stability. Two methods should be distinguished:
- reinforcement which is placed in engineered fill and compacted in layers; flexible strips of galvanised steel are laid on the layers of the fill and bolted to retaining panels at the vertical boundary of the fill
- reinforcement of natural ground or fill by 'soil nailing' can be achieved by driving rods into the ground from the surface

There are many proprietary materials and reinforcing systems and specialist advice is essential.

Figure 1.19 Damage in a cold store caused by frost heave

Unwanted side effects

A danger with fills containing combustible materials is that they may combust spontaneously. This can be caused in a new building when the temperature of the fill rises. Fires of this kind have been put out and voids filled by injecting limestone dust. Alternatively, landfill gases may be collected and removed, and may even be used as auxiliary fuels!

The performance of slotted pipes for removing landfill gas from contaminated land has been examined by BRE investigators. Various properties of the pipes were tested including the friction coefficient, the disruption of flow through the pipes and the static pressure at points along the pipes. The friction coefficient was found to be of little practical use in characterising the performance of the pipes. The pressure variation tests were carried out with the pipe in air, wrapped in geotextiles, and surrounded by fill; the last condition best simulating the conditions on a real landfill site. Results showed that the flow along a slotted pipe diminishes over its length; the rate of flow depends on many factors including the slot dimensions and the permeability of the material surrounding the pipe.

Two types of slotted pipe with different slot distributions were tested to investigate their use in landfill gas extraction. Five of each type were placed in a long trough, each surrounded by one of five different aggregates. Air was sucked out of the pipe by a fan, and pressures along the pipe measured. It was found that the pipe with the smaller open area would be better for use in landfill gas extraction since the pressure in the pipe remained higher along its length. The pressures in the pipes with larger open area were 2–3 times less than those for the pipes with smaller open area. The results showed that the less gas flowed through the pipe, the further it was from the extractor fan; also, that the use of the less open pipe in a borehole would be better as it would draw gas from further down in the well. When the aggregates were compared, it was found that the least permeable aggregate, the 10 mm single-sized, was best for extracting gas. It maintained a high pressure inside the pipe and also allowed gas extraction from further along the pipe.

Health and safety

The newly revised Standard on site investigation, BS 5930[62], contains guidance on procedures for the investigation of contaminated sites. The British Standard for soil testing, BS 1377-3[85], describes laboratory sample tests for some chemicals but not those associated with heavy pollution from industrial processes.

There may be considerable hazards to the health of operatives during a site investigation and the British Drilling Association has published a valuable guide to safe practices on these sites[86].

Contaminants

So far as health considerations are concerned, recommended good practice in the UK has been described by the DoE[73]. A set of trigger values for some of the more harmful contaminants are indicated in Table 1.11. See also DoE Notes, No 42/80[87].

These values should not be interpreted as the maximum permissible concentrations, nor do they define where remedial action is essential. The subject of trigger values is difficult to evaluate for a number of reasons: firstly, there is a wide difference in opinion among experts about the levels that pose a threat to health; secondly, for most elements, the difference in concentration between levels at which beneficial effects or toxic effects occur may be quite small.

Durability

Fill can be formed of many kinds of material having different properties and characteristics; constituent soils and subsoils show a wide variety of characteristics, each having some effect on the durability of building materials used in contact with them. In general, for most sites, materials such as concrete and brick used below ground level should not show undue problems with durability; but materials used in sites with high levels of naturally occurring aggressive agents, especially with those used close to industrial and other residues on contaminated land, are a different matter.

Table 1.11 Contaminants and health

Contaminants	Threshold trigger concentrations (mg/kg air-dried soil)	
	Domestic gardens, horticulture etc	Parks, playing fields etc
Posing hazards to health:		
arsenic	10	40
cadmium	3	15
chromium[A]	25	25
chromium (total)	600	1000
lead	500	2000
mercury	1	20
selenium	3	6
Phytotoxic to any plants but not normally hazardous to human health[CF]:		
Boron (water soluble)[B]	3	3
Copper[DE]	130	130
Nickel[DE]	70	70
Zinc[DE]	300	300

A Hexavalent chromium determined under specific conditions.
B Determined by standard Government method.
C A value of pH assumed.
D Total concentration.
E Phytotoxic effects may be additive.
F Grass is more resistant than other plants.

Some guidance is available on the performance of materials in contaminated land, and the following paragraphs are taken from *Performance of building materials in contaminated land*.

'The durability of construction materials depends primarily on their quality, in terms of both composition and manufacture. For example, high quality materials – such as dense concrete with a high cement content and low water:cement ratio, and high density polyethylene with a high degree of crystallinity are durable in all but the most aggressive conditions. As the quality decreases, the materials are more vulnerable to attack from chemicals and micro-organisms.

In general terms, inorganic chemicals are responsible for the degradation of inorganic materials and organic chemicals for the degradation of organic materials. The exceptions are acids (organic and inorganic) which attack both.

Although both inorganic and organic materials are susceptible to attack by acids, in practice the action of sulfate is probably the major cause of material degradation. In particular, materials that contain a cementitious constituent are very vulnerable to degradation through sulfate attack. The mechanism and severity of sulfate attack is highly cation dependent. Of the sulfates commonly found on contaminated land sites, magnesium sulfate and ammonium sulfate are the most aggressive, because in addition to causing expansion they react with the major cement hydrates causing the cement to lose its binding power. Acid and or chlorides enhance sulfate attack, and an environment which contains a combination of these chemicals can cause severe and rapid degradation of any cement based material.

Plastics and rubbers differ from inorganic materials in that chemicals can affect their performance without causing them to fail mechanically. An example of this type of behaviour is the permeation of plastics pipes. Permeation can present a serious problem, because it can occur at much lower concentrations than are required for degradation. The range of chemicals that can permeate plastics and rubbers is large, but in practice it has been shown that many incidents that occur involve petroleum products.

Apart from degradation by aggressive chemicals, materials are susceptible to attack by and degradation from micro-organisms. A number of different micro-organisms have been implicated in microbial attack, but bacteria are responsible for most microbial corrosion or degradation of materials placed below ground. The more important of these bacteria, with regard to microbial attack, are those related to the sulfur and nitrogen cycles in nature. The extent to which bacteria are involved in the degradation of materials is often underestimated, because as well as being involved in direct reactions which result in the degradation of materials, they can also enhance the rate of chemical attack. Bacteria aid chemical corrosion either by reducing the quality of the material to such an extent that chemical attack can take place, or by metabolising the fragments that result from chemical attack.

At present, the majority of data on the aggressivity of chemicals towards materials are qualitative. This means that professional judgement is needed to appraise the risk presented by an aggressive environment and to select a proper and adequate quality of material. Quantitative data exist in only a few cases.'

Testing

Natural soils may be excavated and placed elsewhere as fill material. In order to optimise the placement and compaction processes, and to minimise any subsequent compression of the fill, laboratory tests are available which attempt to replicate the field processes.

For tests on cohesive fill, the increase in density that can be achieved for a given compaction effort is found to be a function of moisture content; results of a typical series of tests on a clay soil can be plotted as the variation in dry density with moisture content. Results are then used to identify the two parameters of prime interest:
- the maximum dry density
- the corresponding optimum moisture content

Ideally, the fill should be placed at its optimum moisture content, although this is rarely practicable.

Two methods of compaction testing are commonly used: multiple blows with a rammer of fixed weight or, for granular soils, an electrical vibrating hammer. These compaction methods are intended to model the processes that are used in the field to improve fill or loose granular soils and the parameter of prime interest is usually the optimum moisture content which is the moisture content allowing the greatest dry density to be achieved.

The results of compaction tests are normally presented as density as a function of moisture content, and can be compared to theoretical densities for various levels of saturation.

The integrity of all welds in geo-membranes should be tested non-destructively using either approved electric spark or pneumatic (2 kg/cm^2) testing. Selected sections of welds should also be subjected to destructive testing (eg peel and shear tests) by an independent laboratory.

Case study

Slag fill at a housing site causing heave
In 1991, BRE investigators examined a housing site where the ground floor slabs had heaved, and found that the slag fill was inherently unsound. Some of the slag samples analysed did not comply with specifications in that they contained excessive amounts of sulfate and dicalcium silicate. Under wet conditions, dicalcium silicate hydrates and then carbonates when the slag is exposed to air. Reaction between carbonated slag and sulfates under wet, cold conditions results in the formation of a sulfate bearing reaction product called thaumasite. It was the formation of thaumasite (and to a lesser extent ettringite) within confined spaces which led to the severe heave problems evident at this estate and, in turn, required replacement of the defective construction. It was considered that any move to improve the drainage on the estate would probably be detrimental; minimising fluctuations in the water table at its present level may have some benefit in restricting the future rate of heave on this site.

Work on site

Workmanship

Where the major problem with a fill is associated with its loose condition, a simple remedy is to excavate the fill and re-place it in thin layers with adequate compaction. During excavation, unsuitable material can be identified and removed from the site. Where the fill is deep, it may not be practical to excavate the full depth of fill, and it is necessary to determine the required depth of excavation and recompaction. This method, like preloading with a surcharge of fill, involves bulk earthmoving and does not require specialist techniques. The re-placed fill should be 'engineered' with placement carried out to an appropriate specification under controlled conditions. The behaviour of the fill should therefore be well defined and it should be possible to make reliable estimates of settlement under working loads, including long term settlement and bearing capacity. As with preloading it is essentially a whole site treatment.

The method has wide use but there are situations in which it would be difficult to apply.
- Fill handling and storage may be difficult on a small, congested site
- Where there is a high groundwater level, excavation may necessitate dewatering the site; very wet fill will be difficult to handle
- In contaminated ground, special safety techniques and equipment may be required
- Where combustion is occurring, fill may be too hot to handle
- Where some of the fill is biodegradable and hence unsuitable, it may be difficult to find a suitable site to which it can be taken

Normal earthmoving plant is required. The type and quantity of plant selected for a particular job will depend on the following:
- nature of fill material to be excavated
- area and depth of fill to be excavated
- space available for temporary storage of excavated material
- specification for recompaction
- time available for earthmoving

In situ cleaning of contaminated land

Technologies for the in-place cleaning of contaminated soil are not well developed except where the pollution is in the form of light hydrocarbons. There is little experience of their use in the UK and are widely regarded as being expensive.

The techniques available include the following:
- chemical injection to neutralise specific contaminants
- ploughing of the surface soils to aid evaporation of volatile chemicals
- injection of steam to aid leaching and evaporation
- air sparging

Additionally there are very new techniques, under the general title of bio-remediation, involving the injection of micro-organisms to neutralise certain chemicals. Most of these techniques may be regarded as at the experimental stages of development.

Inspection

Comparison of large scale topographic maps made at different dates will allow identification of changes on the site, such as made ground, infilling of hollows or old pits or cellars, removal of vegetation, and demolition of old buildings. Maps at 1:10,560 scale from about 1840 are available, and provide a record of site development over the years.

The positions of old buildings, even if they were demolished many years ago, should be noted. Old basements may hide voids or uncompacted demolition debris. A factory or other works may have left the ground contaminated by toxic waste, presenting a health hazard to construction workers and others using the site, and a source of chemical attack on new foundations.

Chemical attack can result from the natural composition of soils, rock or groundwater in which foundations are to be placed, or from chemical waste. Many British clays contain sulfates. The likelihood of attack from contaminated ground can be determined by looking at topographical maps of different dates to try to detect infilling on the site, and by looking for evidence of old factories or works and trying to assess their potential for pollution (eg gas works have often left the ground contaminated).

The problems to look for are:
◊ the presence of substances that may be toxic to human, animal or plant life
◊ possible pollution of aquifers and watercourses by release of mobile contaminants
◊ attack of chemicals on building materials, including service pipes and cables
◊ contamination of water supplies to buildings by some of the chemicals likely to be present that can pass through plastics pipes without damaging them
◊ foundation problems from the remains of plant and buildings or from the fly tipping of demolition waste
◊ potential combustibility of materials left or deposited on the site
◊ odours

Chapter 1.3

Surface water drainage requirements, water tables and groundwater

For all practical circumstances it is not economic to design for the complete elimination of flooding of surface water drain and sewer systems, rivers and watercourses caused by periods of excessive rainfall. The concept usually employed in making the necessary provision for drainage is that of particular storm frequencies of given intensities and the frequency of flooding which results from them, and just how important it is for building occupants and for society in general to avoid the flooding. Part 2 of BS EN 752[27] recommends, for design purposes and in the absence of any other relevant criteria:
- for rural areas, a storm frequency of once in one year and a flooding frequency in drains and sewer systems of once in 10 years
- for residential areas, once in 2 years (storm frequency) and once in 20 years (flooding frequency)
- for cities and industrial and commercial areas, once in 2 (or 5) years (storm frequency) and once in 30 years (flooding frequency)

In simple terms this means that surcharging of drains might be expected to occur once in 30 years in a city centre. Of course, such flooding may not necessarily be disastrous, depending on whether the area is within the flood plain of a river, and how much rain has fallen in the catchment area upstream, and whether suitable protection has been provided (Figure 1.20).

In calculating the quantities of rainwater run-off, the use of run-off coefficients (Ψ) may be encountered. Essentially these are dimensionless values allocated to particular kinds of surfaces, depending on their nature and geometry, and ranging from 0.0 for fully permeable areas to 0.9 or 1.0 for impermeable areas.

Generally speaking, two kinds of rainfall intensity need to be considered in relation to buildings: rain falling approximately vertically and rain driven by wind. So far as the site is concerned it is the total that needs to be dealt with, particularly if it falls in a relatively short time span.

Intense rainfalls tend to be produced by convectional storms. These do not have the same 'exposed west, sheltered east' pattern of frontal type rain. On the contrary, the main influence of convectional storms is proximity to continental Europe, with the highest incidence in south east England, and the lowest in north west Scotland. Short duration rainstorms, which are those normally used in the design of rainwater disposal systems, vary in intensity for many complex reasons.

Some of the rainfall soaks into the ground and, depending on the ground's characteristics and existing state of saturation, lead to a rise in the level of the water table. If the surface of the ground cannot tolerate more rain, the run-off will need to be disposed of in the same way that drainage and disposal will need to be provided for run-off from roofs and paved areas. Much of the rainfall can be encouraged to percolate into the ground, through carefully designed filter strips and swales, or shallow declivities[28,29]. It may also be necessary to provide flood storage reservoirs to accommodate, on a temporary basis, very large quantities of rainwater which cannot drain away quickly.

Figure 1.20 River flood protection embankment

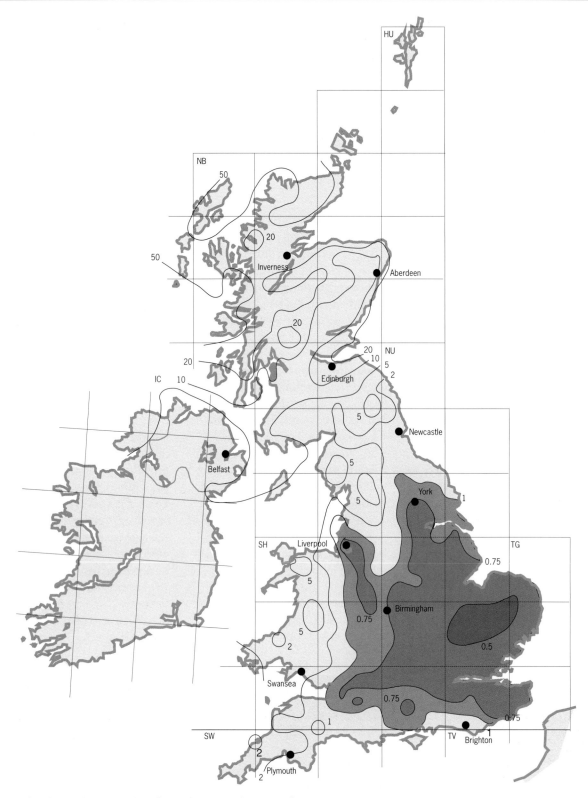

Figure 1.21 A map showing period of years between rainstorms of 75 mm/hr for two minutes (from Meteorological Office data)

This chapter deals with rainfall rates and infiltration characteristics of different soils, while the design of surface water drains and soakaways and flood storage reservoirs are dealt with in Chapter 4.3.

Since it is the cellar or basement which bears the brunt of any inundation of flood water, the aftermath of flooding is dealt with in Chapter 3.2 rather than in this chapter.

1.3 Surface water drainage requirements etc

Existing provisions for rainwater disposal

The majority of surface water drainage provision for existing buildings in the UK will have been designed using either the recommendations of BS 6367[88] or of BS 8301[89].

The quantities usually anticipated have been:
- 50 mm/hr for paved areas
- 75 mm/hr for the roof slopes and flat areas of normal buildings where ponding cannot be tolerated (Figure 1.21)

A different set of criteria for calculating rainwater run-off is recommended for the design of soakaways. A map giving the ratio of 60-minute to two-day rainfalls of a five-year return period for the UK is given in BRE Digest 365[90]. See Chapter 4.3 for simplified descriptions of the design of soakaways.

A simplified method of design for drainage from paved areas is given in BS EN 752-4[91]. Rainfall rates of 140 l/(s.ha) or approximately 50 mm/hr is used for paved areas where limited ponding can be tolerated, or 210 l/(s.ha) or approximately 75 mm/hr where ponding cannot be tolerated.

Water courses and potential for flooding

Water is shown in blue on the old 1:25,000 Ordnance Survey maps. This allows easy recognition of springs, ponds, rivers and other drainage features. Water is also coloured blue on some old six-inches-to-the-mile (1:10,560) maps; old maps may also show changes in the position of water courses when compared with more recent maps. Large scale maps, 1:10,560, 1:10,000, 1:2500 and (in urban areas only) 1:1250, may indicate high groundwater or disrupted drainage by symbols for boggy land.

Buildings built within the flood plain of rivers, and even near smaller watercourses, will be at risk of flooding in certain circumstances (Figures 1.22 and 1.23). The risk for any site can be estimated by the Environmental Agency or the Scottish Environment Protection Agency* or by the Department of the Environment (Northern Ireland) Environmental Protection Division.

There is an interesting difference between Scottish Building Regulation requirements and those in England and Wales concerning flooding. Regulation 16 of the

* The Environment Agency and the Scottish Environment Protection Agency took over responsibility for rivers and their flooding etc in England and Wales from the National Rivers Authority in April 1996, and in Scotland from the River Purification Boards and local authorities from the same date.

Figure 1.23 Flood level depth tablets at Upton-upon-Severn

Building Standards (Scotland) Regulations 1990–94 requires that a site and ground immediately adjoining a building site shall be drained or otherwise treated so as to protect the building and its users from harmful effects caused by flood waters. There is no such requirement for England and Wales.

The potential for flood damage is expected to be an issue of increasing importance in future in all parts of the UK. Currently, river flooding costs £100–200 million of insured losses each year: and the east coast of England is particularly at risk from sea surges and rises in sea level[92].

Increasingly of major importance is the protection of groundwater from pollution. The National Rivers Authority used to require that stringent measures be taken both with new waste disposal sites and with existing ones that pose a threat to water supplies, and this is still understood to be the case under the Environment Agency.

Figure 1.22 The River Severn floods parts of Upton-upon-Severn several times each year

Water tables

There is no inherent reason why buildings on ground having high water tables should not perform satisfactorily. One of the more interesting examples of a high water table is shown in the medieval defensive moat where centuries-old structures show little sign of deterioration (Figure 1.24).

With the possible exception of loose granular materials, the presence of a high water table (eg waterlogged sites) is unlikely to have a significant effect on foundations. However, it can complicate the construction process and may constrain foundation design for that reason. For example, in waterlogged soils, driven piles may be chosen because of the difficulties of excavating below the water table.

One factor that significantly affects the behaviour under load of non-cohesive soils is the level or the flow of water in the soil. Where the strength of the soil has been based on 'dry' conditions, these are given for conditions where the groundwater level is at a depth below the foundation base of not less than the width of the foundations. On the other hand, if the water table is higher than the depth of the foundations, a different value, the submerged value, should be used. The assumed design conditions for lateral confinement and water level must be able to be maintained permanently.

Water tables can vary considerably during the course of the seasons, and this variability must be taken into account in determining, for example, the potential of the soil to accommodate run-off. They may also be altered by changes in the surface of the ground by the incorporation of fill.

Water tables are normally established by the relatively simple means of excavating one or more trial holes. The Scottish Building Regulations, for example, require that the hole should be a minimum of 2 m deep, or a minimum depth of 1.5 m below the invert of any proposed drains, and left for 48 hours before the water table is noted. The height of the water table to be applied in design calculations should be the seasonally highest level encountered[93].

Water flow through the ground

The flow of water exerts a force on the soil, and must be taken into account when, for example, calculating the stability of slopes. Where the flow of water is upwards, the force can exceed the weight of the soil so that there is no contact between particles. For granular soils, the strength is effectively reduced to zero and the soil is described as quick (as in quicksand for example). Quick conditions can occur wherever the upward seepage force exceeds the submerged unit weight of the soil.

A number of other important effects and processes are influenced by water movement and permeability, including the following:
- rate of dissipation of excess porewater pressures and associated primary consolidation of clay soils
- collapse compression of loose, unsaturated natural soils and fill
- liquefaction of some saturated granular soils and fill
- loss of ground due to erosion of fine particles from permeable soils

Uncontrolled seepage can cause the migration of fine particles. Where water flows out of a fill, for example, local instability may occur at the exit point. Where water is flowing through a fill which is susceptible to erosion, the process can be prevented or controlled by protection of the fill with filters which halt the loss of fine material.

Figure 1.24 The medieval moat at Birtsmorton Court, Worcestershire

1.3 Surface water drainage requirements etc

Drainage

Soil behaviour is controlled by the principle of effective stress, and using drainage methods to control porewater pressures is important in many situations. Successful building on some types of ground will be largely influenced by, and contingent upon, appropriate drainage measures. Reduction of pore pressure can increase slope stability, reduce compressibility and reduce the possibility of erosion. There is a wide variety of applications of drainage and a number of aspects need to be considered:
- deep drainage to control groundwater level
- surface drainage to control infiltration of surface water into the ground
- slope drainage to preserve and enhance slope stability
- natural processes of evaporation and transpiration

The following are some of the techniques that might be appropriate in particular circumstances:
- vertical drains to accelerate the consolidation of a preloaded saturated clay soil; sand drains, sand-wicks and land drains have been used
- well points to de-water temporarily
- vacuum well points to accelerate the consolidation of fine grained soils; the cost of electrical power consumption can make this an expensive method
- horizontal drainage layers incorporated in an engineered clay fill during construction to accelerate consolidation

There are two general observations:
- fine grained soils usually acquire a crust of stronger material through the natural process of desiccation; the crust may be quite thin and can give a misleading impression of the condition of the bulk of the soil
- under-drainage installed prior to hydraulic filling may mean that the water table can be lowered substantially and permanently

Consolidation of ground by inundation

Most poorly compacted fill undergoes a reduction in volume when first inundated. This is collapse compression, as described earlier, and represents a major hazard for construction on fill. The phenomenon is further described in Chapter 3. BRE has carried out investigations of collapse compression at several fill sites[94,95,96]. Inundation can occur due to a rising groundwater level or to downward infiltration of surface water.

A main objective in densifying fill by one of the various treatment methods is to eliminate or greatly reduce collapse potential. It might appear that inundation prior to building would be a useful method of ground treatment. However field trials of inundation via shallow surface trenches have shown that there are often practical difficulties in achieving a controlled uniform inundation of the fill and because consistently good results are unlikely to be achieved, the method must be discouraged[83].

Testing

Permeability parameters for cohesive soils can be derived from tests (see Chapter 1.1). For granular soils, there are essentially two tests for determining the permeability of soil to water:
- variable head of water
- constant head of water

Similar tests can be performed either in the field or the laboratory. However, because of the problems of recovering granular samples, laboratory measurements are normally made on recompacted samples, which inevitably involve changes in porosity and particle orientation.

The variable head test is generally used in relatively permeable soils and should be used only in saturated soils since the degree of saturation changes during the test. It is performed by recording, over time, the change in head in a standpipe; the permeability of the soil can then be calculated.

The constant head test is performed by measuring the volume of water which flows, over time, through a cell under a constant head. The flow volume is then used to calculate the coefficient of permeability.

> **Inspection**
>
> Level ground in valleys will probably have a high water table, and the subsoil may be soft or loose and unsuitable for simple foundations. Springs, ponds or rivers may also be a sign of a high groundwater table. The movement of construction plant on such a site could be difficult and the problems of controlling groundwater in excavations should be considered.
>
> The problems to look for are:
> ◊ high water tables
> ◊ surcharging of inadequately designed surface water drains
> ◊ surface flooding of overloaded drainage systems
> ◊ collapse compression of loose fill
> ◊ liquefaction of fill
> ◊ loss of ground due to erosion of fine particles from permeable soils

Chapter 1.4 Site microclimates; windbreaks etc

This chapter describes how microclimate is affected by the geography and topography of a site and its surroundings; and how it can be further influenced by creating a sheltered environment; and, by exploiting the arrangement of buildings and landscape features, give maximum benefit from fine weather and some protection from adverse weather. This can benefit building performance by reducing energy consumption and improving durability, and can make the spaces around buildings more attractive and useful by providing better conditions for outdoor activities.

The primary concern in the UK is to mitigate the cold, wind and wet of the relatively long cool season. The intention should be to enhance, as far as practicable, the heat island effect noticeable in large conurbations, but present to some degree in all built-up areas (Figure 1.25).

General climate of the UK

The main climatic effects on a site depend on its geographical location; for example, proximity to mountains and coasts. Other books in this series have already dealt with particular aspects of the weather and its effects on buildings, particularly wind in *Roofs and roofing*[97] where a map was shown (and reproduced here as Figure 1.26) illustrating the reference basic hourly mean wind speed in metres per second. In the UK, the strongest winds are the prevailing winds. Throughout the country, these prevailing winds come from between the west and south west; the direction does not vary significantly with location. The wind blows from the prevailing direction nearly one and a half times as often as from the opposite point of the compass. The frequency of winds from other directions is between these extremes.

So far as local influences are concerned, the wind is affected by topography, proximity to hills and coasts, and seasonal factors. In the last case the highest extreme winds in the UK are expected in December and January. In the summer months of June and July, extreme winds may be expected to be only about 65% of the winter extremes; for the six month summer period from April to September, normal wind speeds are only expected to be 84% of those in the six month winter period, October to March. Tables of these seasonal factors are published for one month, two month and four month periods starting in each month of the year.

Figure 1.25 The microclimate of a particular site is influenced by many factors, including surrounding development, open spaces and vegetation. A development at London Bridge (above) was tested by BRE at the design stage in model form (below) for the effects of wind

1.4 Site microclimates; windbreaks etc

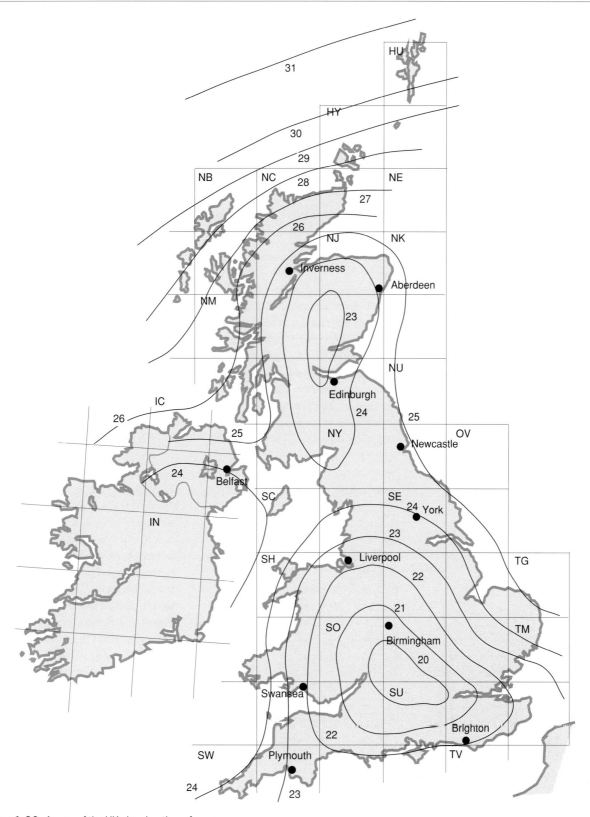

Figure 1.26 A map of the UK showing the reference basic hourly wind speed in metres per second

Temperatures experienced in various parts of the UK were also discussed in *Roofs and roofing*. The variations in air temperature from one part of the country to another are best shown as maps. Examples are printed in Chapter 1.5 of *Roofs and roofing* which illustrate the average minimum and maximum air temperatures for January and July respectively.

Seasonal accumulated temperature difference (ATD, or commonly called degree-day totals) varies by a factor of about 1.6 over the UK between south west and extreme north when values are reduced to sea level. A map of ATD totals is also shown in Chapter 1.5 of *Roofs and roofing* which illustrates just how much more important it is to achieve adequate thermal insulation in the colder northern latitudes of the UK. Totals are considerably increased at higher altitudes. Standard ATD totals give a general indication of the geographic variation in energy needed for space heating, but make only a small fixed allowance for solar gains. Therefore they do not reflect the full potential of the sun to warm buildings and the spaces around them, or of wind induced losses. The use of ATD totals to building-specific base temperatures, as used in the BRE Domestic Energy Model[98], can take account of some of these factors.

Factors influencing microclimate

The microclimate of a site is influenced by factors outside the control of the designer:
- area and local climate
- site surroundings
- shape, form and aspect of ground
- retained existing buildings
- retained infrastructure

and by other factors within the control of the designer:
- form and orientation of new buildings
- site vegetation and tree cover
- walls and other artificial windbreaks
- paving

Some of these factors act as constraints against full optimisation of microclimate, and most need to be considered jointly with other aspects of design. However, if the original decisions about plot subdivision and road layout were taken without reference to their impact on microclimate, the prospects for enhancement of the microclimate to produce a well sheltered design with good solar access may be limited. Ideally, building, landscape and other factors should be considered together as early as possible in the development of any site.

In most sites there is likely to be some compromise between solar access and wind protection. Both microclimatic requirements and potential will vary with location. This reflects, for example, ground form, land values and property type, as well as differences in area climate.

Each site will usually have some microclimatic advantages and disadvantages deriving from its general location and immediate surroundings. Developing a climatically sensitive site plan involves a range of techniques to exploit the advantages and to counter the disadvantages. As the layout of the site and buildings has to be considered as a whole, this is a matter for the initial stages of design; the potential for later intervention or adaptation is limited.

Enhancing the microclimate of a site can achieve a range of aims, both economic and environmental. Reducing the costs of space heating or cooling, together with improved durability, has an obvious, quantifiable attraction. Environmental benefits are not insignificant; they could allow a useful increase in the number of days each year when outdoor conditions are comfortable and a favourable external environment will generally improve user satisfaction.

The principal means by which the microclimate of a site can be improved include:
- allowing maximum solar radiation into the building in winter (low angle sun)
- protecting the building from excessive solar gain in summer (high angle sun)
- protecting the building from external sources of glare
- protecting the building from cold winds
- protecting people in the vicinity from wind turbulence
- protecting the building from driving rain and snow
- protecting the building from cold air drainage and frost hollows at night

It will be rarely possible to achieve a high level of enhancement over the entire area of a site, so there will usually be further choices to be made on its subdivision. For example, recreational areas can be located in the most favoured spots; garaging or refuse storage in the least favoured. Some measures involve little or no additional costs; others (eg earthworks, walls, planting and the extra land they occupy) involve both capital and maintenance costs. Also, building managers need to be made aware of the significance of these features in the site's design if the benefits are to be fully realised.

An improved site microclimate can reduce energy costs for space heating in a number of ways:
- increased air and surface temperatures in the building
- reduced undesigned air change rates
- reduced draughts
- reduced wetting of building fabric

Even if some of these gains are small, the total effect should be significant. For residential developments, average savings in heating costs of at least 5% can be expected (compared to typical site layouts), even in the more sheltered or favourable parts of the British Isles. On exposed sites where wind induced heat losses are greater, substantially larger savings should be realised. Combining microclimatic enhancement with passive solar design can make a major impact on space heating costs, especially during the longer heating season in northern parts of the UK (BRE Digest 350 Part 3[99]).

The balance of factors will differ with circumstances. Shelter from northerly winds would seem the most appropriate in general inland sites with no strong directionality. In other cases (where there is tunnelling, or on or near coasts, or where protection from driving rain is sought) other criteria could apply. In all cases the test should probably be related to how much useful solar gain is blocked by non-northerly wind protection. Where wind speeds are higher, wind protection assumes greater importance and this may justify the use of shelter in sun-blocking situations.

Solar access may be more compatible with other aims since, in many respects, it will complement the desire for sunlight, daylight and view. Wind control may produce conflicts with some visual aspects of design (eg the desire for variety in form, scale and space, and for distant views). Both solar access and wind control design for microclimate raise issues of the rights of, and constraints upon, property developers and neighbours, since one person's building or wind shelter can be another's solar obstruction.

Shelter from the wind

The United Kingdom is subject to relatively high wind speeds with many areas severely exposed by virtue of altitude, topography or proximity to the coast. Most populated areas are in low lying or sheltered locations, but a significant minority are not. In addition, new building tends to be at the edges of built-up areas, and may be on higher, more exposed ground than older property. Although built-up areas can offer substantial protection from the wind, they may contain pockets of exposed land (eg near rivers, playing fields, large roads or railways). The presence of high buildings can also expose nearby low-rise construction to high wind speeds.

In most situations it will benefit both energy economy and environmental comfort to provide as much shelter as possible from the wind during most of the year. During the hottest part of the year, air movement is desirable both for pedestrian comfort in open spaces, and to cool buildings by natural ventilation. It is difficult to fully satisfy these two aims in many parts of the UK since wind directions differ little between winter and summer. A possible solution is to arrange built form and coniferous trees to give shelter from colder but less frequent northerly winds while providing shade from deciduous trees to help counter summer heat.

Factors affecting wind speed near buildings

The wind environment around buildings and the wind pressures on them depend strongly on the roughness of the ground over which the wind passes, and on the extent of the perturbation and redirection of the air flow induced by the buildings. In smooth open countryside, wind speed increases rapidly with height above the ground. Built-up areas offer a much rougher surface to the wind: its speed increases less rapidly with height above the ground, but the flow is more turbulent. However, if the roughness of a built-up area is kept uniform and features inducing local accelerations and ground level turbulence are avoided, it is possible to create a sheltered zone in the first 5 m or so above the ground.

Achieving this kind of shelter requires attention to the form of individual buildings, their arrangement on the site, the use of hard and soft landscape elements, and the provision of wind shelter on any exposed edges of the development. Account may also need to be taken of local topography or the presence of high buildings (BRE Digest 350 Part 3).

The wind exposure of upland sites (eg hill tops and moorland) is usually evident on all but the calmest days. Even if the altitude is insufficient to result in markedly reduced air temperatures, the greater frequency of days when windiness is noticeable affects both external comfort and energy use in buildings. Severe wind exposure will often be revealed in the form of trees and hedges.

The crests of hills and ridges can present particular problems due to the way wind flows over them. The velocity:height profile is compressed, compared with that above level ground, and strong winds can occur very near ground level. For example, around the crest of a 1 in 3 hill, gust speeds only a few metres above the ground will be about one third greater than those at a height of 10 m above a level situation upwind of the hill. Because of this, attempts to establish natural shelter such as trees may have limited success. The exposure of an upland site will depend on the adjoining terrain in different directions. For example, sites with a westerly aspect will generally be more exposed than those with an easterly aspect; a site on an isolated hill will tend to be more exposed than on a hill within a group of hills.

All scales of built development, from outer suburbs to city centres, tend to reduce average wind speeds in the immediate boundary layer. In general, the denser the development, the greater the reduction, but this is accompanied by a proportionate increase in turbulence. Whether this turbulence produces an uncomfortable wind climate at ground level in outdoor spaces depends on the local arrangement and form of buildings: BRE Digest 350 Part 3 suggests ways in which the sensitivity of buildings to wind can be reduced. Existing buildings in the vicinity of a site may produce wind tunnelling or downdraughts that influence the need for wind control and protection.

The edges of built-up areas may present particular requirements for wind protection. Influences include the directions of cold wind and driving rain, and the effects of altitude, topography and any shelter offered by a town. Within built-up areas, large open spaces may increase the wind exposure of adjoining buildings. The way in which ground roughness changes from countryside to suburb is important in determining local wind exposure, and may call for specialised meteorological advice.

Effects of wind control

Reduction of wind speed by wind control should improve the microclimate around the buildings. This can be direct, in terms of reduced mechanical and thermal effects on buildings and people; and indirect, by avoiding the dissipation of external heat gains by mixing with colder air. Wind control implies choosing building design least likely to funnel, concentrate or enhance wind flow patterns near the ground; and using wind sheltering design elements such as courtyard forms, windbreak walls and fences and shelter belts. Without protection, wind speeds at and near ground level can be significantly increased for extended periods in the proximity of tall buildings.

Shelter belts

The inhabitants of the UK's maritime counties need no reminder of the effects of the prevailing wind (Figure 1.27). What is not so frequently seen is the deliberate practice of tree planting to mitigate the worst effects of the wind. An example of the beneficial effects of tree planting is the gardens at Inverewe on the exposed west coast of Scotland. When Osgood Mackenzie planted his first salt-tolerant pine trees on the Inverewe peninsula in 1865 to protect his new house, the spray driven by the Atlantic gales would sweep unhindered over the largely bare rocks on the headland. One hundred and thirty seven years later the site is renowned for its sub-tropical plants. Although this example is perhaps an extreme one, some beneficial effect of sheltering can be achieved for any site.

Landscape design offers many practical benefits through its influence on microclimate. Trees, bushes, walls, fences and ground profiling (eg mounds and banks) can all contribute to wind shelter in addition to their value in providing summer shade. For maximum benefit, landscape elements need to be designed in conjunction with the arrangement of buildings, following many of the principles already described (eg avoiding channelling or funnelling of ground level winds). Vegetation, being permeable to the wind, is less inclined to generate downdraughts than buildings; tall trees, suitably placed, can therefore offer substantial wind protection.

Uses of vegetation divide into:
- major shelter belts to protect the edges of built-up areas or placed at regular intervals within large developments
- smaller-scale planting of trees and bushes to give local protection to buildings or open spaces and to enhance ground roughness generally

Major shelter belts to protect building developments are rare, although they have a long history in agriculture. When fully grown they have the potential to provide a useful degree of wind protection over the entire height of low-rise buildings. However, their effectiveness in early years is more limited, since even quick growing tree species take up to 10 years before giving useful protection. Their establishment therefore calls for a long term landscape planning strategy which extends to the development and maintenance of the plants over the lifetime of the buildings. In some cases advantage might be taken of public or common land to grow shelter belts for community benefit.

Effective wind protection by planted sheltering depends on:
- soil type
- soil moisture availability
- climate
- the pattern of protection sought (eg high protection over a short distance downwind, or moderate protection over a longer distance downwind)

A shelter belt may be designed to grow in several successive stages, with quicker growing species offering early wind protection and acting as nursery stock to protect slower growing trees that will form the eventual belt. As the trees grow taller, infilling at their base with bushes becomes important; this prevents gaps that would channel the wind at low level.

To ensure good and uniform performance, shelter belts need to be integrated into the design of a site or complex, and considered in decisions about road patterns, zoning and solar access. The layout and design of shelter belts should be attuned to local circumstances. In areas with a consistently strong wind from one direction, linear patterns may be appropriate. In other areas, protection may be required from several directions suggesting interlocking patterns that require tortuous passage for the wind.

Local, smaller scale planting may take the form of shelter belts of limited height (6–8 m) that will not block too much solar radiation when placed near buildings. General decorative landscape work can also be exploited; for example, by placing trees and bushes between buildings whose spacing is greater than desirable for maximising wind shelter, and where roads might otherwise create channels for the wind. It may be necessary, though, to compromise on the density of planting close to buildings to avoid excessive obstruction of views or loss of daylight, even where solar access is not affected. The presence of trees will also influence foundation design in many areas. See Chapter 1.5.

Figure 1.27 Trees exposed to frequent winds take on a characteristic habit. This one is on the site boundary of a proposed wind farm in Pembrokeshire

Artificial windbreaks can be used to create instant shelter, either as a permanent solution or as an expedient until plants grow sufficiently to become effective. Solid walls and close fences can provide local protection, but they are inclined to generate excessive turbulence in their wake, rather like flat-roofed buildings. Wind protection over a wider area can be obtained with permeable walls or fences. This mimics the behaviour of planted shelter, and should not be too dense if a large protected area is required. Optimum permeability is generally about 40–50%. If the shelter design enables permeability to be varied, it should decrease from top to bottom (ie the windbreak should be more solid at the base, more open at the top). This is likely to be best for wind control in the 'human' zone, 0–2 m; on the other hand, agricultural windbreaks often have a gap at the base to avoid possible frost damage to plants if cold air is trapped.

The aerodynamic performance of natural and artificial windbreaks can be predicted. It will vary with the form and porosity of the windbreak, and with the speed and turbulence of the incident wind. Given data on the frequency distribution of wind speed and direction, the durations of different wind speeds in the sheltered areas can be estimated. Guidance on assessing wind patterns around high buildings in somewhat similar terms is contained in BRE Digest 141[100].

The case for wind control will normally be judged in terms of its benefits for space heating energy consumption and external comfort. However, the provision of shelter and design to reduce wind sensitivity can also reduce the risk of structural damage from high winds and the degree of weather penetration into buildings caused by driving rain. These objectives will usually need separate consideration in shelter design since the directions of extreme and rain-bearing winds may be different from those of greatest importance to energy use and thermal comfort. Even a permeable shelter belt can offer a useful degree of protection (Figure 1.28).

Various other issues can affect decisions about wind shelter. In heavily built-up areas, the need for air movement to disperse pollutants may outweigh the advantages of shelter. In regions liable to prolonged snow cover, reduced wind speeds in the vicinity of buildings can result in accumulations of deep snow; one way of managing the problem is to provide low, fairly open windbreaks specifically designed to trap snow where it will not form a nuisance. Wind noise is sometimes a problem; reducing wind sensitivity will tend to limit noise, but care should be taken to avoid aeolian effects and mechanical noise (resonances or rattling) in the design of artificial windbreaks (BRE Digest 350 Part 3).

Rain

The drier the fabric of an external wall can be kept, the smaller the amount of heat that will be lost through evaporation of moisture in the wall. Wind sheltering may help to reduce the impact of driving rain; this can have significant effects on the thermal performance of external walls. Hard surfaces that drain and dry more quickly will tend to retain more heat and so have a greater influence on air temperature and radiant exchanges.

See *Walls, windows and doors*[101].

Solar access and air temperature

Solar heat absorbed and re-emitted by external surfaces exerts a warming effect on the external spaces around buildings: average air temperatures will be higher, the daily range of air temperature reduced and more re-emitted radiant heat available. Any such warming will reduce space heating energy demands on the buildings, depending largely on inside/outside temperature differences and radiant exchanges. Maximising the benefits of solar heat requires good solar access to external spaces and surfaces, and attention to the thermal properties of external building and landscape materials (BRE Digest 350 Part 2[102]).

Good solar access is also a basic requirement in passive solar design, dealt with in Chapter 2.6 of *Building services*[20], so as to maximise useful solar gains within buildings. The impact of microclimate improvement on building energy use will tend to be greatest for small low-rise buildings having a large surface-to-volume ratio and occupying the most easily sheltered areas nearest to ground level.

The presence of high buildings can degrade the local climate on adjoining sites. Tall buildings cast extensive shadows which partly block solar access to a wide area of adjacent land. As well as their detrimental effects on external comfort, high buildings may significantly increase the costs of space heating in surrounding low-rise buildings.

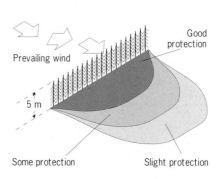

Figure 1.28 It is possible to calculate areas of protection behind a permeable windbreak (eg a hedge or belt of trees)

Frost hollows

The movement of air up or down a slope, even a modest one, can be important: especially for sites at or near the bottom. This effect is due to buoyancy. Air warmed by the ground on a calm sunny day will rise up a slope (anabatic flow), while air cooled by the ground on a calm clear night will drift down it (katabatic flow). Typical speeds for these flows in the UK are 1–2 m/s, depending on the extent and steepness of the slope. Katabatic flows are more important for site development: they render hollows and valley floors colder than locations part way up the sides, and, especially, increase the severity and persistence of frosts where cold air is trapped – the so-called frost hollow. It is important to recognise where the flows are likely to occur, to avoid creating cold air traps at unsuitable points in the layout of buildings or landscape features.

Heat islands

Towns and cities can show marked increases in temperature (the heat island effect) both in winter and summer: especially in summer. Ideally, climatically sensitive site design needs to exploit temperature rises in the winter but counter those in the summer. The effect of large conurbations on space heating degree-day totals can be significant – in central London estimated to be some 10% lower than in the suburbs. Effects such as reduced wetting and long-wave radiation losses, and, conversely, reduced solar gains, also influence the fabric heat balance of buildings in towns.

Solar access and protection

In the UK, making the most of winter solar warmth into and around buildings is desirable for a large part of the year. The range is from about seven months (mid-September to mid-April) in southern England to 10 months (August to May) in the north of Scotland. During these periods, solar gains benefit all types of buildings, but internal gains are put to maximum advantage in passive solar designs where they can significantly reduce heating demand, especially towards the beginning and end of the heating season. See Chapter 2.6 of *Building services*.

Good solar access is needed for building façades and external features in areas where, and at times when, solar gains will be of most benefit. This does not mean that no shade should be provided but rather that it should be arranged so that useful solar gains are not unduly reduced. For direct solar gains (bright sunshine), the shadows cast by buildings, trees, other landscape features and, where applicable, the terrain, need to be analysed for their effects on internal and external solar gains. Gains from diffuse solar radiation (from the sky rather than direct from the sun) are also affected by obstructions, reducing progressively as the area of sky 'seen' by a surface is obscured.

The mechanisms for internal and external solar gains are as follows.
- Internal solar gains comprise the radiant heat received in rooms or spaces after losses from radiation passing through windows. The gains can reduce heating demand in buildings both at the time they are received and subsequently due to heat stored in the building fabric (thermal mass). However, only some solar gains are useful since at times they will exceed heat demand and lead to overheating. The extent of the gains will depend on the areas, positions and types of windows, as well as on site layout factors. The significance of overshadowing will vary between windows, and especially between façades and storeys.

Table 1.12 Transparencies of tree crowns to solar radiation

Common name	Botanical name	Transparency (% radiation passing)	
		Full leaf	Bare branch
Sycamore	Acer pseudoplatanus	25	65
Silver maple	Acer saccharinum	15	65
Horse chestnut	Aesculus hippocastanum	10	60
European birch	Betula pendula	20	60
European beech	Fagus sylvatica	10	80*
European ash	Fraxinus excelsior	15	55
Locust	Gleditsia spp	30	80
European oak	Quercus robur	20	70
Lime	Tilia cordata	10	60
Elm	Ulmus spp	15	65

* The beech retains dead leaves for most of the winter, reaching bare branch condition briefly before new leaf growth in the spring.

Notes:

These data apply to individual tree crowns. Multi-row belts or blocks of trees let virtually no radiation through when in leaf, and very little when in bare-branch condition. Most of the data are based on measurement of light, but can be used for solar radiation generally.

The values are averages from a range of sources. They must therefore be treated with caution, noting that in any case there will be considerable divergence in the transparencies of individual trees, especially in summer.

- External solar gains comprise the radiant heat received in external spaces. In cold conditions human comfort is increased directly by radiant heat including that reflected from nearby surfaces. The spaces around buildings will be more effectively warmed if the gains can be absorbed and stored in external thermal mass. This is typically provided by masonry external skins to buildings and by hard landscape features (eg pavings and freestanding walls). In most cases, a space sufficiently sheltered to benefit fully from solar gains will receive direct sunshine for only part of the day in winter. The arrangement and use of spaces should therefore reflect whether they are warmed in the morning or afternoon.

While solar access is more simply assessed in terms of shading of direct solar radiation, this may not be fully representative, especially for internal solar gains. Averaged over all weather conditions in the UK, a substantial amount of energy is available from diffuse solar radiation. This is particularly significant for points shaded from direct sunlight for substantial parts of the day (eg north-facing façades and open areas adjacent to them). On the other hand, large obstructions may give some benefit by reducing long-wave radiation loss at night.

Daylighting

Daylight availability is a further element of microclimate, affecting the amount of energy needed for artificial lighting, particularly in domestic buildings. Daylight design techniques include methods for quantifying the effects of external obstructions and the reflectivity of external surfaces (BRE Digests 309[103] and 310[104]). These are similar in concept to methods for assessing passive solar availability to buildings, but apply equally to any façade rather than just those receiving direct sunlight. Daylighting needs might conflict with microclimate design if dark-surfaced materials are used to absorb solar radiation.

Shading effect of trees

Trees are a very important influence on availability and protection against solar radiation (Table 1.12). Conifers offer similar densities in both winter and summer; but it is convenient that deciduous trees offer different characteristics, with bare branches in winter allowing desirable low level solar gains, and full leaf conditions in summer providing protection from undesirable solar gains.

The implication of the above discussion is that, ignoring other factors, coniferous trees should be planted to the north of buildings, and deciduous trees to the south.

Noise

Much of the assumed noise attenuation property of vegetation is psychological rather than real. However, the sound generated by air currents and the rustling of leaves can mask other noises.

Industrial pollution

Vegetation can absorb dust and microscopic particulate matter. Most pollution tolerant species concentrate the particles in leaves or bark and then lose this buildup when the leaves fall in autumn or bark is shed. London plane trees are perhaps the classic example of this latter characteristic.

Inspection

The problems to look for are:
◊ unusual vegetation patterns
◊ frost hollows
◊ proximity of tall buildings giving rise to increased wind speeds
◊ coniferous trees on the south of buildings
◊ deciduous trees on the north of buildings

Chapter 1.5 The effects of vegetation on the ground

It is generally accepted that trees in the vicinity of buildings are a positive feature that improves the impact of buildings on the environment, both in visual and environmental terms. Trees can assist with the visual integration of buildings into the wider landscape by softening hard edges and screening specific views while framing others. By acting as carbon sinks, trees can reduce the effect of carbon dioxide emissions. However, there are also detrimental effects to consider, particularly the effects which vegetation may have on the ground, and affecting, in turn, the buildings sited on that ground.

The positions of trees or hedges will normally be of concern to building professionals, especially when they are close to existing or proposed new buildings, and even more so when the site is underlain by shrinkable clay (Figure 1.29). The growth of trees already on site may lead to further settlement of shallow foundations, while the removal of trees and hedges may cause heave. The magnitude of these movements will depend on the size, type and location of the trees, the plasticity of the clay and the groundwater level. Unstable slopes are sometimes tree covered because the disruption of subsoil drainage by ground movements makes them boggy and therefore difficult to cultivate.

Soil shrinkage caused by trees taking water from the ground can result in foundation subsidence. Soil swelling caused by the ground recovering moisture following tree removal can result in foundation heave. BRE Digest 298[105] gives simple design guidance on minimising these effects in clay soils and points to some dangers in current foundation practice.

See *Subsidence damage to domestic buildings: lessons learned and questions remaining*[42].

HAPM has indicated that of the sites founded on shrinkable clay that they inspected in the early 1990s, some 1 in 4 were considered not to have complied with published guidance with regard to existing trees, and some 1 in 5 did not take adequate precautions when trees were to be removed.

Characteristic details

Site vegetation
Aerial photography provides a permanent record of site vegetation at a particular point in time. For a given site, photography cover may be available about every 10 years from 1946. Aerial photography specialists can estimate the height of trees or bushes from stereo pairs of photographs. Large scale topographical maps will show the positions of hedge lines, woodland and, occasionally, of isolated trees.

It has been common practice for many years, enshrined in building regulations and before then in local authority building byelaws, to ensure the removal of vegetation which might be of potential harm to the foundations of a building. *'The site of any building, other than an excepted building, shall be effectively cleared of turf and other vegetable matter'* (Regulation C2 (1) The Building Regulations 1965[25]). This was normally accomplished by removing topsoil, though not all potentially detrimental root systems were removed by this process.

Figure 1.29 Trees are often considered a valuable amenity to be protected, but they may affect the stability of buildings if they are too close to them

1.5 The effects of vegetation on the ground

Main performance requirements and defects

Desiccation
For the purposes of this book, the word desiccation means any reduction in the soil's moisture content caused by evaporation or transpiration of water through the roots of plants. The effect of this reduction is to increase suction in the soil water and, therefore, the stress acting on the solid soil particles, bringing about a volume reduction as mentioned in Chapter 1.1. If soil is removed from the ground and placed in a warm and dry environment, the suctions produced by evaporation would increase progressively until practically all the moisture was removed from the sample. In the ground, however, where there is a constant supply of water from surrounding soil, the suctions reach limiting values and the soil is able to achieve a new equilibrium at a lower moisture content. Therefore, a soil described as desiccated may in fact have a moisture content that is reduced by only 1–2% below its undesiccated condition.

Desiccation can reduce the moisture content of firm shrinkable clays to values below the plastic limit. This makes the soil much stiffer, and gives it a dry and friable appearance. Near the surface, desiccation often causes the ground to crack. In highly shrinkable soils, cracks 25 mm wide and 0.75 m deep are not uncommon during dry summer months.

Determining if a clay is desiccated is usually done by measuring water content in samples taken from two (or more) boreholes, one in the test area, the other a control borehole well away from any desiccating influences[105]. Sometimes determining the desiccation proves impossible to achieve, either because there is no part of the property that is outside the influence of trees or because the characteristics of the soil (eg based on visual appearance or plastic and liquid limits) vary significantly over short distances.

The physical process of desiccation involves removing water from the soil by external forces. The process creates suction in the remaining soil water; in other words it goes into a state of tension, bringing the soil particles closer together (shrinkage) and increasing the forces between them (the soil becomes stronger and stiffer). However, other factors than desiccation may also bring the particles together with similar changes in the soil strength and stiffness. It follows, therefore, that the most reliable way of determining desiccation is to measure a suction value in the soil water.

The suctions in the soil can be measured by a number of different techniques. The method that is most commonly used in the UK is the filter paper test[106]. In this test, samples of the soil are placed in intimate contact with pieces of filter paper, or other porous material; the whole assembly is sealed using, for example, cling film, and left to come to equilibrium. The filter paper is then removed and weighed before and after drying in an oven to determine its moisture content. For a particular porous material, there is a unique relation between the moisture content and the pore suction. So, provided the material has been calibrated, the measured moisture content can be converted directly to a suction.

When using a single suction profile, it is important to appreciate that a sample of saturated, and therefore undesiccated, over-consolidated clay will register a suction value. This is because the ground stresses confining the sample in the ground will, upon the sample being released from them, generate equal but opposite stresses in the soil water.

Light vegetation and small scrub
In grass covered areas, the effects of evaporation and transpiration by light vegetation are largely confined to the uppermost 1–1.5 m in firm, shrinkable clay soils.

Trees and hedges
Trees are an essential feature of many sites, urban as well as rural. As noted, they are allies in the struggle against global warming, for they lock up large quantities of carbon dioxide. But there is a downside too. Trees, and to a lesser extent hedges and large shrubs, can extract moisture from depths greater than 1.5 m. For example, a large oak tree may desiccate the soil to 6 m or more. For high-plasticity clays, which tend to have very low permeabilities, rainfall during winter months cannot fully replenish the moisture removed by large trees during the summer. Consequently, a zone of permanently desiccated soil develops under the tree. If a building is too near this zone, it may suffer as a result (Figure 1.30, and Figure 1.31 on page 72).

Figure 1.30 A cracked corner of a house near a Lombardy poplar tree

1 The basic features of sites

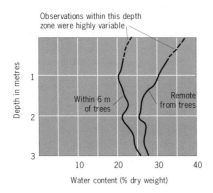

Figure 1.31 Soil moisture content profiles adjacent to and remote from trees in heavy clay, indicating desiccation. The time range of observations was from August to January, and the mean observation month was December for both curves. The observations were taken from 25 different borings

Figure 1.32 The safe distance of a tree from a building founded on shrinkable clay

As the tree grows, the desiccated zone increases in depth and width, producing more subsidence which is likely to affect any nearby structures. The extent of the desiccated soil depends on the moisture demand of the tree. In general, broad leaf trees have a greater moisture demand than evergreens. Because of their large size, oak, elm, willow and poplar are notorious for having caused damage. However, these trees are not the ones which nowadays most commonly cause damage to housing because other trees with lower moisture demands, notably plane, lime and ash, are more frequently planted in close proximity to buildings. Using information collected by the Royal Botanic Gardens during the 1970s, and from its own observations, BRE has suggested that the trees most likely to cause damage, in descending order of threat, are as listed in Table 1.13[105].

For each type of tree, Table 1.13 gives the distance between the tree and the building within which 75% of the reported cases of damage occurred. As a rule of thumb, it would appear that damage can usually be avoided by ensuring that the tree is no closer to the foundations than its mature height (Figure 1.32). For the less thirsty trees, this figure can be reduced to half the mature height. There are two reservations about this generalisation. First, it takes no account of the shrinkage potential of the soil at the depth of the foundations. Second, it is the leaf area of the tree rather than its height that ultimately determines its moisture demand. The rules should therefore be treated with caution. There is room for different views on the use of generalised categories of tree height as one of the criteria, and one source argues that each case should be treated on its own merits[107].

Desiccation in the surface soil varies seasonally, being greatest towards the end of the summer and smallest in late winter or early spring. The magnitude of the corresponding movements will reflect the magnitude of the changes in desiccation; an unusually dry year will cause greater ground movements than an average year.

Precautions on tree and vegetation removal

The removal of low growing vegetation is relatively straight forward; however the felling and grubbing out of mature trees can be more difficult and hazardous. Appendix A of BS 3998[108] should be complied with together with the Forestry and Arboriculture Training and Safety Council Safety Guides. Trees should be taken down in small sections to avoid damage to adjacent trees and vegetation if these are being retained.

When a tree is felled or dies, or excavation for, say, new foundations severs a substantial part of the root system, the removal of water ceases and the suction in the desiccated soil begins to draw moisture from undesiccated clay at the extremity of the zone and from any free water entering from cracks above. The low permeability of very plastic soil like

Table 1.13 Risk of damage by different varieties of tree			
Name	Maximum height (*H*) of tree (m)	Separation between tree and building for 75 % of cases (m)	Minimum recommended separation in shrinkable clay (multiple of *H*)
Oak	16–23	13	1
Poplar	24	15	1
Lime	16–24	8	0.5
Common ash	23	10	0.5
Plane	25–30	7.5	0.5
Willow	15	11	1
Elm	20–25	12	0.5
Hawthorn	10	7	0.5
Maple and sycamore	17–24	9	0.5
Cherry and plum	8	6	1
Beech	20	9	0.5
Birch	12–14	7	0.5
White beam and rowan	8–12	7	1
Cypress	18–25	3.5	0.5

London Clay makes the rehydration extremely slow. One case was recorded where elm tree removal resulted in swelling that lasted for more than 25 years and caused building heave of 160 mm.

On clay soils where trees are removed before construction begins, precautions should be taken to minimise risk to the building's footings as a result of the swelling of the clay soil following tree removal.

- A compressible layer should be applied only to the side of the trench nearest to the former site of the tree before placing concrete. If applied to the further side, or to both sides, the resistance to lateral displacement of the foundations will be reduced
- A slip plane layer should be applied on the side of the trench fill, so that upward swelling of the clay is less likely to lift the foundations. It should not extend below the zone of swelling since this would reduce the resistance to uplift provided by the stable soil below the swelling zone
- Care must be taken to accommodate swelling soil beneath ground beams and ground supported floor slabs (BRE Digest 298)

If it is known that trees will be planted in shrinkable clay soils after construction of the building, consideration should be given to ensuring either a safe separation distance or to deepening the foundations to reach a zone that will not be desiccated by the tree roots. If rows of trees are to be planted, the separation distance or the foundation depth should be further increased.

The removal of trees arrests the de-watering process and the clay returns to its natural water content and volume. If the tree is younger than the building then the cracks its presence caused will recover. If the tree is older than the building, or is removed before construction starts, the heave that accompanies reversion of the clay to its original volume can occur comparatively rapidly and is usually more damaging than clay soil shrinkage[12].

In Chapter 4.2 of NHBC *Standards*[109] a comprehensive set of guidance is given for the design of low-rise building foundations near trees.

Figure 1.33 Shrinkage cracks in a road and pavement on London Clay indicating desiccation by trees

Precautions on alteration of pavings round trees

Although relatively impermeable coverings such as asphalt or concrete may help to reduce evaporation, they will also encourage run-off which may in turn force trees and other large vegetation to extend their root systems to maintain their supply of moisture. Consequently, where there is shrinkable clay, consideration should be given to the long term changes in desiccation and associated ground movements that might occur as a result of surrounding large trees with extensive areas of concrete or asphalt (Figure 1.33). Where there is a risk of causing damage to nearby buildings, more permeable alternatives such as pressed concrete slabs with open joints could be used or grading the coverings, so allowing the water to enter the ground round the trees.

Improving surface water drainage may also reduce the supply of water to established trees and thereby trigger some long term movements as equilibrium in the ground is adjusted. Where a grassed area is being replaced by a patio or similar feature, one possibility is to provide a soakaway in a position where existing trees can exploit the supply of water rather than channelling the run-off into a drain. In some circumstances, it may be beneficial to improve the supply of water to the soil under the foundations by grading

---

Case study

Measurements of desiccation on a site in the London area

A London Clay site having a very high shrinkage potential was monitored over a period of seven years. The site was partly covered in grass, but part was approximately 5 m away from some large poplar trees.

Measurements confirmed that ground movements in the grass-covered area were largely confined to the top metre of soil, although the unusually dry weather of 1989 and 1990 produced movements of 6 and 13 mm respectively at a depth of 1 m. The movements in the vicinity of the poplar trees were considerably larger and, at a depth of 1 m for example, exceeded 35 mm even in an average year such as 1988. However, despite the fact that water content and suction profiles indicated that the soil was desiccated to about 4.5 m, only very small movements were recorded at a depth of 2 m during 1988; in other words, in an average year the desiccation between 2 and 4.5 m appeared to be relatively insensitive to seasonal variations.

Nevertheless, during 1989 and 1990 measurable ground movements were recorded at depths of 2, 3 and even 4 m indicating that prolonged periods of dry weather can have an effect, including on deep seated desiccation. Also, it was evident that a lot of the moisture extracted from below 2 m was not replenished during a single winter cycle, so that two consecutive dry summers cumulatively affected the degree of desiccation in the deeper soil.

Some 1.5 m deep concrete pads positioned an average of about 5 m from the poplars showed the effects of cumulative moisture losses which produced a ratcheting effect on the pads, resulting in settlements of more than 50 mm over a three year period. These movements were in addition to any long term subsidence associated with the growth of the trees and would be capable of causing serious damage in most buildings, particularly if only part of the building were affected, maximising differential movements and therefore distortion and damage.

Monitoring
Volume change of clay soils due to changes in moisture content is a common cause of damage to low-rise buildings. The growth of trees in these soils desiccates the clay, so creating greater volume change. For a number of years, BRE investigators examined the behaviour of various types of low-rise building foundations at a shrinkable clay site in the Home Counties, and some examples from these studies are described in the adjacent case studies.

Use of planting to stabilise slopes
Slope stabilisation can be enhanced with the judicious use of ground cover and shrub planting, particularly if done in conjunction with bio-engineering techniques. However, vegetation can contribute both to stability and instability. For example, grass grown on suitably prepared slopes encourages run-off during storms, helps prevent scour, and physically binds the soil together; but a layer of topsoil bound together by grass can itself be unstable and slide like a carpet on underlying gravel or weathered rock. Although the physical strength of roots is not generally thought to be a key factor, the desiccation caused by shrubs and trees tends to improve stability by reducing the moisture content and increasing the strength of the surface soil. However, trees may become a liability when they get very large as their weight and the action of wind on them both contribute to localised instability. Moreover, on some jointed rocks, the roots may open up the joints in the rock, allowing water to flow in and eventually detach blocks.

Remedies for damage
Subsidence damage to domestic buildings: lessons learned and questions remaining[42] lists four possible courses of action depending on a number of factors, including the severity of the damage:
- doing nothing
- acting against the offending trees
- stabilising the building
- repairing the building

Doing nothing would be appropriate where the damage is minor, will not worsen, is patently caused by vegetation in extreme weather conditions, and there is no pressing need for action. Action against the offending tree is covered in the next section, 'Work on site'. Stabilising the building is covered in 'Remedial work' in Chapter 2.1. So far as repairs are concerned, there are a number of techniques available for repairing the building, depending on its construction; for example, brickwork can be reinforced against further cracking by stainless steel reinforcement bars inserted into slots cut into bed joints, or be strengthened by 'stitching' with short lengths of reinforced concrete sections.

See also Chapters 2.1 and 2.2 in *Walls, windows and doors*[101].

Case study
Monitoring movement in shrinkable clay soils
Measurements were made in an area originally covered by some large poplar trees. Surface movements at points amongst the trees and in open ground showed that 35 mm of vertical movement can occur in open ground between summer and winter and 90 mm amongst trees.

Following detailed observations of exceptional seasonal effects during the prolonged dry weather of 1989 and 1990, all of the trees in the vicinity of the experimental foundations were felled in September 1990. In the following five years the ground continued to swell as moisture returned slowly to the desiccated soil.

In another case, regular precise survey levelling, carried out around the perimeter walls of a bungalow built on shrinkable clay, showed that there was a seasonal fluctuation of vertical movement of 47 mm closest to adjacent trees coupled with progressive settlement. Lateral displacement of 66 mm occurred in the foundations below the DPC together with cracking of the external and internal walls. It was concluded that the damage had been caused by a combination of ground shrinkage due to water abstraction by trees on the outside of the foundations, particularly by an oak tree 9 m from the bungalow, and swelling of the clay on the inside of the foundations due to tree roots being severed during foundation construction.

Case study
Long term heave of a building founded on clay soil after tree removal
Trees were felled on a clay site two months before a three-storey office block was built on strip foundations. Heave was expected. A precise levelling system was installed during construction and movements monitored for 27 years. The maximum heave recorded was 22 mm reached 10 years after construction; other parts of the building heaved less but took up to 17 years to reach the maximum movement.

---

the patio towards a flower bed or gravel surround to the house. However, there is a danger that this may actually encourage the growth of roots in this area, creating potential problems if there is a prolonged period of dry weather.

1.5 The effects of vegetation on the ground

Work on site

Surveys
The existing vegetation can be a good indicator of existing ground conditions, showing areas of poor drainage, low or high pH, and soil fertility. A relatively detailed inspection of existing plant cover of a site is worth undertaking, particularly if trees are present, to establish what is worth preserving.

Storage and materials handling
Topsoil should be stored in mounds not higher than 2 m, and away from working areas. The area of topsoil should not be used for storing construction materials or be compacted by building plant.

Supervision of critical features
Trees are easily damaged during construction operations and positive protection measures should be put in place to prevent casual damage. Temporary fencing can be erected to protect not only the tree but the soil area around it.

Workmanship
Where damage to buildings is attributable to clay shrinkage, and it is reasonable to believe the shrinkage has been exacerbated by the presence of trees, there are four ways in which the effect of the tree can be mitigated, any of which may provide a cost effective remedy:
- tree removal
- tree reduction
- root pruning
- root barriers

Tree removal
Removing the tree altogether is likely to have the greatest and most immediate effect on the levels of desiccation in the soil. This course of action should be safe if the trees are no older than any part of the building since any consequent heave can do no more than return the foundations to their original level. In these circumstances there is no advantage in a staged reduction in the size of the tree, and the tree should be completely removed at the earliest opportunity.

If a tree is removed, water will return slowly to the soil causing it to swell, and the process will continue until the porewater suctions that desiccation induces have dissipated to equilibrium values. The time scale over which these changes take place depends largely on the permeability of the soil, although the extent and magnitude of the desiccation, and the availability of free water are also important factors. There is one well documented case history which reports measurable movements continuing 25 years after the removal of some large elm trees; however, the heave was 90% complete within 10–15 years of tree removal[110].

The time taken for the soil to recover mainly depends on the permeability of the soil. In heavy clays, such as London Clay, it may take tens of years for the ground to come to equilibrium; in more permeable soils, or soils containing permeable layers, full recovery may be achieved in one winter. Depending, then, on the soil conditions, the degree of damage and the circumstances of the owners, tree removal may or may not be an acceptable solution.

Where a tree is younger than the property, its removal can, at most, only return the foundations to their original position and should not, therefore, cause significant damage.

Where the tree is older than the house, or perhaps more commonly there are newer additions to the house, it may not be advisable to remove the tree altogether because of the danger of inducing damaging heave. There is a very real danger of damage resulting from the removal of the tree. In such cases, it may be prudent to investigate the state of desiccation in the ground and estimate the heave potential in the soil adjacent to the foundations, using the procedure described in Chapter 2, before deciding on whether or not the tree can be removed.

A more common cause of damage, however, is the removal of trees prior to construction which can lead to problems unless appropriate measures are taken to isolate the foundations from the effects of the swelling soil. Problems of this type normally manifest themselves within 10 years of construction, and the past presence of trees can usually be confirmed by reference to local records, building control submissions or aerial photographs.

Tree reduction
Where it is unsafe to remove the tree altogether and the damage is relatively minor, consideration can be given to some form of pruning (eg crown thinning or crown reduction). Pollarding is often mistakenly specified because the height of the tree is the parameter most commonly quoted as being a measure of its likelihood to cause damage. In fact, since the vast majority of the moisture extracted by the roots passes straight up through the tree and is transpired, the leaf area is a more relevant parameter. Indeed, pollarding of the sort often specified can severely damage the tree. Pruning to thin or generally reduce the crown is, therefore, generally preferable to pollarding.

Pruning a tree may reduce its ability to extract water, though it must be borne in mind that inappropriate pruning can irreparably damage some trees, while insufficient pruning may have little effect on water extraction. In addition, pruning may simply encourage rapid regrowth of the tree to something like its former state. Different trees react in different ways to pruning; arboricultural advice should be sought if considering pruning.

The available knowledge of the effects of different forms of reduction on the water demand of trees is very limited, and research is in progress to rectify this[42].

Root pruning

The usual way of doing this is to excavate a trench between the tree and the damaged property to a depth sufficient to intercept and cut most of the roots; it is important, however, that the trench is positioned far enough away from the tree to ensure that its stability is not jeopardised. In the fullness of time, the tree will almost certainly grow new roots to replace those that are cut; however, in the short term there will be some recovery as the degree of desiccation in the soil under the foundations reduces and, in cases where the damage has only appeared in a period of exceptionally dry weather, a return to a normal weather pattern may be all that is needed to prevent further damage occurring.

Root barriers

Root barriers are essentially a variant of root pruning. However, instead of simply backfilling the trench with soil, the trench is either filled with concrete or lined with an impermeable layer to form a permanent barrier to the roots. Little is known about the effectiveness of root barriers in preventing root activity beneath building foundations. They are usually fairly expensive and disruptive to install. There is considerable uncertainty about how extensive a root barrier should be, both laterally and in depth. It must be remembered that if a barrier is effective, the outcome will be the same as if the trees were removed; therefore, barriers should only be considered where tree removal would not cause unacceptable heave. Root barriers require careful consideration and detailing at the ground surface, where roots could potentially grow over the barrier, and around any services passing through the barrier.

Inserting root barriers close to trees can be dangerous. If a significant part of the root system is severed, the tree may lose lateral stability and fall. Even if rapid instability is not caused, the tree may slowly die and become unstable at a later date. Similarly, the potential for undermining the building foundations while constructing the barrier must be considered.

Supervision of critical features

Where trees are to remain in spite of the damage they have caused to buildings, and remedial measures such as underpinning are undertaken, it will be necessary to take precautions to protect the tree.

Where large, mature trees are present, it is not uncommon for the soil to be significantly desiccated to a depth of 6 m or more. However, in these cases, it is likely that the levels of desiccation below about 3 m will remain at an approximately constant level as long as the tree remains in place. The design of an underpinning system to withstand the ground movements associated with the removal of the tree will be radically different, therefore, from one designed simply to provide an adequate foundation while the tree remains.

The lifespan of most common, large deciduous trees such as oak, willow, ash, sycamore, horse chestnut is in excess of 100 years; therefore, it may not be necessary to design against the eventuality of the tree being removed unless there is good reason to believe the tree is diseased or nearing the end of its life. It follows that, wherever possible, nearby trees should be preserved and particular attention should be paid to avoiding damage to the trees and their roots during any underpinning operations[111].

Inspection

Where cracking is occurring in the structure, monitoring may be necessary. A description of monitoring techniques is given in Chapter 2.1 and in *Cracking in buildings*[12].

The problems to look for are:
◊ construction starts too soon after tree removal
◊ new trees planted too close to existing buildings
◊ inappropriate pollarding
◊ dead trees (killed during construction) close to new buildings

Chapter 2 Foundations

Foundation design normally consists of three steps.
- Calculation of the total loads acting on the foundations. This calculation should include dead loads (ie the weight of the walls, roof and contents of the building), live loads (ie wind and snow loading), and any bending moments acting at the base of columns or walls
- Estimation of whether the pressure acting below a minimum width foundation of, say, 0.5 m is likely to exceed the pressure that the ground can safely sustain without excessive deformation (ie the allowable bearing pressure). If it is, then wider or deeper foundations will be needed
- Determination of a minimum depth for the foundations. This may be based on the requirements for the structure (eg the need to accommodate a basement or underfloor heating) or the need to found at a depth where the soil is suitably stable (eg below the surface layer which is assumed to be affected by changes in moisture and temperature)

This second chapter deals with all kinds of foundations, including strip and pad foundations (Figure 2.1), and the more commonly encountered kinds of piled foundations; that is to say, both bored and driven in various materials and various configurations.

Strip foundations are, as the term implies, narrow foundations placed under loadbearing walls in order to spread the load over the ground. They are dealt with in Chapters 2.2 and 2.3.

Piles are relatively long and slender foundation elements that are used either to transmit the applied loads through a surface layer of soft or unstable soil to an underlying competent stratum; or, by their length and frictional resistance (known as skin friction) when placed or driven into otherwise unsuitable ground, to transmit the applied loads without appreciable movement. They are often the most economical way of supporting very heavy loads and provide a way of resisting uplift and lateral forces. Piles are extremely useful for limiting the settlements of large structures on stiff clay; equally, small piles can be used to protect houses and light industrial buildings from shrinkage and swelling in the surface layer of clay soils.

Because of the wide range of uses for piles and the variety of soils encountered, piles of many different types and methods of installation have been developed. There are, however, two basic types:
- cast-in-place (or replacement) piles
- driven (or displacement) piles

Cast piles consist of concrete which is placed into a preformed hole before the insertion of reinforcement, or which is continuously pumped into the void created as a continuous flight auger is retracted. (A continuous flight auger is one in which the screw of the auger is as long as the pile depth.) Piles can have bases that are flat, pointed or bulbous, and shafts that

Figure 2.1 Adequate foundations are crucial to the stability of even the smallest of buildings. Here, a sloping site of disturbed ground has given rise to a number of awkward problems for the contractor

are vertical or raked. They are dealt with in Chapter 2.4. Driven piles are fabricated from timber, steel or reinforced concrete, and are also dealt with in Chapter 2.4.

In addition to piles used for holding up buildings, piles may also be used for holding down buildings, as in the case of buoyant structures; for example, those with very deep basements and light superstructure.

Chapter 2.1   General points on foundations

The foundations of a building transmit the loads acting within and on the structure into the ground, and, at the same time, transmit the effects of any ground movements back into the structure. Some foundation movement is inevitable since soil and even rock will deform under loads applied by and through foundations. Ground movements may also occur as a result of processes that are unconnected with the applied loads (eg changes in moisture content). For the most part it is quite impractical to design foundations to be totally immovable throughout the life of the building; therefore, a certain amount of movement has to be allowed for. The essence of successful foundation design is to ensure that the level of movement transmitted to the superstructure is acceptable, that the consequent distortions never exceed tolerable levels, and, not least, that the cost of the foundations is as low as possible while meeting these criteria.

This chapter examines how much foundation movement is likely to be tolerable and the methods that are commonly adopted for designing foundations to ensure that the movements do not exceed these limits. The different types of foundations that are commonly used and how the selection of the foundations is influenced by the prevailing ground conditions are described in later chapters.

The chapter also deals with ground improvement techniques; that is to say, techniques used to densify poor ground so that economical foundations can be safely used.

It is useful to draw a nominal distinction between shallow foundations and deep foundations, even though this distinction is not always apparent in practice. BS 8004[68] defines shallow foundations as being less than 3 m below ground level. It points out, however, that shallow foundations where the depth-to-breadth ratio is high may need to be designed as deep foundations; and, similarly, deep foundations with small depth-to-breadth ratios may need to be designed as shallow foundations. Examples of shallow foundations are strip footings, pads and rafts, described in Chapters 2.2 and 2.3. Examples of deep foundations are piles and piers, described in Chapter 2.4.

This present chapter also deals with underpinning, since it may be needed in a situation demanding remedial action whatever the type of foundation – corbelled brick, concrete strip or pad, and ground beam.

Existing buildings of whatever age, if they have been subjected to differential movements in the past, may already have been underpinned. In Victorian times, underpinning was commonly undertaken as part of remedial or refurbishment work on major buildings. The architect Thomas Graham Jackson (1835–1924), for example, undertook several such assignments during his professional life, including Winchester Cathedral, Holy Trinity Church, Coventry, All Saints Church, Stamford and Hatfield Church, Doncaster[112].

At Winchester, the ground was particularly wet, containing a layer of peat some 2 m thick, and the fabric had settled considerably in spite of the existence of short oak piles laid in Norman times and the trunks of beech trees laid horizontally in the twelfth century. In 1906 the underpinning was founded on the gravel bed some 5 m below ground level, the water table being around 2 m from the surface. Consequently, at the suggestion of the engineer, Francis Fox, a diver was employed to excavate for and to place the mass concrete. Iron rods and cramps were used for reinforcement.

Characteristic details

Foundations in common use

For soils with adequate bearing capacity, the types of foundations in common use have usually been strip for loadbearing walls and pads for the columns of framed structures (described in detail in Chapter 2.3). In soft soils, common foundation types have consisted of rafts to spread loads (dealt with in *Floors and flooring*[66]), and piles to take loads down to ground levels where the risk of movement is minimal (Chapter 2.4).

According to HAPM, of over 2000 housing sites they inspected during the early 1990s, nearly three quarters were of the traditional concrete strip or trench fill, some 1 in 6 piled, 1 in 11 rafts, and 1 in 25 vibro columns (described later in this chapter). There is no comparable information on foundations used for other building types.

While the foundations for most lightly loaded structures can be designed on the basis of experience and using Section 1E of Approved Document A to the Building Regulations for England and Wales, more heavily loaded foundations need to be designed in accordance with accepted practice, as is described in BS 8004. Further information is also contained in BRE Digest 318[113].

Factors governing the choice of foundations

The factors governing the choice of foundations and their subsequent performance are ground conditions, cost and ease of construction. In many cases foundations have been selected at an early stage in the design of a project, with little attention being paid to the actual ground conditions. Many foundation problems could have been identified and avoided during the early stages of projects, and costs could have been reduced by increased emphasis on desk study and site investigation work. Other factors which have an important influence on foundation design include project organisation, site supervision and building regulations. The most common causes of difficulty encountered by the designers and builders interviewed during a survey carried out by BRE investigators in the mid-1980s[114] were:
- building on fill
- shrinkable clay sites
- ground treatment such as vibro replacement

More recent information indicates that the situation has changed little since that time.

Even at the end of the twentieth century, foundation design was largely carried out using either theoretical (very rare for low-rise structures) or empirical design procedures based on soil properties determined from laboratory and in situ test methods. The designer then uses the results of both the laboratory and in situ tests in the choice of soil properties to be assumed for foundation design. Since the design procedures themselves are based either on empiricism, or on theoretical behaviour of idealised soils, the whole process involves a high level of uncertainty. This is compensated for in the high factors of safety applied. It is anticipated that the forthcoming Eurocode 7 will introduce new procedures, though still with a strong element of empiricism, which will inevitably have an effect on UK design procedures.

Where an inadequate soil under a proposed building is underlain at a depth of less than around 30 m by a stronger ground such as rock or gravel, it may be economic to support the structure on piles. There is a great variety of patent and other types of bearing piles, and it is important to be sure that the right type of pile is used; and the total carrying capacity of the group of piles supporting a building as well as the carrying capacity of individual piles must be known within reasonable limits. It is always necessary to establish whether a piled foundation is indeed the solution, for in some conditions it may weaken the ground unduly.

Improvement techniques for poor ground

Some sites carrying existing buildings may have been subjected to improvement before construction began, affecting subsequent behaviour of the foundations and structure.

Where site investigation has indicated that the ground has poor load carrying properties and that significant differential movements could occur over the area of a building, improving the ground by using an appropriate ground treatment method prior to development of the site can form a major element in providing an economic and technically adequate solution to the foundation problems. There has been growing interest in ground treatment techniques in the UK since the scarcity of good building land has led to construction on ground hitherto regarded as unsuitable for development. General information and guidance is provided in this chapter, but it should be recognised that on sites where ground treatment is required, the services of a suitably experienced geotechnical engineer will usually be essential.

The selected treatment technique must be able to remedy the relevant inadequate characteristics of the ground, whether it is for load carrying or other parameter. This means that the deficiencies in the ground properties must be correctly diagnosed before the appropriate treatment method can be specified.

Techniques which are used to increase the density of non-engineered fill include:
- compaction by surface impact loading (dynamic compaction or rapid impact compaction)
- deep vibratory techniques (vibro)
- preloading with a surcharge of fill
- excavating and refilling in thin layers with adequate compaction

Geotechnical problems with soft, saturated clay are associated with the low undrained shear strength and the high compressibility of the clay. The very low permeability of the clay is one of the controlling factors in determining the usefulness of any treatment method. Some methods used on fill can also be used on soft clay; for example, vibrated stone columns or preloading. However, the objective of the treatment and the mechanisms of soil behaviour which are utilised in the treatment may be quite different.

Dynamic compaction

Dynamic compaction of fill is effected by repeated impacts of a heavy weight onto the ground surface. The dropped weight is one of the simplest and most basic methods of compacting loose fill (Figure 1.8 in Chapter 1.2). A typical application would be one in which the energy input per blow is obtained by dropping a 15 tonne weight from a height of 20 m using a large crane. A greater energy input per blow requires special lifting equipment and, unless a very large site requires treatment, it would probably not be economic to use greater impact loads. Typical total energy input is of the order of 2000 kNm/m² although there are considerable variations depending on the type of fill and the degree of improvement required[115].

Certain limitations of the method should be recognised.
- High mobilisation costs associated with the large crane required to drop the weight usually mean that areas smaller than 5000 m² cannot be treated economically
- It is inadvisable to use the method close to an existing structure due to the potentially damaging vibrations caused by the impact of the weight onto the surface of the fill. Minimum distances of 30 m have been quoted, but much depends on individual circumstances such as whether the adjacent building is also on the filled ground or is built on natural ground
- Flying debris constitutes a hazard for personnel, vehicles and structures close to the impact point. Some shielding to catch the debris may be required
- With many fill sites it will be necessary to provide a granular blanket to form a working platform for the crane. The cost of this granular material may be substantial in relation to the total cost of the treatment

Rapid impact compaction
A mobile rapid impact ground compactor initially developed for repairing explosion damage to military airfield runways has been adapted for use in building and civil engineering applications (Figure 2.2). The compactor comprises a modified hydraulic piling hammer acting on an interchangeable articulating compacting foot. A 7 tonne weight falls 1.2 m onto the foot at a frequency of about 40 blows per minute. The compaction process should include a minimum of two treatment phases, with high energy primary compaction using abutting compaction points to improve the ground at depth and a low energy secondary treatment to ensure adequate compaction of the material up to the ground level. The standard 1.5 m diameter compacting foot is very effective at punching into the ground to give the maximum depth of treatment (Figure 2.3). A larger foot should be used for a lighter secondary treatment.

The technique produces compaction of the fill by impact loading. It is similar in principle to dynamic compaction using a drop weight but the following differences should be noted:
- the energy per impact is much lower
- the number of blows per unit time is much greater
- the foot that transmits the energy to the fill remains in contact with the fill

The rapid impact compactor can improve the engineering properties of a range of fill and natural sandy soils. Trials carried out by BRE show this to be a promising technique for the improvement of miscellaneous fill of a generally granular nature up to depths of about 4 m.

The process produces virtually no flying debris and could work safely, with normal precautions, in close proximity to other site operations or structures, including roads and railways, where vibration and noise alone may not be of great importance. However, vibration and noise may make the method unsuitable for some urban sites, though measures can be taken to reduce noise emissions.

Vibro treatment
The various related ground improvement techniques collectively termed vibro have been extensively used to improve foundation conditions for low rise developments in the UK. The BRE research in this area has placed particular emphasis on field monitoring at selected sites, where the effectiveness of vibro techniques can be assessed in the light of monitored field performance.

Figure 2.2 A rapid impact compactor

Figure 2.3 Close-up of the footprint of the rapid impact compactor

Vibro stone columns

A large proportion of current building projects in the UK are carried out on poor or marginal ground. Vibro treatment has been frequently used in cases of filled or poor ground on which reinforced strip footings are to be founded. BRE investigators have carried out field trials to study these two connected aspects of vibro ground treatment[116].

In essence, a vibro stone column is formed by using a large poker vibrator to penetrate the ground and form in its wake a continuous and dense column of suitable and durable broken stone from the required maximum depth of penetration at the competent stratum up to the ground surface. The soil conditions must, of course, be appropriate. Either top feed or bottom feed may be used, and the stone is fed into the column either in a dry or wet state, depending on circumstances and the particular design. The column must not be contaminated with spoil.

The basic tool is a cylindrical poker which contains an eccentric weight in its bottom section. Rotation of the weight results in vibrations in a horizontal plane being transmitted to the soil as the poker penetrates into the ground. The long cylindrical hole formed by the poker is backfilled with granular material to form the stone columns which stiffen the soil (Figure 2.4).

A loose sand can be compacted with no necessity to add any stone material, although in this situation the level of the ground surface will be lowered. This process is rarely appropriate in the UK.

In most soils stone columns are used. The treatment is generally intended to stiffen and densify the ground, to seek out weak spots and to reduce differential settlement. Soils can be effectively compacted if the percentage of silt size particles is smaller than about 15%; otherwise the ground is improved solely by the presence of the relatively stiff stone columns.

Either compressed air (dry method) or water (wet method) is used to remove any loose debris and assist the penetration of the poker. Water is also used to support the sides of the hole in softer or more unstable soils. In the UK the dry method is predominantly used. This is principally because on small sites there can be environmental objections to the quantity of water used in the wet method and to its disposal.

Treatment typically comprises the placement of lines of stone columns beneath loadbearing walls or on a grid pattern beneath rafts. The geometry of the proposed structure constrains the spacing of the columns in plan. For small structures such as low-rise housing, there is little scope for variation in treatment pattern; treatment points along each length of footing are generally at about 2 m centres with additional treatment points under groundbearing slabs if slabs are deemed to be suitable[117]. In some cases suspended ground floor slabs may be required. As it is not necessary to treat the whole site in a uniform way, the method is well adapted to traditional housing and light industrial units.

Where vibro stone columns are used in a granular non-engineered fill it will often be possible to densify the fill significantly. However, the installation of stone columns in a soft saturated clay is unlikely to improve the clay; any benefit will be associated with the presence of the relatively stiff stone columns.

Well over half the applications of vibro in the UK are in fill, with some work in soft clay. If the fill contains discrete zones of organic matter or any other substance that may decrease in volume with time due to chemical reaction or solution, these may have to be removed and replaced with granular fill. Significant volumes of unacceptable material render a fill unsuitable for vibro treatment. The reduction in volume of the fill would lead to a loss of lateral support for the stone columns and consequent settlement. For natural soft ground, a lower limit of undrained shear strength of 15–20 kPa is often quoted. If the dry process is employed, an undrained shear strength in excess of 30 kPa is required to ensure that the hole created by the vibrating poker stays open during backfilling with stone. The amount of improvement that can be achieved in natural soft ground may be quite limited. This means that on this type of ground the method is often more appropriate to low-rise industrial buildings than settlement-sensitive housing.

Extensive buried obstructions can seriously impair a treatment programme if not identified and removed ahead of the works. A firm or desiccated crust may need to be pre-bored to avoid impeding the penetration of the vibrating poker.

Stone columns could form paths for water to penetrate into untreated fill at depth; this could have serious consequences in a loose, partially saturated fill susceptible to collapse compression on inundation. Such fill

Figure 2.4 Vibro stone column ground improvement

2.1 General points on foundations

can include opencast mining backfill, mining wastes and chalk.

The possibility that vibration will cause settlement of adjacent structures and buried services needs to be carefully considered; limiting distances of 2–5 m have been quoted. Although the noise level is relatively low, there may still be environmental problems.

It is usual for there to be extensive records kept of the treatment process including dates, plant types, depths, quantities of stone in each column, ground heave and vibrator details. Load and penetration testing may be appropriate (Figure 2.5).

Pre-loading

Pre-loading with a surcharge of soil can be used on both fill and soft clay (Figure 2.6). In the former application the density is increased by reducing the air voids and most of the compression occurs as the surcharge is placed. In the latter application preloading compresses the soft clay as water is slowly squeezed out of the voids. The speed at which consolidation occurs is reduced by the low permeability of the clay. It will usually be necessary to install vertical drains to achieve consolidation within a reasonable period (Figure 1.7 in Chapter 1.1).

The superior load carrying characteristics of many natural soils are due to over-consolidation from preloading during their geological history. This over-consolidation has made them stiffer under applied loads than comparable normally-consolidated soils would be. In a similar way load carrying characteristics can be improved by temporary preloading with a surcharge of fill prior to construction.

The method can be used on two different types of soil:
- uncompacted fill with large air voids. Where the fill is in a loose unsaturated state, compression will occur mainly as the surcharge is placed and consequently there is no need to leave the surcharge in position for an extended period

- soft, saturated, natural clay soils. Consolidation may take a considerable time and it may be necessary to install vertical drains to speed up the rate of consolidation, controlling the rate of placement of the surcharge to ensure that the preloading does not cause instability due to excess porewater pressures in the fill

Consolidation times are a function of the square of the length of the drainage path and horizontal permeability is usually much greater than vertical permeability. Therefore the quickest possible drainage is achieved using vertical drains, as closely spaced as practicable and economical. It was once common for vertical drains of sand 200–500 mm in diameter, and spaced at 1.5–6 m centres, to be installed using both displacement and non-displacement techniques. Prefabricated land drains are now widely used. Generally these are 100 mm wide and a few millimetres thick with plastics cores wrapped in a paper or fabric filter.

Treatment by preloading does not require specialist equipment or skills unless vertical drains are required. Pre-loading with a surcharge of fill requires only normal earthmoving machines, and the appropriate type of plant for a particular job will depend on the following:
- the quantity of surcharge fill to be moved
- the haul distance
- the ability of the surcharge fill to withstand site traffic

The following must be specified:
- the type of fill to be used for the surcharge
- the extent in plan, height and slope angle of the surcharge
- any restriction on the rate of placing of the surcharge to maintain stability (usually only a problem where a saturated low permeability fill is preloaded)
- the period the surcharge is to be left in position

The height of the surcharge should be calculated in relation to the properties of the uncompacted fill that it is necessary to improve[118,119]. This is simple where settlement caused by the weight of buildings will be the major problem, but more difficult where other causes of settlement are perceived to be major hazards.

Figure 2.5 A simple loading test in progress to assess the effectiveness of vibro treatment

Figure 2.6 Surcharging poor ground prior to the development of a site

There are some factors which limit the practical usefulness of the method:
- A relatively large area is needed if preloading with a surcharge of fill is to be practical; it is essentially a whole site treatment
- the cost depends on the haul distance and consequently a local supply of fill is usually required

Chemical stabilisation

The behaviour of soft, natural soils (and of fill) may be improved by additives which may either modify soil behaviour by physico-chemical processes or cement the soil together. Chemical stabilisation may have one or more of the following objectives:
- to improve load carrying properties (eg improve bearing capacity, reduce compressibility or increase resistance to water softening)
- to reduce permeability
- to stabilise chemical contaminants

Cement, lime, bitumen and various chemicals may be mixed with the ground in the following situations:
- during placement of fill in layers
- by some in situ mixing process (eg lime columns, as used extensively in Scandinavia)
- by injection into the ground

Table 2.1 Empirical skin friction values[120]	
Type of soil	Skin friction (kN/m²)
Silt and soft clay	7–30
Very stiff clay	50–200
Loose sand	12–36
Dense sand	33–67
Dense gravel	50–100

Main performance requirements and defects

Structural requirements

For tall buildings and others involving heavy loads on the foundations, the choice of foundation design and the bearing capacity of the soils will be critical to subsequent performance. These buildings are most likely to have had the benefit of geotechnical and structural engineering expertise. On the other hand, most foundation problems for lightly loaded buildings are unrelated to applied loadings from the structure, and prescriptive requirements tend to predominate.

Bearing capacity of shallow foundations

The ultimate bearing capacity of shallow foundations is normally determined by postulating a failure mechanism in the soil and calculating the force required to overcome the shear strength of the soil acting along the failure planes. The more realistic the mechanism, the lower (and more precise) the calculated bearing capacity.

Bearing capacity of deep foundations

For deep foundations it is necessary to consider the bearing capacity as two separate components: skin friction acting on the vertical faces or shafts of the foundations, and base resistance acting on the undersides of the foundations.

For deep foundations other than piles, such as piers, the contribution of skin friction will depend on the degree of contact between the soil and the foundations. These foundations are often used where there is a surface layer of soft or erodible soil; in these circumstances, it would be inappropriate to rely on skin friction. In other situations – for example, where a shaft in competent ground is backfilled with concrete to form a pier – it may be appropriate to allow some skin friction, although the value will depend on the degree of loosening and swelling that occurred while the excavation was open. In most instances, precise calculations are unwarranted and it is usual practice to adopt rule-of-thumb values such as those given in Table 2.1.

Negative skin friction
Where a deep foundation passes through a layer of fill or soft soil, there is a risk that the surface soil will settle in relation to the foundation due to a variety of processes (eg consolidation under fill used to raise ground levels, groundwater lowering, disturbance of the soil caused by pile driving). Any skin friction associated with this settlement will induce a downward loading on the foundation – a process known as negative skin friction or downdrag.

Building regulations

Most foundations of buildings built in the last half of the twentieth century will have complied with the requirements of the Model Byelaws, and building regulations. The Building Regulations state that *'The building shall be so constructed that ...... loads are ...... transmitted to the ground (a) safely and (b) without causing ...... deformation of the building as will impair the stability of any other building'*. Requirement A2, Ground movement, states that *'The building shall be so constructed that movements of the subsoil caused by swelling, shrinking or freezing will not impair the stability of any part of the building'*.

Compliance with the Regulations is deemed to have been satisfied if the guidance contained in Approved Documents is followed. These Approved Documents call upon a host of widely recognised documents, chiefly the Codes and Standards published by the British Standards Institution.

Eurocode 7

Few existing structures, other than experimental ones, have been designed under new rules. Nevertheless, it is to be expected that Eurocodes, particularly Eurocode 7[121], will come to play an increasing role in the design of foundations, as with other aspects of the structure of buildings.

2.1 General points on foundations

Factors of safety, permissible stress and limit state design

The structural Eurocodes, particularly for geotechnics, have introduced into the UK new methods of performing design calculations. Traditional British geotechnical design has featured the use of a 'lumped' factor of safety to ensure that the resistance offered by a foundation, say, is sufficiently greater than the forces attempting to disturb it. Another way of describing this approach is using the expression 'permissible stress' design: the stresses acting in structural elements (or the ground) are limited to permissible levels by the application of a factor of safety. The permissible level of stress ensures not only that the element will not break but also that deformation is acceptably small. In structural engineering, the permissible stress design approach was replaced in the 1980s by the so-called limit state design method in which limiting conditions, such as ultimate failure of a structural member (ultimate limit state or ULS), are explicitly and separately avoided by limiting the stresses in the element through the use not of lumped factors but partial factors; also to be avoided is a serviceability limit state (SLS) at which unacceptable deformations occur. In the lumped factor method, values of applied load and element resistance are assumed to be approximately average or perhaps moderately conservative values, and a single factor is used to reduce the loads (or increase the resistance) by an amount derived from experience of the past behaviour of similar designs.

However, the partial factor method separates the lumped factor into some of its component parts. For example, the uncertainty of the load is expressed through the use of a partial factor for the load, while the uncertainty of the strength of the structural element is expressed through a partial factor applied to the material strength.

In principle, it is more logical to use partial factors rather than lumped factors because the former attempt to deal explicitly with uncertainty in unconnected parameter values and separately for different limit states, whereas the lumped factor approach embodies all uncertainty into one large factor and cannot explicitly cater for the different limit states. However, most UK experience of calibrating design methods to observations of the performance of structures or analysis of why they have failed is embodied in the lumped factor values used for so long. Nevertheless, a great deal of work has been undertaken to ensure that the new design methods in the Eurocodes have been calibrated to previous practice so that new designs will be sufficiently safe. Not surprisingly there remains a reluctance to abandon a tried-and-tested method of design until sufficient experience has been gained of the use of the partial factor design method.

Depth of influence of loaded areas

The determination of the depth of influence of a loaded area at ground surface is one of the basic problems in geotechnics and has important applications. The depth of influence of a foundation load has to be evaluated in order that the settlement which will be caused by the loading can be calculated. The depth of influence also needs to be assessed when designing a scheme of ground improvement by preloading with a surcharge of fill[122].

Tolerable settlement

There are a number of criteria that can be applied to determining how much foundation movement a building can tolerate; these include:

- visual appearance
- serviceability
- structural integrity

For the majority of buildings, allowable foundation movements are likely to be limited by their effect on visual appearance and serviceability rather than the integrity or stability of the structure. Serviceability is often the prime consideration for industrial buildings and will depend on the function of the building, the reaction of the user and owner, and economic factors such as value, insurance cover and the importance of prime cost. In other cases for non-domestic buildings, the level of tolerable movement will be dictated by a particular function of the building or one its services; for example, the proper operation of overhead cranes, lifts, precision machinery, or drains. On the other hand, for domestic buildings, visual appearance will probably take precedence, and while expectations and perceptions will vary between individuals, it is generally desirable to avoid cracking of walls and cladding materials, and distortion, the obvious signs of which include sticking windows and doors.

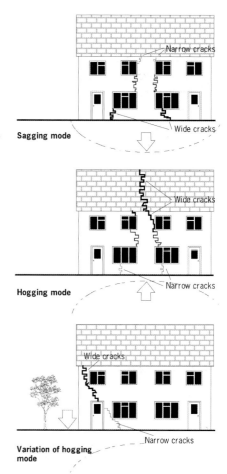

Figure 2.7 Crack patterns associated with different modes of distortion

Differential movement or settlement

If the building and its foundations are designed so that they move all together, to the same extent and at the same rate, then all that is affected are the connections of the building to its external services; these topics are dealt with in appropriate later chapters. However, if the foundations of different parts of the building move independently of each other (differential movement), there will inevitably be some distress on the fabric.

HAPM says that, of the sites they inspected during the early 1990s, some 1 in 6 sites where movement joints should have been provided made no provision to accommodate differential movement.

How much building distortion and, therefore, foundation movement is tolerable depends on a wide range of factors; in particular, on the length-to-height ratio of the wall and the mode of deformation (ie whether the wall is hogging or sagging as shown in Figure 2.7 on page 85). Distortion can be defined in a variety of ways, although it has been concluded that the most meaningful parameter in the context of cracking is deflection ratio. This parameter can be defined as the maximum vertical deviation from original construction divided by the length of the affected portion of the wall. Foundation movement, on the other hand, is normally quantified in terms of differential settlement which is the (maximum) vertical displacement of one part of the building with respect to another[123].

In practice, it is difficult to predict differential settlement, let alone deflection ratio. Engineers have tended, therefore, to depend instead on correlations between measurements of maximum total settlement and manifest damage. Proposed design limits for total settlement[124] are summarised in Table 2.2.

The tighter limits for foundations on sand are attributable to two factors:
- because of its greater permeability, settlement is likely to take place more rapidly on sand without allowing any creep or stress redistribution in the structure
- granular soils tend to be less homogeneous than clay soils, thereby tending to reduce the difference between differential settlements and total settlements

While many buildings will tolerate differential settlement of the order of half the average total settlement, it is a widely held view that the average masonry two-storey dwelling will not tolerate differential settlement exceeding about 25 mm. Considerable work remains to be done to improve prediction of building response to settlement. However, with the wide diversity of structural form and materials used, the likelihood of such improvement remains small, and, consequently, the understandable attitude of most designers is to try to eliminate all settlement.

Groundborne vibration

It has long been alleged that groundborne vibration (eg from road or rail) can cause damage to buildings, but it is notoriously difficult to find conclusive evidence that this has happened. Authenticated cases of damage to buildings by vibration are very rare, and it is far more likely that damage has more underlying causes which are simply triggered by groundborne vibrations.

To quote from *Cracking in buildings* [12]: '*The belief that buildings can be damaged by, for example, traffic vibration rests to a large extent on the remarkable ability of the human frame to detect vibrations of very small amplitude. The resulting subjective impression of heavy vibration may then be reinforced by other factors, such as windows rattling in the air currents created by a vehicle passing in close proximity. Therefore surveyors should not be unduly influenced by occupants' subjective impression of vibration: the true cause of any cracking ascribed to vibration is virtually certain to lie elsewhere*'.

A computer database was set up to store a substantial amount of data relating to cases of groundborne vibration and structural damage in the UK. The data were analysed to investigate a number of parameters including values of peak particle velocity (ppv) and frequency of ground vibration. Vibration cases where damage was alleged or verified were analysed separately as well as with non-damage cases. Frequency of vibration, severity of damage, ppv, structural types affected and vibration sources were analysed for the damage cases. Relationships between damage, foundation types and soil types were also considered. The results of the analyses do not indicate any threshold value of ground vibration ppv for damage. Similarity, no single structural type, foundation type or soil type can be positively identified as being the most susceptible to damage.

Case study

Investigation of alleged damage due to blasting

BRE investigated complaints of cracking damage to a building which had been attributed by the occupants to blasting at a nearby opencast coal pit. A survey of damage noted narrow cracks in the mortar joints of exterior stonework, and there was also slight damage to interior plaster. No evidence was found of increasing severity of damage over time. The overall degree of damage was considered to be slight and it was concluded that the crack damage had resulted from a variety of causes which included minor foundation movements, and thermal and moisture movements of building components. None of the damage seen in the survey could reasonably be attributed to blasting, or indeed from other groundborne vibrations.

Table 2.2 Limits for maximum settlement

	Clay (mm)	Sand (mm)
Isolated foundations	65	40
Rafts	65–100	40–65

2.1 General points on foundations

Thermal insulation

Many reinforced concrete rafts, and most cellars and basements, will have been built without the benefit of thermal insulation; but for those built since building regulations were amended in 1992, it will have been necessary to incorporate thermal insulation. This will usually take the form of polystyrene boards, but polyurethane foams may also have been used either for void filling or for thermal insulation.

Thermal insulation for floors is covered in Chapter 1.3 of *Floors and flooring*, and for cellars and basements in Chapter 3.3 of this book.

Diagnosis of problems

Questions which need to be addressed

Foundations rarely fail in the sense that the superstructure or parts of it collapse. Although materials may degrade with time – for example, concrete may be attacked by chemicals in the groundwater and steel may corrode – it is rare for these processes to affect the foundations to the point when they are no longer capable of supporting the building. What is usually meant by failure, therefore, is that they have failed to meet their primary function which is to carry the superstructure loads safely to the ground, with no unacceptable movements or damaging deformation of the building. The essence of successful foundation design is to ensure that the movements transmitted to the superstructure never exceed tolerable levels.

When investigating a damaged building to determine if the foundations are at fault, there are three basic sequential questions which need to be addressed:
- have the foundations moved?
- what potential is there for further movement to occur?
- if further movements are likely, what can be done to reduce their potential for causing further damage?

To answer these questions it is essential to identify as accurately as possible what has caused the damage by examining as much evidence as possible. However, the scope of the investigation is likely to be constrained by considerations of cost. The essence of a good survey of a damaged building is, therefore, to identify at an early stage whether or not remedial works are required so that decisions can be made on the cost effectiveness of further investigations.

Any material deforms under applied loading and soil is no exception. For granular soils, such as sands and gravels, the deformation is largely elastic (ie recoverable) and almost instantaneous. For clays, however, there is an additional long term component associated with the gradual squeezing out of porewater (ie consolidation); in low-permeability clays this process can take many years. Furthermore, more compressible soils such as soft clays and fill tend to creep under the applied load; this process can continue throughout the life of the building, albeit at a decreasing rate. Different parts of the building may settle by varying amounts due to either variations in the applied foundation load or variations in the properties of the underlying soil; it is this differential settlement that can lead to cracking. Provided, though, that the allowable bearing pressure of the soil is not exceeded (Chapter 1.1), the settlements should be small, often no more than a few millimetres. Any associated damage should show during the first few years after construction and should be easily repairable.

Settlement tends to be a serious problem only where the allowable bearing pressure is exceeded and this can normally be attributed to poor or inappropriate foundation design, or the existence of a unidentified soft layer (eg peat) at shallow depth.

Subsidence

The factors giving rise to subsidence include:
- clay shrinkage
- erosion of soils
- lowering groundwater levels
- rising groundwater levels
- compression of fill
- collapse of mine workings and other cavities
- construction processes
- vibrations

These factors are now discussed in turn.

The effects of evaporation and moisture extraction by vegetation on clay soils have been described in Chapter 1.5. These processes can cause damage to any buildings that are not founded at a sufficient depth to protect them from the volume changes that occur in the surface soil, the damage being most likely to occur during prolonged periods of dry weather. Since domestic buildings insurance was extended to include damage caused by subsidence, there have been several unusually dry periods in south east England: 1975–96, 1989–90 and 1995–96. The trend in the level of insurance claims shows that there were sharp rises following these dry periods, particularly following 1989–90 when the effects of two consecutive dry summers were exacerbated by a dry intervening winter. This coincidence of damage with dry weather and analysis of claims information indicate that clay shrinkage is by far the commonest cause of damage to low-rise buildings.

However, the results of recent and earlier BRE research have confirmed that, even during prolonged periods of dry weather, the effects of evaporation are largely confined to the surface metre of soil. Hence, the types of building that are likely to be susceptible to movement as a result of clay shrinkage are largely restricted to:
- those buildings that do not conform to the current British Standard requirement of a minimum foundation depth of 0.9 m in clay soils, where there are no nearby trees or other large vegetation
- those buildings near to trees or other large vegetation and whose foundations do not conform to current NHBC recommendations[125], either because they were built before 1969 (when NHBC recommendations were first published) or because their circumstances have changed since they were first constructed. Often this change in circumstances is brought about by the owners planting potentially large trees too near the foundations, or by constructing an extension, on shallow foundations, too close to large vegetation

While surface erosion by wind and rain can remove soil from around buildings, this process rarely leads to an adverse effect on foundations. Internal erosion can, however, cause damage to buildings. Percolation of water through granular soils can wash out fine particles or dissolve soluble minerals, creating voids or reducing the density of the remaining soil and thereby producing subsidence. Most problems are caused by fractured drains or water pipes.

Where the lateral stresses acting on clay soils are reduced, for example adjacent to a cutting, excavation or steep slope, the clay will tend to absorb water and soften. This process will tend to reduce the bearing capacity of the soil and may, in extreme cases, lead to an increased possibility of foundation loads inducing a shear failure in the soil. More commonly this process may produce some limited movements as it allows soil to move towards a loosely backfilled trench, such as a deep drain. This softening will occur wherever the soil is below the water table for at least part of the year, which in the UK means that most excavations that are close to and go below foundation level are likely to induce some limited ground movements. Having softened, the soil would be largely unaffected by a sudden influx of water; for example, from a leaking drain. However, it is a common practice to lay drains in a bed of granular material and, in these circumstances, leaking water may wash out fine particles allowing further movement of the already softened clay.

Another way that a leaking drain can have an influence on foundation movements in clay soils is where there are trees nearby. In these cases, a long term leak could reduce the state of desiccation in the soil, producing recovery, and lead to local heave.

Water abstraction or general changes in groundwater regime can cause certain soils (eg loose sands, silts and peat) to reduce in volume and to subside. This is a common problem in fen areas where lightweight buildings on rafts have been adversely affected by drainage schemes.

Rising groundwater levels are a recognised problem in certain cities (notably London, Birmingham and Liverpool) where cessation of pumping of water from underlying aquifers, formerly extracted for industrial use, is allowing groundwater to rise to levels not seen for a century[126,127]. Elsewhere in the UK, ending of pumping for regional drainage in former coal mining areas may lead to a rapid rise in groundwater levels[128]. Groundwater rising into unstable room-and-pillar mine workings may trigger local surface subsidence.

In clay strata overlying aquifers, the effect is to cause swelling which may result in vertical heave in buildings, particularly those founded partly on deep and partly on shallow foundations. Long piles may fail in tension. In both clays and granular soils there may be a reduction in bearing capacity of end-bearing piles.

For both causes of rising groundwater there is a risk of seepage into basements and attack on foundations and buried services from corrosive (eg sulfate-rich) groundwater. In the case of rising coal mine waters there can be a temporary hazard from increasing methane gas emissions.

Unless it is placed under carefully controlled conditions, fill tends to be far more compressible than most natural soils (Chapter 1.2). Therefore, where the presence of a small amount of fill goes undetected during construction (eg when building on the edge of an old backfilled pit), there is an increased likelihood of damage as a result of excessive differential settlement.

Even where the presence of the fill is recognised, special care is needed because of the tendency of these materials to compress slowly with time, either as a result of self-weight consolidation or from degradation of any organic component. Fills may also be sensitive to a change in the groundwater level: some may settle dramatically the first time they are inundated – the collapse compression process (Chapter 1.2).

As described in Chapter 1.1, shallow mine workings are prone to collapse causing localised subsidence of the overlying ground surface as the void migrates upwards. A similar phenomenon can occur in limestone and chalk deposits which contain natural cavities or solution features formed by percolating water. Deeper abandoned mines can cause subsidence over wider areas as groups of pillars, left to support the roof, collapse progressively. Modern, longwall mining, where deliberate roof collapse follows coal extraction, can produce a wave of subsidence that may distort and crack any buildings in its path. However, the movements are often of relatively short duration and, in many cases, the buildings will be left in a stable

2.1 General points on foundations

and serviceable condition. Problems are most likely to occur towards the periphery of the subsidence area where permanent differential movements are greatest and in areas containing faulted rock; these faults may be reactivated by the mining subsidence, locally causing large differential movements at the surface.

All construction work, and in particular deep excavation and tunnelling, generates deformations in the ground that can affect nearby buildings. There is a tendency for the surrounding soil to move towards the excavation or tunnel, the magnitude of the movements depending largely on the method of construction and the properties of the soil. A common design rule-of-thumb is that no movement occurs at distances further than three times the depth of the excavation. The effects of excavation are well known, especially in urban redevelopment schemes. Careful monitoring of adjacent buildings is frequently employed to give warning of undue disturbance and to control the excavation process.

As previously noted, there is little evidence of groundborne vibration causing damage to buildings other than in extreme cases such as explosions, earthquakes or driving piles close to a building. Vibration may, however, exacerbate existing damage (ie encourage existing cracks to extend and widen) and could conceivably result in slight compaction of some soils (eg loose sands or poorly compacted fill).

Heave
The four processes that give rise to heave are:
- removal of overburden
- swelling of desiccated clays
- frost heave
- chemical reaction

These processes are now discussed in turn.

Just as application of foundation load causes settlement, the removal of load due, for example, to excavation for a basement or general lowering of ground levels, will produce some recovery or heave in the soil. In granular soils it will occur instantly; in cohesive soils it may take place over a period of many years. BRE has monitored the continuing upward movement of a Northern Line Underground tunnel beneath the Shell Centre basement on the South Bank in London since its excavation in 1953[35]. Unless large quantities of soil are being removed, the movements are unlikely to be more than a few millimetres. It may, however, be appropriate to design foundations to allow the underlying soil to swell without lifting any of the building; for example, by providing a void under the floor slab of a piled building.

Desiccation is essentially a reversible process. It has been already noted in Chapter 1.5 that reversal can occur if a tree is removed.

Prolonged periods of freezing weather can cause surface layers of chalky and silty soils to expand, resulting in heave damage to buildings on very shallow foundations. However, in the UK, it is generally accepted that this process is unlikely to affect any buildings founded below 0.5 m. Where damage has occurred, it has frequently been to new, part-completed buildings in highly exposed conditions. Special consideration may need to be given to refrigerated buildings used for cold storage.

Most chemical reactions that affect foundations are expansive and therefore could result in heave. The commonest form is sulfate attack of concrete[33]. The lifting of floor slabs supported on sulfate-rich fill material was, until relatively recently a common problem. Figure 2.8 shows how sulfate attack of the underside of a concrete slab can cause it to arch and crack. Fill materials that may be particularly high in sulfates include burnt colliery shale, brick rubble with adhering plaster and some industrial and mine wastes.

Rarely, sulfate attack can weaken foundation elements such as piles and result in their failure. Sulfate attack occurs only where there is water. A high or fluctuating water table will normally saturate the whole of the foundations and cause general uplift. However, where the source of the water is internal leakage or external run-off, the damage may be concentrated in the wettest areas.

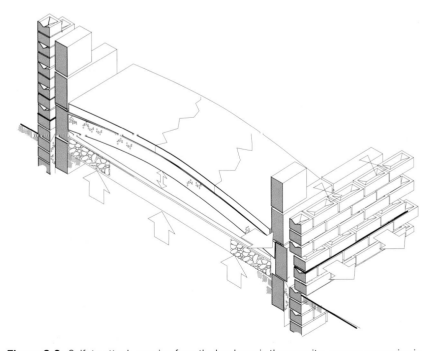

Figure 2.8 Sulfate attack, coming from the hardcore in the oversite, causes expansion in the concrete slab, leading, in turn, to the slab doming, and displacing blocks and bricks in adjacent walls

Landslip

Landslip occurs in one of two ways:
- as a sudden, bodily movement of a large mass of soil sliding on a shear plane (slope failure)
- as a gradual movement downhill of a surface layer (slope creep)

Many incidents of landslides are slow movements of old, reactivated slope failures.

In the UK it is uncommon for buildings to be sited on slopes that are naturally unstable; there are, however, exceptions such as on the Isle of Wight, where hundreds of houses are situated on the Under Cliff, which is a complex of old landslides which are slowly moving. Evidence of previous movements should normally be detected during the site investigation (Chapter 1.1) and the location avoided or steps taken to improve its stability (as described in this chapter). There is a possibility of decreasing the stability of existing slopes by excavating near the toe of the slope, or by loading the crest with a building or fill material, or by making local alterations to the natural slope or the groundwater regime.

Of greater relevance to most buildings is the potential for relatively shallow clay slopes (say 7–12°) to migrate slowly downhill. There is very little information on this phenomenon. It is thought that damage caused by slope creep is commonly (and wrongly) diagnosed as clay shrinkage; however, it is possible that the two phenomena are related. Desiccation cracks that open up during periods of exceptionally dry weather may not close completely, causing a net movement down the hill. Similarly, slopes can exacerbate clay shrinkage by increasing run-off, thereby reducing the amount of rainfall infiltration into the ground.

Investigating damage

Despite the fact that there has been a sevenfold increase in the level of claims for heave and subsidence damage to domestic properties, there has, until recently, been a dearth of guidance on how best to investigate claims and select appropriate remedial measures. This section, which is based both on observations of foundation behaviour made by BRE over the last 50 years and on involvement in a number of recent claims, outlines the key elements of a cost effective investigation.

The initial survey of the damaged property should include a sketch showing the position, width and taper of all internal and external cracks[34]. There are two very good reasons for doing this: first, it provides an objective record against which future damage can be compared, and second it facilitates the identification of the mode of distortion (ie the way in which the foundation movements are affecting the building as a whole). This is important, as it may highlight damage that is being caused, for example, by rotation of a wall and may, therefore, be well removed from the area that is subsiding or heaving. In other words, the location of the damage does not necessarily correspond to the location of the foundation movement.

It is obviously important to establish the history of the damage, and information supplied by the owner or occupier should be corroborated wherever possible by examining the surfaces of cracks for dirt deposits indicative of age, looking for evidence of past repairs etc. The size, position and species of all trees close enough to have an effect on the foundations should also be noted.

Where there is exposed brickwork on the outside of the building, a simple survey should be performed using a portable water level[129] or an optical level to determine whether there are any significant level changes along brick courses. Alternatively, where the external walls are rendered, an indication of previous movement may be obtained by measuring the verticality of walls.

In some cases, other types of distortion survey are appropriate[130]. For example, where heave is suspected, it may be appropriate to measure levels on the floor slabs as well as the external walls. Ideally the distortion survey should form part of the initial inspection: it takes about 1 or 2 hours for one person to check brick course levels around a detached house using a water level.

The local geology should be checked by reference to the relevant 1:50,000 scale Drift (or Drift and Solid) geological map (or, in some areas, larger scale maps). Again, it may be appropriate to check old maps, records and aerial photographs for the location of pits, streams, trees etc. It may also be possible to obtain copies of the plans for the original building work and any extensions; these can then be used to identify the depth of foundations, heave precautions, floor slab details etc and save on the cost and disruption of trial pits. There is always, though, a probability of significant differences between the plans and the actual construction.

Having decided that some foundation movement has occurred, it is generally desirable to conduct a physical investigation for confirmation. Since physical investigations are relatively expensive, it is important that they are properly targeted. At least one trial pit should be excavated early in the investigation and, in straightforward cases, this may be the only ground investigation that is required. Trial pits are needed to confirm the depth and condition of the foundations; they are also used to obtain soil samples for index property tests, which are then used to establish the soil's volume change potential (Chapter 1.1). A hand auger can be used below the base of the pit to obtain deeper samples for an indication of the depth of any desiccation (Figure 1.13 in Chapter 1.1).

There are a number of possible methods of forming boreholes that might be used in the investigation of building damage. Light percussion drilling (sometimes incorrectly known as shell and auger drilling) is occasionally used. This is substantially more expensive than trial pitting and should, therefore, be carefully specified. Typical examples of where its use might be warranted are:
- where removal of a large mature tree is being considered, and it is important accurately to determine the depth and degree of desiccation
- where a building is suspected of being on a backfilled pit and it is important to confirm the geology

When investigating shrinkable clays, samples of undesiccated clay should be obtained from a control borehole for comparison wherever possible.

If only disturbed samples are required to a maximum depth of about 6 m, a hand auger might be used. Driven tube window samplers are being used increasingly; in clay soils, these can form boreholes to a maximum depth of about 8 m relatively rapidly.

The most usual manifestation of failure is cracking of the fabric. However, there are a number of other causes of cracking which first need to be investigated.

Cracking

BRE Digest 251[34] presents a classification of visible damage to walls with particular reference to ease of repair of plaster and brickwork or masonry (Table 2.3). Crack width is just one factor in assessing category of damage and should not be used on its own.

Possible causes of cracking

Foundation movement is only one of many processes that can cause distortion and cracking in buildings. The first step in investigating a damaged building is, therefore, to try to identify any damage that can be attributed to causes other than foundation movement. Some of the more common processes that affect buildings are listed below:

Thermal effects

Varying amounts of expansion and contraction in different building materials produce strains that cause cracking. Flat roofs and long south facing walls without expansion joints are particularly vulnerable. Dramatic thermal effects can be generated by fires; for example, an expanding steel roof truss can punch through the masonry supporting its ends.

Creep

Over long periods, some building materials deform slowly under load, and these very slow movements may eventually lead to cracking. An extreme example of creep is roof spread, where prolonged loading of inadequately tied pitched roofs can result in appreciable lateral loading being transferred to the walls.

Moisture changes

Many building materials such as calcium silicate (sand-lime) bricks, lightweight concrete products, plain concrete slabs, some plasters, rendering and timber, shrink as they lose moisture. This shrinkage occurs normally during the first few years after construction but can be triggered in older buildings by the installation of central heating. In contrast, new, ordinary clay bricks expand as they absorb moisture. The damage associated with these processes is usually fairly minor and unlikely to be a continuing problem.

Table 2.3 Classification of visible damage to walls

Category of damage	Description of typical damage (*Ease of repair in italic*)
0	Hairline cracks of less than about 0.1 mm which are classed as negligible. *No action required*
1	Fine cracks which *can be treated easily using normal decoration*. Damage generally restricted to internal wall finishes; cracks rarely visible in external brickwork. Typical crack widths up to 1 mm
2	*Cracks easily filled. Recurrent cracks can be masked by suitable linings.* Cracks not necessarily visible externally; *some external repointing may be required to ensure weathertightness*. Doors and windows may stick slightly and *require easing and adjusting*. Typical crack widths up to 5 mm
3	Cracks which *require some opening up and can be patched by a mason. Repointing of external brickwork and possibly a small amount of brickwork to be replaced*. Doors and windows sticking. Service pipes may fracture. Weathertightness often impaired. Typical crack widths are 5–15 mm, or several of, say, 3 mm
4	Extensive damage which *requires breaking-out and replacing sections of walls*, especially over doors and windows. Windows and door frames distorted, floor sloping noticeably*. Walls leaning or bulging noticeably*; some loss of bearing in beams. Service pipes disrupted. Typical crack widths are 15–25 mm, but also depends on number of cracks
5	Structural damage which *requires a major repair job, involving partial or complete rebuilding*. Beams lose bearing, walls lean badly and require shoring. Windows broken with distortion. Danger of instability. Typical crack widths are greater than 25 mm, but depends on number of cracks

* Local deviation of slope, from the horizontal or vertical, of more than 1 in 100 will normally be clearly visible. Overall deviations in excess of 1 in 150 will lead to problems (eg unstable furniture)

Over-stressing
Local crushing of masonry may occur where loading is increased by, for example, structural alterations. Increased floor loadings (eg storage of heavy items in a loft) can cause ceilings to sag and crack.

Frost attack
Freezing and thawing of absorbed water can cause cracking and spalling in porous materials such as fired clay products, natural stones, and weaker concrete materials.

Chemical attack
Aqueous solutions of sulfates can attack cement and hydraulic lime mortars and concrete blocks causing them to expand. The problem is restricted to exposed walls and slabs that remain wet for prolonged periods. Carbonation can affect porous concrete and calcium silicate products causing them to shrink. Some concrete aggregates may react with alkali present in the hardened cement causing the concrete to expand and crack – a process known as alkali silica reaction.

Corrosion
Corroded wall ties can lead to the failure of cavity brickwork. In addition, the expansion associated with the oxidation of steel fixings and reinforcement buried in porous building materials can cause movement, spalling or cracking of the materials in contact with the metal.

More detailed descriptions of cracking processes, and how they can be recognised and remedied, are given in *Cracking in buildings*[12], *Why do buildings crack?*[131], and in *Roofs and roofing*[97], *Floors and flooring*[66], and *Walls, windows and doors*[101].

Coefficients of linear thermal expansion, and of reversible and irreversible moisture movements, for common building materials are given in Tables 1.1, 1.3 and 1.4 of *Walls, windows and doors*.

Cracks caused by foundation movement
Most of the processes listed above tend to produce a profusion of small, widely distributed cracks. However, shrinking and swelling of bricks can cause local cracking around windows and at junctions of walls, though it is rare for these cracks to exceed 5 mm in width and they seldom extend through the dampproof course (DPC). Foundation movement, on the other hand, tends to produce a few isolated cracks which are often continuous through the DPC and frequently more than 5 mm wide. Because foundation movements tend to affect large sections of wall, the cracking usually shows itself at weak points such as window openings and doors; moreover, the pattern and taper of the cracking can be associated with a particular mode of distortion[34].

Other indications of foundation movement are sticking doors and windows, displaced roof joists, gaps in roof tiles and disrupted services. The best way of confirming that the foundations have moved is to measure the out-of-level along a course of bricks or to measure the verticality of external walls. Damage is unlikely to have occurred unless there has been at least 25 mm of differential movement (40 mm in older buildings)[124], and this should be distinguishable from any construction tolerances which should be less than about 20 mm for a typical pair of semi-detached houses. Building errors are usually randomly distributed whereas foundation movement is indicated by a trend in level change.

Likely indicators of foundation movement include:
- a few isolated cracks show at weak points in the structure
- cracks taper from top to bottom
- cracking is continuous through the DPC
- crack width often exceeds 5 mm
- cracking occurs internally and externally at the same locations
- sticking doors and windows
- disrupted drains and services
- out-of-level or out-of-plumb walls

'Clay soils, in the top metre or so, undergo marked seasonal volume changes. Buildings founded within this zone move bodily up and down with those volume changes, the amount of movement depending in part on the depth of the foundations. Provided that all parts of the building are founded at the same level and the soil is uniform, no damage necessarily ensues. In practice it is desirable to limit the amount of such bodily movement in a building, in part to avoid damage to drainage systems.

Seasonal wetting and drying and the influence of minor vegetation produce negligible volume change in clay soils at 0.9 m, and this is widely accepted (possibly rounded to 1 m) as the minimum depth for foundations on clay soils. Significant volume changes at depths greater than this are associated with the de-watering effect of the root systems of major vegetation, principally trees. The crucial significance of this is that the volume change is localised and the corresponding movements at foundation level differ at various points under the foundations. It is differential foundation movements that crack buildings'[12].

Diagnosis
It is sometimes difficult to identify the cause of damage with confidence, especially where the damage is relatively minor. By the application of a basic knowledge of how buildings and the ground behave, it should be possible to narrow the choice further and to make some useful statements about the likelihood of the remaining causes. If the damage is severe enough or the case complex enough to warrant further investigations, these can then be properly targeted.

While an exhaustive account of all the deductive processes that can be applied to the information obtained during an investigation is beyond the scope of this chapter, the following comments illustrate the basic principles:

2.1 General points on foundations

Type of damage
Several possible causes can be eliminated by establishing whether the damage has been caused by subsidence, heave or landslip. This is often achieved by examining the pattern of damage, aided where necessary by simple measurements of changes in level along brick courses or the verticality of walls. Moreover, certain types of damage are characteristic of particular causes: for example, under-sailing of brickwork below DPC is indicative of heave associated with swelling soil, or landslip if the building is on a slope.

Soil type
The nature of the underlying soil is one of the most important factors in determining the cause of foundation movement. For example, only cohesive soils change volume with variations in their moisture content. Therefore, if the soil is identified as being predominantly granular (eg sand, gravel, chalk, limestone etc), the most common cause of foundation movement, that is seasonal shrinkage and swelling of the surface soil, can be immediately eliminated. On the other hand, if the soil is found to be a clay, erosion can usually be eliminated, although care is needed since many clays contain lenses of coarser material that could be affected. The existence of internal erosion can usually be identified by its effect on general ground levels and the presence of unusual quantities of water.

If the soil is a soft clay or fill, there is a strong possibility of the damage being caused by excessive settlement. This is particularly true if the thickness of the fill or soft layer varies across the site. With fills, there is the added possibility of movements as a result of self-weight compression, degradation or collapse compression, and these can often be recognised by their effect on ground levels generally. The likelihood of a building having been constructed on soft clay or fill is often determined by reference to local knowledge, geological maps, and records and plans showing, for example, the position of old pits. Failing this, a site investigation based on several boreholes is likely to be needed to confirm the nature and variability of the underlying soil.

Foundations
Another obvious factor to be considered is the design of the foundations. For example, any foundations that are more than 0.5 m deep are unlikely to have been adversely affected by frost heave; and any that are more than 1.0 m deep are unlikely to have been affected by clay shrinkage, unless there are trees or excavations nearby. Where the foundations are significantly deeper than 1 m and the soil is a shrinkable clay, trees may have been cleared from the site prior to construction suggesting clay heave as a cause of damage. This possibility should be checked by looking for deep roots in boreholes and obtaining aerial photographs of the area that pre-date construction. Using piles, pads or raft foundations for low-rise, relatively lightweight buildings may indicate that the builder or designer was anticipating difficult ground conditions.

Poor workmanship or inadequate specification may have contributed to the damage. For example, the forces generated by swelling clay can break inadequately reinforced piles or displace insufficiently protected trench fill foundations, and sulfate attack can weaken ground beams and other foundation elements not constructed with sulfate resisting concrete. Poorly prepared trenches can cause strip and trench fill foundations to settle excessively; sometimes this movement will occur during a dry period as soil shrinks away from the side of the concrete, causing movement to be confused with the effects of general clay shrinkage.

In some circumstances the design of the floor slab may also have an important influence. For example, a groundbearing slab significantly increases a building's susceptibility to ground heave, particularly where the internal leaf of cavity walls, as well as internal walls, rest on the slab.

Environment
Local topography can be important. In clay soils, some form of landslip is possible wherever the slope of the ground exceeds about 7°. In addition, localised movements are possible near cuttings and excavations, and these features may also have an influence on groundwater levels and consequently on the degree and extent of any desiccation. Some small movements are possible wherever there are poorly backfilled trenches at or below foundation level, especially where there is a flow of water to wash out any granular backfill material.

The existence of current and previous mining activity and the likelihood of the ground containing natural cavities is normally checked by reference to relevant geological maps and local records.

Water
The presence or absence of water can provide important clues. For example, where groundwater runs freely into a trial pit or borehole excavated in a clay soil, this indicates the existence of layers of coarser grained soil which can have a dramatic effect on the permeability of the soil mass; desiccation levels are, therefore, likely to be able to change far more quickly than they would in a pure clay. Moreover, where there is free water for any length of time there can be no desiccation; the discovery of water in a borehole should, therefore, be considered carefully when deciding on the likelihood of clay shrinkage or swelling as possible explanations for the manifest damage.

Generally speaking, chemical attack is only possible where the foundations are inundated and chemical processes having contributed to movement is confirmed or denied by testing the groundwater for aggressive compounds such as sulfates. Chemical testing can also help identify the source of the water; for example high levels of organic matter may indicate leakage from foul water drainage.

The possibility of changes in the groundwater regime having induced ground movements should also be considered. Rising water tables can cause increased loading on retaining walls, basements, and other watertight elements, leading to collapse compression in fill, while a general lowering of the water table can lead to settlement of loose sands, silts, peat and soft clays with a relatively high organic content. The pumping and de-watering operations that often accompany large excavations may have a local effect on groundwater levels.

Trees

The effects of trees on the soil is described in Chapter 1.5.

Where the soil is a shrinkable clay, trees and other large vegetation obviously play a key role in influencing ground movements. A large broad leaf tree such as an oak can have a significant effect on desiccation levels to a depth of 6 m and over an area of more than 1000 m^2. Wherever damage is attributable to clay shrinkage, the position of any nearby trees should be noted.

Although cases of trees causing damage in non-shrinkable soils such as sands, gravels and chalk are rare, the possibility should be considered where large trees are extremely close to the foundations.

Composite effects

In many cases damage cannot be attributed to a single cause and there may be several contributory factors. For instance, a property may be more susceptible to damage as a result of clay shrinkage because it settled differentially during the first few years following construction. Similarly, damage to a house built on an unstable clay slope may be exacerbated as the result of seasonal volume changes in the surface soil. In making a correct diagnosis it is, therefore, important to gauge whether the manifest damage and distortion are consistent with the movement attributed to a cause. Where a building is thought to have been damaged by clay shrinkage, the question that must be answered is: 'are the measured pattern and magnitude of the foundation movements compatible with the depths of the foundations, the properties of the soil, and the size, type and position of any nearby trees?' If the answer is no, then there is reason to suspect the contribution of other processes and further investigations and monitoring are likely to be needed to identify them.

Monitoring

Monitoring is a powerful tool. It can help to confirm the cause of damage and assist in deciding an appropriate course of action. Nevertheless, it should not be seen as a substitute for a proper and rational investigation and, in general, its use as an aid to diagnosis should be restricted to cases where the cause of the damage is not obvious. However, having taken some measures to mitigate the cause of the damage, such as removing a tree, monitoring takes on a completely different role; it is then being used to assess whether the measures taken are likely to provide a satisfactory long term solution, or whether further remedial measures, such as underpinning, are ultimately required.

Types of monitoring

There are three ways in which monitoring can be used to aid the investigation of a damaged property:

To establish cause of damage

Where the cause of the damage is not obvious, monitoring can be a very cost effective diagnostic tool. For example, where a house is founded on a slope, monitoring of vertical foundation movements will help distinguish clay shrinkage from landslip. Measurements are likely to be needed for at least a year before any long term subsidence or heave can be distinguished from seasonal movements. This may rule out monitoring as a diagnostic tool in cases where the damage is serious enough to make it unreasonable to postpone repairs.

To measure rate of movement

Where the cause of the damage is self-evident, monitoring can be used to establish whether the damage is progressive and, if so, the rate of progression. This might be carried out to determine an appropriate course of remedial action. Additionally, it can be a useful technique where it is known that the damage has been caused by a process that has a limited duration, such as heave following removal of a tree. In cases where the damage is attributable to a persistent effect, such as seasonal shrinkage and swelling of a clay soil, monitoring of this type is likely to be of limited benefit.

To check effectiveness of remedial action

Where sensible action has been taken to mitigate the cause of the damage (eg removal or pruning of nearby trees), monitoring can be used to gauge the effectiveness of the remedy. A decision can then be taken at the end of the monitoring period about whether further measures such as underpinning are required to stabilise the foundations. This proactive form of monitoring is generally preferable to the other varieties as it is associated with damage prevention and helps to avoid prolonging the duration of the investigation unnecessarily.

Monitoring techniques

Where a decision has been made to monitor a building thought to have been damaged by ground movements, there are two choices: to monitor the damage or to monitor the movements of the building. Also it is sometimes useful to measure the movements of the ground; for example, where landslip is thought to be a contributory factor, measurements of lateral ground movement are likely to be more helpful than measurements made on the structure. The various techniques that can be applied to monitoring damaged buildings are described briefly in the following sections.

2.1 General points on foundations

Crack width monitoring
The most common and simplest way of monitoring subsidence damage is to measure changes in the width of existing cracks. There are several ways in which this can be done:

Steel ruler
Provided sufficient care is taken, crack widths can be measured to the nearest millimetre using a steel ruler. However, because the readings tend to be subjective and it is difficult to ensure that the crack is measured at the same point each time, this form of measurement is normally used only for recording the state of damage during the initial investigation.

Magnified graticule
Internal cracks in plaster or other smooth finishes can be monitored by measuring the offset between two pencil marks using a magnifying glass fitted with a graticule. With care, movements can be measured to a resolution of 0.1 mm.

Plastics tell-tales
One of the most popular systems is shown in Figure 2.9. It consists of two overlapping plates screwed to the wall, one marked with a cursor, the other with a scale marked in millimetre intervals. The two plates are mounted on opposite sides of the crack so that the cursor is initially in line with the centre of the scale; any subsequent shear or normal movement of the crack can then be resolved to the nearest millimetre by recording the position of the cursor with respect to the scale. The advantage of this system is that a reading can be taken at any time by anyone, including the occupiers, without the need for additional measuring equipment or an original zero reading. The disadvantages are that the plates are relatively obtrusive, vulnerable to vandalism, and have only a limited accuracy. Some tell-tales are provided with integral Demec points (see the next column); this greatly improves their accuracy.

Glass tell-tales
Cementing glass strips across cracks used to be a popular method of detecting progressive movement. However, the strips are vulnerable to vandalism and, when they crack, they give no indication of the magnitude and direction of the movement.

Demec points
The distance between two special stainless steel discs can be measured very accurately using a demountable mechanical strain meter or Demec gauge. This system, which was developed for measuring strain in concrete and masonry, has a resolution of 0.002 mm and will, therefore, respond to changes in temperature and moisture in brickwork which are of no relevance to determining whether or not the foundations are moving. Demec gauges require a flat surface; they are not, therefore, suitable for measuring cracks at corners. Their main disadvantage, however, is their limited range; in cases where significant movements are occurring, it may be necessary to install replacement discs at intervals to preserve the continuity of the readings. Demec points can also be used in conjunction with vernier or digital callipers. As described below, callipers provide sufficient resolution and range for crack monitoring; therefore, provided sufficient care is taken over the readings, an overall accuracy of ± 0.1 mm should be achievable.

Brass screws
This technique involves the installation of small brass screws either side of the crack and measuring the distance between them using a calliper (Figure 2.10). This system has the advantages of being simple, robust, relatively unobtrusive and, by using the callipers in different modes, capable of measuring cracks in corners and other awkward positions, as shown in Figure 2.11 (on page 96). Moreover, if three screws are arranged in a right-angle triangle, it is possible to resolve both normal and shear crack movements. Using a digital calliper, a resolution of 0.01 mm is obtainable and an overall accuracy of better than ± 0.1 mm should be easily achieved.

Figure 2.9 Crack monitoring using a plastics tell-tale

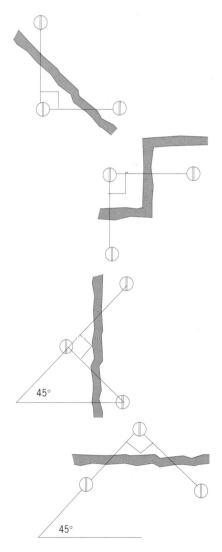

Figure 2.10 Crack monitoring by accurately measuring the distances apart of three screws inserted into the fabric

Figure 2.11 Crack measurement using brass screws

Displacement transducer
Although there are obvious cost penalties in using electronic devices to monitor crack widths, there may be circumstances when these more sophisticated techniques are justified. For example, it may be necessary to monitor continuously in order to activate an alarm in the event of sudden movement, or the crack may be in an inaccessible place, such as a railway tunnel. Two commonly used devices are the linear variable displacement transformer (LVDT) and the potentiometric displacement transducer. Both of these devices can be read manually using a hand-held unit or automatically using a data logger, and can easily provide an accuracy of ± 0.1 mm.

Foundation movement monitoring
However accurately crack widths are measured, only the symptoms and not the cause are being monitored. Therefore the results of monitoring crack widths alone can be ambiguous; a crack may form for one reason and progress for another. For example, previous distortions that are unconnected with the cause of the cracking may lead to further movements as locked-in stresses re-distribute themselves throughout the building, while the foundations remain stationary. In the majority of cases, it is preferable to measure the movements of the foundations as well as the crack widths. There are two ways this can be done:
- periodic level monitoring
- measuring changes in the inclination of elements of the structure

Periodic level monitoring
Vertical movements are usually monitored using precise levelling. To find out what is happening, rates of movement must be determined; these may be only a few millimetres a month. Consequently, the portable water level gauge used for level surveys is not sufficiently accurate for this purpose, nor, in most circumstances is normal site surveying equipment. In rare cases where a continuous check on building movements is needed, for example where an historic building is being affected by nearby excavation work, electronic tilt sensors or precision water gauges with electronic pressure sensors may be used.

A precision optical level has a resolution of 0.01 mm and, for a single building, it is normally possible to achieve an overall accuracy of better than 0.5 mm (ie the closing error of the survey should be no more than 0.5 mm). For measurements of this accuracy to be meaningful, it is essential that monitoring points of some form are attached to the building.

Wherever possible, level measurements should be referred to a fixed datum. For most domestic applications, a stormwater drain or similar feature is usually sufficiently stable for this purpose. However, where there are no deep drains or where absolute accuracy is imperative, a dedicated deep datum must be installed to a suitable depth.

Monitoring is normally required for a period of at least 12 months to distinguish seasonal movements from long term subsidence or heave; ideally, readings should be taken every month, but in practice every two months is usually adequate.

Changes in the inclination of elements
The alternative way of monitoring foundation movement is to measure changes in the inclination of elements of the structure. While this technique is limited to the detection of differential movements, it is more amenable to the use of electronic sensors and a check can be kept on absolute movement by periodic surveys using a precision level.

Although sensitive electronic tilt meters or inclinometers have been available for several decades, their cost has restricted their use to demountable devices. However, the development of relatively low-cost electrolevels of the type shown in Figure 2.12 has meant that it is now practical to semi-permanently install enough sensors on a structure to allow its movement to be monitored continuously. These sensors are sensitive enough to measure extremely small movements; typically, an electrolevel with a range of ± 3° mounted on a 1.5 m long beam can resolve relative movements of the ends of the beam to ± 0.1 mm.

A precision water gauge may be used for continuous monitoring of vertical movements. This device consists of two interconnected water-filled chambers. Floats in these chambers continuously monitor any changes in level between the two locations to an accuracy of ± 0.2 mm. This technique was successfully used to measure subsidence of the Mansion House in the City of London during tunnelling for the Docklands Light Railway. A cheaper variant now uses electronic pressure sensors to record changes in level.

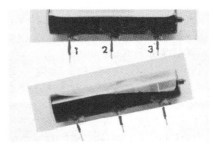

Figure 2.12 Electrolevels for measuring changes in the inclination of building elements

2.1 General points on foundations

Ground movement monitoring
Techniques similar to those described for measuring foundation movement can be applied to the measurement of ground movements. Optical levelling can measure the vertical movement of the ground surface or of rods anchored at various depths. Where relatively large movements are taking place, extensometers will measure the movements at various depths in the ground. Demountable inclinometers lowered down flexible plastics tubes can be used to profile lateral ground movements; trains of electrolevels can be deployed in horizontal, vertical or inclined tubes to monitor profiles of ground movement.

Remedial work

Where it has been confirmed that damage is attributable to foundation movement and there is a likelihood of further movement a decision has to be made on what action is needed to prevent further damage occurring. One solution is to underpin the foundations. Underpinning is often very disruptive and expensive, however, and should only be pursued once due consideration has been given to the alternatives which include mitigating the cause of the movements, repairing or strengthening the superstructure, or stabilising the soil.

Having correctly identified the cause of the movement, it may be possible to take steps to remove it altogether or to mitigate its influence on the foundations of the damaged building. Typical techniques are:
- repairing a leaking drain (Chapters 4.1 and 4.2)
- removing a tree causing clay shrinkage (Chapter 1.5)
- installing a retaining wall to prevent movement towards an open excavation (Chapter 5.1)
- using gabion walls to stabilise a slope (Chapter 5.1)

Unfortunately, these approaches cannot always be relied upon to prevent all further movement since a change may have occurred in the ground that is not necessarily reversed by removing the cause. In the case of leaking drains or nearby excavations, cavities may have formed in the ground which take some years of consolidation and collapse before completing their effects on the foundations. Nevertheless, in many cases, it may be possible to mitigate the cause at modest cost and then monitor the damage to gauge whether this action is likely to provide a satisfactory long term solution.

Repairing or strengthening the superstructure
Where the cause of the damage is confirmed as a process that is largely finished (eg initial settlement) or is likely to occur only rarely (eg clay shrinkage during a period of exceptionally dry weather), it should be possible to prevent further damage by repairing or strengthening the superstructure. The techniques that can be employed include: tie rods, strapping, resin bonding of brickwork and cavities, brick stitching, mortar bed reinforcement and corseting. Details of most of these techniques can be found in any good textbook on structural repair such as *The repair and maintenance of houses*[132]. Corseting, however, is a relatively new technique, which merits a few words of explanation.

Corseting consists of casting a reinforced concrete beam around the perimeter of the building, usually at or below ground level and connecting it to the brickwork by means of dowels. This stiffens the building at foundation level, and helps it bridge local areas of subsidence. It can therefore reduce the differential settlement associated with variations in the compressibility of underlying fill or soft soil. Corseting has also been applied to buildings suffering damage from slope creep movements.

A patented system using a post-tensioned beam is Hoopsafe. This system does not use dowels; instead, prior to concreting, the beams are tensioned to compress the structure contained within them[133].

Soil stabilisation
Grouting
Certain types of unstable ground will be improved by injecting grout to fill voids, reduce permeability and increase strength. Cement grouts can permeate silty fine sands with permeabilities down to about 5×10^{-4} m/s, while chemical grouts can permeate finer grained soils with permeabilities as low as 1×10^{-6} m/s but become increasingly expensive as higher permeation is required. The various grouting techniques available include:
- permeation grouting where grout is used to fill voids with little change to the soil volume and structure
- compaction or intrusion grouting where a stiff grout is used to compress soil by creating bulbs or lenses of grout
- fracture grouting where the soil is fragmented by hydraulic fracturing and the fragments are coated and compressed by the grout but not permeated by it
- jet grouting where the soil is mixed in place with a stabilising grout under very high nozzle pressures; soft, fine grained soils may also be partially removed prior to jet grouting by means of air or water jetting techniques. Jet grouting can be carried out in most soil types

Compaction grouting or fracture grouting can be used to correct or prevent differential displacements of structures. These processes are referred to as grout jacking (where it is carried out after the displacement has occurred) or compensation grouting (where it is used to prevent differential displacements occurring, especially during tunnelling operations).

Chemical treatments

Certain chemicals can be used to reduce the shrinkage potential of clay soils. However, owing to the extremely low permeability of most shrinkable clays, mechanical mixing is normally required to introduce the chemicals effectively into the soil mass. This severely limits the usefulness of chemical additives, although some proprietary treatments are available. This technique seems to have been used only rarely in the UK.

Underpinning

Where existing foundations are inadequate, they may be improved by underpinning, which involves either providing new foundations or, more usually, extending the existing foundations downwards to reach stiffer or more stable ground. Underpinning, therefore, has a variety of applications, including increasing the bearing capacity of foundations prior to refurbishment work and preventing foundations being undermined by nearby excavations. The most common use, however, is as a remedy for damage caused by foundation movement, especially where the differential movement is excessive and progressive.

Underpinning is often a difficult operation and not without risk. There are occasional collapses. A newspaper article, *The day our house fell down*[134], reported the collapse of a house in west London in the late summer of 1998 while it was being underpinned. Underpinning had been advised after cracks appeared following a dry summer and monitoring had taken place for two years.

As noted in Chapter 1.5, where large mature trees are present, it is not uncommon for certain soils to be significantly desiccated to a depth of 6 m or more. In these cases, though, it is likely that the levels of desiccation below about 3 m will remain at an approximately constant level as long as the trees remain in place. The design of an underpinning system to withstand the ground movements associated with the removal of a tree will be radically different, therefore, from one designed simply to provide adequate foundations while the tree remains. For example, where the soil is desiccated to 6 m, the former requirement might necessitate a pile-and-beam system with pile lengths in excess of 12 m and extensive heave protection, while the latter could probably be achieved using mass concrete to a depth of 2.5 m under external walls. The cost differential between these two options for a large house could be substantial.

The use of underpinning as a remedy for subsidence and heave damage has increased significantly since the early 1970s when cover for these defects was added to domestic building insurance policies. However, authoritative reports[135,107] have concluded that the majority of remedial underpinning work has not been technically justified, and may be at least partially responsible for the growing reluctance amongst insurers to accept underpinning solutions. Nevertheless, in cases where there is relatively severe damage or there is nothing else that can be done to mitigate the cause of the damage, underpinning is often the correct solution and should be specified without delay. In other cases, where the cause of the damage can be mitigated, alternative measures such as those described in the preceding section may be more advisable and more cost effective.

There are four main underpinning techniques that can be applied to low-rise buildings:
- mass concrete
- pier-and-beam
- pile-and-beam and piled raft
- mini-piling

Mass concrete underpinning

Mass concrete underpinning is often referred to as traditional underpinning since the principle has been in use for centuries. In the past, low labour costs and the absence of ready-mix concrete meant that traditional underpinning was constructed in brickwork. Now, mass concrete is invariably employed.

Mass concrete underpinning consists of extending existing strip or pad foundations downwards to reach a stratum that is stable and of sufficient strength. The underpinning is carried out in a series of bays as shown in Figure 2.13. The width of each bay is determined by the ability of the walls to span the gap created; for most low-rise buildings with competent brick or stonework, this is likely to be in the range 1.0–1.4 m. Where there is any doubt about the wall's ability to span the bay, the wall should be needled and the load transferred to temporary supports bearing on the ground.

Only the bays with the same number are excavated at any one time, so that no more than 20–25 % of the wall is left unsupported. The mass concrete is cast into the bay to leave a 75–150 mm gap between the

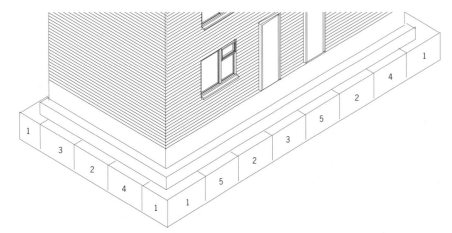

Figure 2.13 Sequence of operations for mass concrete underpinning. The bays are numbered to indicate a typical sequence of excavating, concreting and pinning up

concrete and the underside of the existing foundations; having allowed the concrete to cure for a minimum of 24 hours, this gap is then pinned up by ramming in a concrete with a maximum aggregate size of 10 mm and just enough water to enable the mixture to remain in a ball when squeezed in the hand.

Having completed a series of bays, work can commence on the next series of bays, although it is normal practice to allow at least 24 hours between pinning up and the excavation of an adjacent bay. This procedure is repeated until the bays become contiguous, forming a strip under the entire section of building being underpinned.

A variant of mass concrete underpinning is staggered or hit-and-miss underpinning. Instead of forming a continuous strip beneath an existing footing, the underpinning remains as discrete bays or piers. The span between piers is determined by the strength of the existing footings. It would not normally be used in cases where foundation loads are relatively high or where footings are shallow, insubstantial or cracked as a result of excessive ground movement. Using hit-and-miss underpinning is, therefore, somewhat limited, but it can be an economical solution in favourable circumstances and is especially suitable where existing foundations are reinforced.

In low-rise buildings, the use of mass concrete underpinning tends to become uneconomical at depths between 2 and 2.5 m because of the high labour costs involved in hand excavation. For larger buildings, however, where the applied foundation pressures are higher, the greater costs of excavating to depths of considerably more than 2.5 m can be justified. Mass concrete is also used to greater depths in refurbishment and redevelopment works, where the underpinning may also be acting as a retaining wall; in these circumstances, it should never be less than 0.6 m thick.

The presence of a high water table in the ground may preclude the use of mass concrete underpinning, particularly if the ground through which excavation is to be made is loose or otherwise unstable. In cases where continuous underpinning is employed and it is founded on a sloping impermeable stratum, it may be prudent to provide paths to allow groundwater to pass through the foundations.

Pier-and-beam underpinning
This system of underpinning was introduced shortly after the 1939–45 war and is now widely used in similar or amended forms. It consists of a reinforced concrete beam formed at or above footing level spanning between mass concrete piers founded at appropriate depths.

Except in very unusual circumstances, the beam is constructed first. Small areas of brickwork below DPC are broken out and sacrificial props, known as stools are fixed in the wall. The stools are usually manufactured in steel and have precast concrete top and bottom plates. They are normally supported on the existing footings or are inserted in the wall above the footings. If the footings are too weak or too shallow, individual pads resting directly on the ground may be used for support. Once the stools are in place, the remaining brickwork, or existing footings, is removed. Reinforcement is then threaded through the stools and formwork fixed to the sides to allow the beam to be cast in situ; the stools are thus incorporated into the beam. If the underpinning has been provided to arrest heave in desiccated clays, the underside of the beam will require protection from the heaving soil; this is usually provided by a proprietary compressible or collapsible board material.

After a suitable period, typically 24 hours, the beam is pinned up to the existing brickwork above it. Isolated piers are then excavated to an appropriate depth, concreted and pinned up to the underside of the beam or original footings.

Pier-and-beam underpinning can be used satisfactorily in most ground conditions. Because the beam is constructed first, the building and the operatives are at far less risk than with mass concrete underpinning. However, while it is possible to construct piers to depths of 8 m or more, the high labour costs associated with hand excavation tend to restrict the maximum depth to about 5 m. High water tables in poor ground increase costs substantially, particularly if shields have to be used; and running sands and silts may prove to be impossible to excavate without endangering the structure. Alternative methods of underpinning should, therefore, be considered in these ground conditions.

Pier-and-beam underpinning is particularly applicable where significant lateral movement of the ground is anticipated. Being relatively massive in comparison with piles, piers are able to withstand large lateral forces, and in the event of their movement, lateral forces are not necessarily transferred to the building because there is no structural connection between the piers and the footings or beam.

Pile-and-beam and piled raft underpinning
These pile based systems are similar in concept to pier-and-beam underpinning but have advantages where there is no suitable bearing stratum available within a depth economical for hand excavation of piers (say 4 or 5 m) or where it is necessary for the underpinning to pass through loose or waterbearing strata.

Where beams are being used, these are normally constructed in a manner similar to that for pier-and-beam underpinning, except that the beams are extended at corners and intersections to form caps for attaching to the pile heads. Intermediate supports are formed by pairs of piles using needle capping beams or, where internal access is restricted, using cantilever pile caps.

Pile diameters usually vary between 150 and 400 mm for low-rise buildings, although smaller diameters may be used on very lightly loaded structures. Because of

Figure 2.14 Figures 2.14 – 2.18 show a sequence of events for inserting piles in a house built on shrinkable clay. In this first picture, a mechanical auger is boring the ground before insertion of a pile. The whole of the suspended timber floor, sleeper walls and original concrete slab have been removed

Figure 2.15 A pile adjacent to the separating wall. The DPC is being indicated. The original foundations were on piers of brickwork founded at approximately 2 m depth, bridged by brick arches spanning 2.5 m. One of these arches can be seen behind the pile reinforcement

Figure 2.16 Deformable packing being inserted between the piles to carry the reinforced concrete floor slab. On each side of the photograph, jacks support the internal partitions

Figure 2.17 Reinforcement in place before casting the floor slab

Figure 2.18 A section of an external one-brick wall removed to accommodate the toe of the floor slab

access limitations and the proximity of vulnerable structures, the piles are usually excavated by auger and cast in situ; however, the smaller piles may be driven.

Alternatively, where external access is restricted, the existing floor slab can be removed and the piles installed inside the property (Figures 2.14 and 2.15). The piles are then connected using a reinforced concrete raft that is keyed to the external walls below ground level by removing small sections of brickwork in turn (Figures 2.16, 2.17 and 2.18).

Pile-and-beam underpinning can be used satisfactorily in virtually all circumstances but is generally most economical at depths in excess of 4–6 m. It is usually preferable to form the ground beams first since vibration and loss of ground during piling may have serious effects on the foundations of a severely damaged building. This is not always attractive to the contractor because of the awkward detailing involved in forming the pile caps after completion of the beams.

The piled raft variant is particularly cost effective where the underpinning is being designed to withstand soil swelling as this often requires existing floor slabs to be removed to enable a suspended floor to be installed.

2.1 General points on foundations

Mini-piling underpinning
In this method of underpinning, loads are transferred directly from the structure to the piles, either by needles or cantilevered pile caps, or by placing the piles directly through the existing footings. In low-rise buildings the pile supports would normally be at 1 m centres or less. Because of the small spans between piles, pile loads are small and small diameter piles or mini-piles are invariably used. Typically, these piles have diameters ranging from 75–150 mm and are formed by either filling bottom-driven steel or plastics casings with cement grout, or are augered and cast in situ. The technique adopted depends largely on the ground conditions. Guidance on the design, supervision and approval of remedial works and new foundations for low-rise buildings using mini-piles is given in BRE Digest 313[136].

Mini-piling underpinning differs from most others in that it depends to a degree on the inherent strength of the existing structure. This fact may be significant when considering older properties or properties in a bad structural condition. Mini-piling systems fall into four broad categories.

Raked piling
In this method, holes are cored at an angle through the existing foundations and a raking pile is formed and cast directly into the footings. The success of the system depends on the ability of the existing footings to withstand the lateral, vertical and torsional forces imposed by the alternately positioned piles (Figure 2.19). A rake angle of 10° ensures that for most common foundation depths the piles pass through the underside of the footings close to the centre. Where footing depths around a building vary, the angle of rake should be adjusted accordingly, but should not exceed 15° to the vertical. Care should be taken with the larger rake angles to ensure that each pile passes through the middle third of the base of the footings. Raking piles are normally installed at spacings between 0.7 and 1 m, depending on the condition of the existing footings and the soil. These piles are also varied to suit the plan and features of the building, and are raked alternately in opposite directions to minimise bending moments.

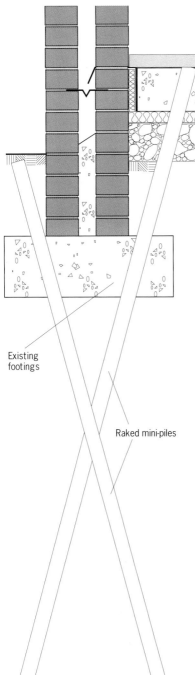

Figure 2.19 Raked mini-piles installed through existing footings

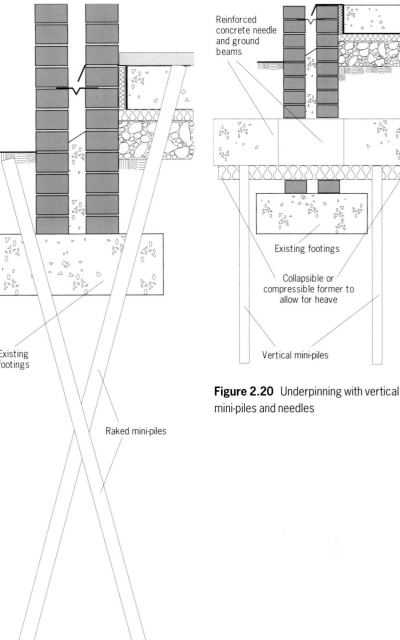

Figure 2.20 Underpinning with vertical mini-piles and needles

Needle piling
In this system, the wall to be underpinned is supported by short needles at approximately 1 m centres carried on pairs of piles installed either both sides of the wall or on one side only in a cantilever configuration (Figure 2.20). In cases of shrinkable clays it will be necessary to provide for movement beneath the needles and ground beams.

Types of mini-pile[136]

The majority of small diameter mini-piles are formed of driven steel casings filled with cement grout and provided with nominal reinforcement by a single central bar. In clay soils, where the sides of an unsupported borehole will not collapse, casings are not normally necessary, and augered cast-in-situ grout or concrete piles can be used. Augered piles, in nominal diameters from 100–450 mm, can be formed with continuous-flight augers in unstable ground without the need for casings. Because augered mini-piles are not normally cased, good quality, highly workable concrete or grout is required to protect the reinforcement from corrosion. This is particularly important for the smaller-diameter piles, where the cover is minimal, and for all piles in soils where acid attack on the concrete is possible.

For the larger pile sizes, hollow stem, continuous-flight augers are available. These are particularly useful in soils where an unsupported borehole is liable to collapse because the grout or concrete can be pumped down the auger stem while it is being withdrawn.

For loads up to 20 kN, mini-piles are typically 60 mm diameter and for loads up to 60 kN they are around 100 mm diameter. Mini-piles larger than about 150 mm diameter are usually excavated by auger.

In floor slab applications where small diameter, driven mini-piles shorter than 2.5 m are adequate, plastics casings have proved satisfactory. These are bottom driven through a mandrel resting on an oversize steel baseplate.

Where soil heave is likely, augered piles can be isolated from the swelling soil by cardboard or plastics tubes inserted in the borehole (Figure 2.21).

Figure 2.21 Cardboard tubes for insertion in boreholes to isolate piles from swelling of soil

Floor slab piling
Subsiding ground floor slabs and rafts can be stabilised by coring through the slab and inserting mini-piles to a suitable depth. The head of the pile is then cast into the slab. Spacing of the piles depends on the strength of the floor slab.

Pali radice (root piles)
These constitute a proprietary form of mini-piling that uses a high-strength grout injected into small diameter holes to form the pile. The grout is usually placed by means of a tremie pipe through which compressed air is fed to ensure that the grout is forced into intimate contact with the soil as the tube is withdrawn. Small diameter (100 mm) *pali radice* are reinforced with a single bar, while a cage or tube is used for larger diameters (250–300 mm).

The use of mini-piling is often restricted by the small diameter of the piles and their ability to withstand horizontal forces. Their main applications, therefore, are where destabilising forces are acting vertically and adequate lateral support is afforded to the slender pile by the surrounding ground.

Because of the slenderness of these mini-piles, their use to remedy clay shrinkage or heave movements is not normally recommended since horizontal forces may be transmitted unless precautionary measures are taken.

Mini-piling is comparatively cheap, having a low labour content, and has the advantage, where driven piles are being used, of being a relatively clean and disturbance-free operation. The majority of mini-piling contracts for underpinning low-rise buildings are small and the work is frequently completed in three or four days. The operatives carrying out the work are likely to have only a rudimentary understanding of the factors affecting pile bearing capacity, structural continuity, concrete strength, metal corrosion and rising damp. Much of the work will be in difficult situations with limited headroom, in corners of rooms or where there are underground services. Therefore, a higher level of supervision should be maintained for mini-piling works than is usual for other types of underpinning.

Structural walls underpinned with mini-piles should normally be isolated from moisture rising up the piles by the DPC in the walls, but dampproof membranes in floors punctured by the installation of mini-piles need to be repaired. Bituminous materials applied to the pile heads cannot be relied upon to prevent moisture penetrating the floor screed above the membrane. If rising damp might be a problem, floor finishes and screeds should be removed completely and a new continuous dampproof membrane laid over floor slab and pile heads[136].

See also BRE Digest 245[137] and Chapter 3.1 in *Floors and flooring*[66].

Special care is needed when forming *pali radice* in loose soils or below the water table to ensure that the grout is continuous and sound. The application of *pali radice* to remedial underpinning is very limited in the UK.

The design of remedial mini-piling[136]

Wherever possible, mini-piles should be designed on the same principles as larger piles in similar soils. A pile derives its support from friction on the shaft and bearing forces at the base. The ultimate bearing capacity Q_u is given by:

$$Q_u = Q_s + Q_b$$

where Q_s is the maximum resistance to sliding developed on the shaft Q_b is the load which causes shearing of the soil at the base.

The working load is given by dividing Q_u by a safety factor which is usually between 2½ and 3, depending on the degree of confidence in the ground investigation and anticipated quality of workmanship.

In firm clays, Q_s will almost certainly be the larger component of the ultimate bearing capacity. For mini-piles used to underpin structural walls or to support structural walls in new construction in these soils, the value of Q_s should be determined by soil mechanics methods or by test loading. No contribution from shaft friction should be included when the mini-piles are installed through soft clays and loose granular soils, or fills likely to consolidate or compact under their own weight, or superimposed floor loadings. The friction developed on many smooth plastics casings is generally less than that developed on steel and concrete, and the Q_s term should be ignored when plastics casings are used unless loading tests demonstrate that adequate friction is mobilised by the shaft.

If soft clays and loose fills are subjected to surcharge or are undergoing consolidation, they can apply downdrag forces to the pile shaft as they settle around it. It may then be necessary to assign a negative value to Q_s for that part of the shaft within the settling soil.

Simple design charts for determining the allowable loads on piles in clay from the undrained shear strength, C_u, of the clay are available.

Calculations of the maximum shaft friction, Q_s, and base resistance, Q_b, for piles driven into granular soils require measurements of soil properties. However, if the characteristics of the hammer have already been determined, it is possible to obtain an approximate value of the sum of Q_s and Q_b for driven mini-piles in granular soils during the driving process, as discussed below.

When a drop hammer is used to drive a pile, the resistance of the pile to penetration is related to the weight of the hammer, the height of the drop and the movement of the pile under each blow. As long as the weight of the hammer is sufficient to drive the pile through non-loadbearing strata at an acceptable rate and is comparable with the weight of the pile, a satisfactory resistance in the bearing stratum is normally reached when at least 10 blows are necessary to cause a penetration of 25 mm.

When a vibratory hammer is used to drive a pile, the resistance to penetration depends on the energy delivered to the pile and the pile movement in a manner similar to that for drop hammers. Most mini-piling contractors presume that adequate resistance has been obtained if the pile penetration is 10 mm or less during a period of 10 seconds, irrespective of the power of the hammer and the size and weight of the pile. This should be checked over a penetration of at least 100 mm.

In silty clays and fine sands the bearing capacity of a pile can fall with time owing to movements of water between the soil particles. Test piles and randomly selected working piles should be re-driven after a period of rest to ensure that the required capacity has been maintained.

Forces carried by mini-piles

Vertical mini-piles, accurately aligned beneath loadbearing walls in new construction or used systematically to support settling floor slabs in existing buildings, are unlikely to be exposed to forces other than axial compression. However, in addition to the axial forces, bending moments will be applied, due to eccentric loading, when piles are raked or placed vertically on each side of the walls. Clay soil swelling can cause lateral as well as vertical forces and may, therefore, subject the piles to bending and axial tension. Raking piles should not be used where heave is likely to occur.

Because the slenderness ratios (pile length to pile diameter) of mini-piles can be high, their loadbearing capacities may be limited in some cases more by their strength as columns than by the support of the soil.

Buckling

Mini-piles are frequently longer in relation to their diameters than most reinforced concrete piles of larger diameters and the possibility should be considered of buckling due to lack of straightness, deviation from the true alignment or small eccentricities of loading. However, it is known that even very soft soils offer considerable resistance to buckling. Nevertheless, in view of the lack of information available it seems reasonable to guard against the possibility of buckling by limiting mini-pile lengths to about 75 times their diameter so that the length of the smallest common mini-pile (of around 65 mm diameter) should not normally exceed 5 m.

Reinforcement

Steel casings are not necessary to provide compressive strength to axially loaded piles unless the loads are large. However, they may be essential in providing tensile strength in piles subjected to bending. Steel casings may, therefore, need protection from corrosion in natural soils containing acid groundwater and in aggressive fills. Some protection can be provided by using galvanised steel or by applying a protective coating. Alternatively, the thickness of the casing section can be increased to provide sacrificial steel.

Most mini-pile contractors installing piles less than 100 mm diameter provide nominal additional reinforcement in the form of a single central bar; this should be galvanised and held centrally by wire spiders. The central bar is valueless in resisting bending but it can be useful in three ways.

- It can be used with a vibrator to compact the concrete or cement grout inside the casing
- In new construction, or where footings or floor slabs have been broken out, it can be bent over at the pile head for connecting into new ground beams or floor
- It provides additional tensile resistance when the piles are restraining a building on heaving ground

Some authorities prefer to dispense with reinforcement in small piles for settling floor slabs because of the difficulty in placing the concrete. Where the underpinning system is such that large eccentric forces have to be carried, conventional cage reinforcement must be designed to BS 8004, and the foundations and the pile diameters made large enough for adequate cover to be provided. For these situations, the bending resistance of the casing should be ignored but it may be assumed that corrosion of the casing will not affect the frictional resistance to axial forces.

Structural connections

Where raking piles are to be installed through footings, the thickness and concrete quality of the footings must be sufficient for adequate bond to be developed. If this is not the case, wall underpinning using vertical piles and needles, and ground beams, should be considered. To achieve good mechanical coupling between underpinning piles and the footings of loadbearing walls, mini-pile casings must be cut off, or driving continued so that the tops of the casings finish below the underside of the footings. Some contractors use an internal tube cutter. Supervisors should ensure that no casings are left to penetrate the footings because the bond between the pile and the footings would then be provided only by a thin annulus of grout around the casing. This applies also to the installation of mini-piles through settling floor slabs. The concrete or grout is brought through the drilled hole in the floor or footings and finished flush with the surface.

Designing underpinning
Wherever underpinning is used it is important that it is correctly specified; poorly designed underpinning may create more problems than it solves. For example, partial underpinning is often acceptable, but special care is needed where the underpinning is terminated. Similar special care is needed where a semi-detached or terraced property is being underpinned. As is the case with foundations for new buildings, it is generally uneconomical, and often unwise, to design the underpinning not to move at all; the governing criterion should be to limit movements to an acceptable (ie non-damaging) level. However, it can be argued that a damaged building may be more vulnerable to further foundation movement than a new building, in which case it would be appropriate to be more conservative in selecting the required depth for the underpinning than would be required for a new building in similar circumstances.

The load applied to the soil by mass concrete underpinning and the piers of pier-and-beam systems must not exceed the allowable bearing pressure of the soil. Where there is no evidence of the building having been damaged by excessive settlement, this figure may be based on standard values but in other cases it must be calculated from site specific data.

It is possible to use jacks to compensate for future movements. However, it is rare for jacking to be used for domestic applications.

To install mass concrete or pier-and-beam underpinning it is usually necessary to excavate an access pit on the outside of the wall before removing the soil from beneath the original footings. It follows that the line of action of the wall loading will normally be eccentric to the centre line of the finished underpin, inducing an overturning moment. In more heavily loaded applications, this eccentric loading may lead to the allowable bearing pressure of the soil being exceeded on the inside edge of the underpin. This condition may be avoided by adopting the middle-third rule: the width of the underpin is selected to ensure that the line of action of the resultant load passes through the middle third of the base.

Normal pile design criteria as described in Chapter 2.4 should be applied to the design of pile-and-beam, pile-raft and mini-piling systems. As with all forms of deeper underpinning, appropriate soil parameters must be determined from a proper ground investigation, usually incorporating boreholes. Where driven mini-piles are used, they are usually driven to a specified resistance and it is sometimes argued that a detailed ground investigation is not, therefore, needed. It is possible, however, that a set can be achieved on a buried obstruction and this can be a major problem when using this type of pile on fill sites.

Particular attention should be paid to the uplift, downdrag and lateral loading that may be induced by future movements in the unstable ground. Guidance on the design of laterally loaded piles is given in a CIRIA report[138] which indicates that most building foundation piles can provide considerable resistance to lateral forces. In pile-and-beam systems it may be possible to improve lateral restraint further by extending transitional beams into the unaffected parts of the property.

Where underpinning is being specified as a remedy for ground movements associated with fill, a layer of soft soil, internal erosion, collapse of an underground cavity, or landslip, the selection of a suitable depth for the underpinning is a fairly straightforward operation; the underpinning must be taken to a depth where the ground is stiff enough and stable enough to provide foundations that will be unaffected by these processes. For piled systems, this implies that the length of the pile may have to be approximately twice the depth of the unstable soil to ensure that there is sufficient resistance to downdrag or uplift forces. Similarly, where underpinning is being specified as a remedy for clay swelling following tree removal, the foundations must be taken below the depth of desiccated soil for a sufficient distance to provide adequate resistance to uplift forces.

Where underpinning is being undertaken following damage to buildings caused by trees, and where the tree is to remain, there are three approaches that are commonly adopted to select the correct depth for the underpinning.

NHBC Standards
A common misconception is that the tables in Chapter 4.2 of the NHBC Standards give depths of desiccation. What they give, in fact, are recommended depths for new house foundations which are intended to prevent the movements transmitted to the superstructure reaching intolerable levels. In practice, these depths can be considerably less than the depth of desiccation; for example, the maximum foundation depth recommended by the NHBC is 3.5 m, while it is generally accepted that significant desiccation is often found at depths of 6 m and more near large trees[139]. Designing underpinning to these NHBC guidelines could, therefore, produce foundations that continue to move. The level of movement should be within the bounds that are tolerable for new buildings, say a maximum of 20 mm, and this may be acceptable provided the structure has not been unduly weakened by the previous foundation movements. However, if this method were to be adopted, it would be prudent in most cases to reduce the anticipated movements by increasing the recommended foundation depths given in the NHBC tables.

Last visible root
Most underpinning is subject to building control: the approach adopted by some local authorities is to select the required depth on site according to local experience. The usual way of doing this is to require the excavation for the underpinning to go a certain distance, for example 1 m, below the depth of the last visible root. Clearly, this can lead to

2.1 General points on foundations

Underpinning unstable domestic floor slabs[136]

Domestic floor slabs can settle owing to unsatisfactory or poorly compacted fills or to the compression or shrinkage of natural soils. Where ground floor slabs have been cast on permanent shuttering there is sometimes a risk that the bearings of the permanent shuttering may have deteriorated, or hardcore may have been insufficiently compacted, with a consequent risk of settlement at these points.

See also Chapter 3.1 in *Floors and flooring*, and BRE Digests 251 and 276.

These settlements are revealed by gaps opening between skirting boards and floor finishes and, in more severe cases, by cracks in internal partitions founded on the slab, damage to floor finishes or distortion of door frames (Figure 2.22). In many of these cases, settlement is either insignificant or not continuing and slab treatment is unnecessary. If slab movement on poorly compacted fill does continue, it may sometimes be stopped by careful grouting beneath the floors without the need for piling. In more severe cases, it will be necessary to install mini-piles through the floors, or to break out badly damaged floors and replace them on mini-piles or better quality fills. It will also be necessary to break out cast-in-situ slabs when soil is heaving owing to the removal of trees or to the use beneath the slabs of expansive fills, such as burnt colliery shales. Where mini-piles support reconstructed slabs, they can be sleeved and the slabs cast on compressible or void-forming material to absorb any further tendency for the soil to heave.

Where mini-piles are to support settling floor slabs, the thickness of the slab and the depth of fill above the natural soil should be checked during the investigation of the suitability of the ground for the use of mini-piles. The suitability of applying grouting to the fill and to loose natural soils and gaps beneath the floor slab should be determined. Notes should be made of the positions of all underground services that might be affected by piling.

Before a settling floor slab is underpinned, any voids (either directly beneath it or in poorly compacted hardcore supporting the slab) must be grouted through holes at about 1 m spacing in the slab. Checks should be made that the grout has not penetrated the inner skin of cavity wall construction and risen in the cavity above the DPC; this can lead to serious problems with mould growth. Mini-piles under a floor slab should normally be placed at centre-to-centre spacings not exceeding 1.5 m with maximum distances from slab edges of 300 mm (Figure 2.23). Where the slab carries partitions, the pile layout should be designed to take account of the increased loading. If pile spacings are large or the slab thickness significantly less than 100 mm, the permissible concrete stresses may be exceeded at slab-pile joints, even with casings cut off at the correct depth. In these situations it may be necessary to break out the slab and replace it with a thicker one. Additional slab support should be provided by forming inverted conical cavities immediately beneath the slab at each pile position before the pile concrete is poured. Where a floor slab is settling so much that it has cracked, mini-piles should be installed each side of the crack at a distance from it not greater than 300 mm.

Mini-piling is particularly appropriate for floor slab restoration where grouting of poor fill alone cannot be used because of the presence of a high water table in the fill, or where grouting of weak cohesive soils close to the underside of the slab will be ineffective. Many piling rigs can operate where headroom is as little as 2 m and piles can usually be installed within about 40 mm of existing walls.

A novel use of mini-piling is for upgrading existing floors of light industrial buildings when increased load capacities are required.

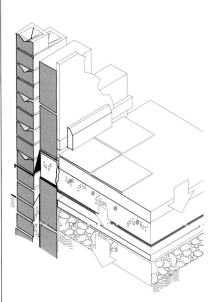

Figure 2.22 Settlement may show by a gap opening up at the floor/wall interface. The gap may be concealed by a skirting or floor covering

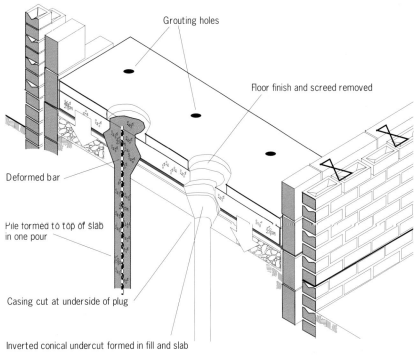

Figure 2.23 Mini-piling a settling floor slab

difficulties on site if the required depth is significantly greater than the depth anticipated at the time of quoting for the works. In addition, there are three potential pitfalls with this approach:
- trees can extract water through very fine fibrous roots which may be difficult to identify in an excavation
- more roots, or the zone of desiccation below a large tree may extend several metres beyond the last visible root
- the presence of roots does not necessarily indicate that the ground is desiccated

Nevertheless, this approach is likely to give a depth that is not dissimilar to that based on NHBC recommendations and may help identify situations where root activity is unusually vigorous. It would, however, be imprudent to specify any underpinning depths based on the last visible root that are less than those recommended in Chapter 4.2 of the NHBC *Standards*[125].

Desiccation profiles
Profiles of moisture content or suction can be used to quantify the extent and the magnitude of the desiccation, as described in Chapter 1.5. Some judgement is still required, however, to distinguish areas where the level of desiccation varies seasonally from those where it is relatively constant. As a rule-of-thumb, in a soil with a high potential for volume change, it can be assumed that the depth of seasonal variation is about 1m plus half the remaining depth of desiccation. However, in a soil with a low potential for volume change, movements are more likely to be almost entirely seasonal. If adopting this approach it would not be prudent to specify any depths that are shallower than those recommended in Chapter 4.2 of the NHBC Standards.

Where a tree responsible for damage to a building is not yet fully mature, there are two options:
- remove the tree and design the underpinning to withstand the consequent heave
- do nothing to the tree and make provision in the design of the underpinning for the increasing levels of desiccation produced as the tree grows. In the latter option, consideration should also be given to the appropriateness of designing for the tree's future removal or death, on the assumption that the building will last longer than the tree

Designing underpinning against heave
Special precautions are needed where the underpinning is being designed to protect the structure from the effects of soil swelling following tree removal. First the underpin has to be founded at a depth where the movements produced by the change in desiccation are tolerable. Where large deciduous trees are involved, a ground investigation incorporating boreholes is likely to be required to provide the necessary information. The procedure is essentially similar to that for new foundations, and it may be acceptable to use the same allowable movement (ie 20 mm) in cases where damage has been caused by subsidence prior to the removal of the tree on the basis that any future movement is tending to reverse existing distortions. Equally, where existing damage is attributable to ongoing heave, it would be prudent to use a smaller value for the allowable movement, say 10 mm. In less extreme cases, for example where there are small deciduous trees or conifers, it may be more cost effective to make some conservative assumptions regarding the depth of desiccation and to use this as the required depth for the underpinning.

Where a pile based system is being used, the length of the pile should be selected with due consideration to the expected uplift forces generated by the swelling soil. As a rule-of-thumb, however, the required length of pile is likely to be approximately twice the depth required for mass concrete or pier-and-beam underpinning. More detailed guidance, leading to a shorter pile, is available[140].

In addition to selecting a depth to ensure adequate stability, it will be necessary to protect the underpin and the existing foundations from the lateral and uplift forces generated by the swelling soil. The vertical faces of mass concrete, piers, ground beams and original footings can be protected by layers of low density expanded polystyrene or other suitably compressible materials. It is usually most important to ensure that this material is placed on the inside of the foundations where the soil is likely to be more confined and, therefore, capable of generating greater pressures. Even where trees are left in place, the underpinning is likely to cut through roots and cause the soil under the building to swell.

Where a pile-and-beam or pier-and-beam system is being used, uplift forces can be reduced by providing a compressible layer on the underside of the beam replacing the original footings. Alternatively, where the beam has been built into the wall, it may be preferable to remove the original footings altogether and create a void under the beam.

Swelling will also affect floor slabs, and if significant movements are anticipated, it may be necessary to replace any existing groundbearing floor slabs with suspended floors.

2.1 General points on foundations

Partial underpinning
It is usually unnecessary to underpin the whole of a structure that is only partly affected by ground movement. An underpinning scheme that does not include all loadbearing walls is normally referred to as partial underpinning. In some cases, the underpinning will be restricted to one side of the property; in others, it may be the internal walls that are left in their original condition. Similar considerations apply to the underpinning of a semi-detached or terraced house where it may not be possible to extend the remedial scheme to a neighbouring property.

Provided the scheme is properly engineered, there is no reason why partial underpinning should not be satisfactory. However, special care must be taken where the affected property is founded on shrinkable clay. All buildings founded on clay soils, even those with foundations complying with current guidelines, will move up and down as a result of seasonal moisture changes in the surface soil. Nevertheless, the occupants are likely to be unaware of these movements because the whole house is moving as a unit and differential foundation movements are small. However, there is a danger that, if part of the building is underpinned, the levels of differential movement in the remaining part will be increased, because the stability of the underpinned section will have been radically improved.

One way of avoiding damage as a result of partial underpinning is to extend the underpin under the unaffected part of the building and to step up progressively to avoid creating a sudden change in the support conditions. This is most easily achieved with mass concrete underpinning where the depth of each bay can easily be varied; the usual practice is to decrease the depth of underpinning in 0.3 m steps over bays of about 1 m width until it merges with the original foundations. This approach can also be applied to pier-and-beam and pile-and-beam systems, although coarser steps are likely to be needed because of the greater distance between piles or piers. Moreover, there is also a practical minimum depth to which piles can be installed. For mini-piling systems, it should be possible to use progressively shorter piles at the extremities of the underpinning, although this is not common practice.

Alternatively, the likelihood of damage can be reduced by ensuring that the depth of the underpinning is not over specified, so that the underpinned section of the building continues to move in sympathy with the rest of the structure. This may increase the risk of existing cracks widening, but may provide a more cost effective solution than, for example, having to underpin all internal loadbearing walls.

Testing
It is important that any ground treatment is carried out to an appropriate specification under controlled conditions, with adequate testing to ensure that the required quality is achieved. The specification may describe the method of treatment in detail or may define the required performance of fill. There are problems with both approaches. Many fill sites are heterogeneous and it may be necessary substantially to modify the treatment (for example by adjusting the input energy in dynamic compaction) as the work progresses and the extent of the required improvement becomes clearer. A performance specification will only be satisfactory if some adequate test can be carried out after treatment to determine whether or not the required performance has been achieved. Supervisory personnel should be on site during ground treatment, not just the operatives. This will increase costs but is considered to be essential to ensure adequate quality control.

Testing will be required to meet the following objectives:
- investigating the properties of the ground prior to treatment
- assessing the degree of improvement effected by ground treatment. (Testing may be carried out to give a comparison of properties before and after treatment)
- assessing the effectiveness of treatment. (Some testing may be carried out during treatment)
- confirming that, for example, vibration and noise levels are acceptable. (Environmental testing may be needed during treatment)
- confirming that a specified improvement has been achieved. (Testing may be carried out after treatment)
- determining the load carrying characteristics of the treated ground (by field load tests)
- monitoring long term movement following treatment

Various types of testing may be required:
- laboratory tests on recompacted samples at the design stage to predict field behaviour of recompacted fill
- control testing as the fill is placed and compacted (eg for density, moisture content, and particle size distribution for clay fill)
- load tests during filling and, especially, when filling has been completed, to confirm predicted behaviour
- in situ penetration tests, such as those described in Chapter 11, to provide a comparison of treated and untreated ground; in heterogeneous fill the tests may not be practicable and could be misleading in fill with very large particles
- geophysical measurements may sometimes prove useful in indicating improvement
- buildings built on the treated fill should be monitored, where possible, for long term settlement

The most common form of testing for vibro is the plate loading test. A 600 mm diameter plate is placed on top of a treatment column and the load versus deformation behaviour is determined during a relatively quick loading and unloading cycle. The load is applied by a hydraulic jack using the weight of a vehicle or crane as reaction. This type of test is normally carried out as a routine control procedure. It may give some indication of workmanship and uniformity but the results of the test cannot be used for design or to predict the long term movements of structures which are founded on a large number of columns spaced at varying intervals.

Inspection

The problems to look for with unstable concrete slab floors include:
◊ gaps opening below skirtings caused by settlement of hardcore or by subsidence
◊ sticking of doors caused by doming of slabs as a result of sulfate attack, clay heave or expansion of hardcore
◊ cracking of slabs from settlement of hardcore
◊ ducts below floors collapsing

The problems to look for in remedial underpinning using mini-piles include:
◊ piles not long enough to reach an appropriate bearing stratum
◊ piles not long enough to reach levels at which shrinkage or heave is unlikely
◊ further movements due to consolidation of compressible soils at depth
◊ piles too short to prevent settlement due to adjoining deep excavations
◊ raked piles used under poor quality existing foundations
◊ no provision for heave under needles or ground beams in expansive clay soils

Chapter 2.2

Old brick and stone footings

Before the introduction of concrete strip foundations for masonry walls, it was common practice to provide masonry footings wider than the wall above in order to spread the imposed load on the soil. Where the ground was of poor bearing capacity, with high water tables leading to boggy areas, there may even have been brushwood faggots or fascines placed beneath the footings in the hope of improving bearing capacity.

As well as describing the lower courses of corbelled brickwork, this chapter deals with masonry below DPC level – particularly its durability.

Characteristic details

Basic description
In Georgian and Victorian times the foundations of small buildings tended to be very shallow in depth, often no more than half a dozen or so courses of bricks. This brickwork below ground level was often stepped in order to spread the load, as shown in Figure 2.24. Before the widespread introduction of concrete in footings towards the end of the nineteenth century, unless the walls were founded on rock, progressively wider courses of bonded masonry were absolutely necessary to spread loads over adequate widths of subsoil. It was common practice to use as many bricks as possible laid as headers, that is to say, normal to the line of the wall so that the corbelling action was maximised. Where the wall was of stone, the same principle would apply, with the maximum use of through stones (Figure 2.25).

Local authority byelaws in the last years of the nineteenth century required the course above the concrete to be twice the width of the wall to be carried, with successive courses above that reducing by quarter brick width each side until the required thickness of wall was achieved. In modern construction, this stepped profile has been dispensed with as a simple mass concrete strip adequately distributes the load.

Main performance requirements and defects

Strength and stability
There is an important difference between the behaviour of brickwork built in lime mortar, and brickwork built in cement mortar. Masonry footings which used lime mortars are much more tolerant of movement than are those built with cement mortars, the ductile nature of the lime mortar allowing the bricks to move slightly relative to each other when under stress. In consequence, footings of buildings built before the 1930s, before lime mortars began to be superseded by cement mortars, are much less likely to crack when the soils on which they are founded shrink or heave.

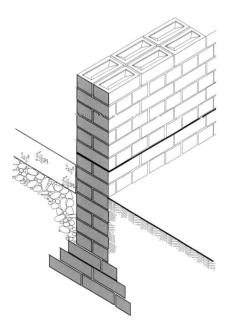

Figure 2.24 Maximum use of brick headers in footings gives the best loadbearing qualities

Figure 2.25 Typical stone footings widely used until the end of the nineteenth century

Figure 2.26 A narrow diameter hole has been excavated in a confined space alongside a one brick solid external wall to establish the depth and width of what turned out to be brick footings. Both depth and width had to be established by probing with a long crowbar. The footings were on shrinkable clay, and were unusually deep for an Edwardian house. Tree roots are visible in the sides of the excavation

Figure 2.27 A variety of contaminants has obviously been in contact with this 100 year old clay brickwork below DPC level, but only minor damage has resulted. There is also evidence of previous alterations and some repointing

Where alterations are to be made to an old building, it is crucial to expose and measure any masonry footings, and to carry out an assessment of the loadbearing capacity of the soil on which they are carried (Figure 2.26).

Durability

It should be noted that discussion of this topic in this chapter applies only to those materials used below ground level. Different considerations apply to materials used above ground level, and reference should be made to the appropriate parts of *Walls, windows and doors*[101].

Only a few influences are known to reduce the normally indefinite life of materials recommended for construction of foundations, the most significant of these being sulfate attack and frost. Mortar for brick and blockwork can have very variable durability. Breakdown may be hastened by the use of unsuitable, poor quality or incorrectly prepared materials. The relevant codes give extensive guidance on the preferred specifications for given circumstances. Local building control departments should have information on the presence of aggressive soil or groundwater in particular areas. Where any doubts exist, samples should be taken for analysis.

Washing out mortar from footings or fines from the ground beneath foundations can result from leaking water mains or drains.

Clay bricks

So far as the bricks themselves are concerned, there is a wide range of performance. Engineering bricks are highly durable, the inherently good resistance of these materials being due principally to their low porosity. The deterioration of brick depends on the nature of the contamination to which it is exposed, and can occur as a result of either a chemical interaction with the ceramic materials – leading to the dissolution of the glassy phase, which in some cases can constitute as much as 60% of the brick – or a physical expansion mechanism due to the crystallisation of salts within the brick pores[77]; this is potentially a more serious problem (Figure 2.27). The salts are transported by water to the interior of the brick and can derive from the external environment or from the rehydration of the soluble phase of the brick. The most deleterious salts are those that are readily hydratable (eg sulfates of sodium and calcium).

The site of crystallisation is determined by the dynamic balance between the rate of evaporation of water from the brick surface and the rate of solute migration to the site of crystal growth. If the rate of solute migration is faster then the rate of evaporation, then crystallisation occurs on the surface of the brick (efflorescence). Although efflorescence is unsightly it is not harmful. If the rate of solute migration is slow, or if the salt is relatively insoluble, crystallisation will take place within the pores of the brick. This is usually referred to as sub-florescence or cryptoflorescence, and it can result in delamination, flaking and spalling[77].

A more detailed explanation of the effect of crystallisation of soluble salts on clay bricks is to be found in Chapter 2.2 of *Walls, windows and doors*.

Calcium silicate bricks

Calcium silicate bricks are resistant to attack by most sulfate salts in the soil and groundwater. The durability of calcium silicate bricks under salt crystallisation attack can be attributed to a combination of their very low soluble salt content, and their low porosity and coarse pore structure. However, calcium silicate bricks may be attacked by high concentrations of magnesium and ammonium sulfate. They may also suffer severe deterioration if they are impregnated by strong salt solutions, such as calcium chloride or sodium chloride, and then subjected to frost[77].

2.2 Old brick and stone footings

Mortars

Mortar is subjected to the same agents of deterioration as is concrete. One of the main causes of the deterioration of mortar is sulfate attack. The chemical reactions of sulfate with the hydrated constituents of the cements used in mortars vary with the type of sulfate, the access of atmospheric carbon dioxide and the ambient temperatures. The reaction between sulfate and the hydrated tricalcium aluminate in Portland cement results in the expansive formation of ettringite. At low temperatures the deteriorated sample may also contain a calcium silicate carbonate sulfate mineral (thaumasite) which, like ettringite, causes a swelling of the original material[77].

See also the section, 'Durability', in Chapter 2.3.

Remedial work on site

Inspection

Other than at exposed edges of concrete rafts and ring beams, most foundations are difficult to inspect and the condition of the superstructure will frequently have to be used as an indicator of whether there has been material failure in the foundations.

If there is evidence of foundation movement, two initial questions must be answered. First, does the pattern of movement indicate a failure of the foundations? And second, has movement now ceased? Where a foundation failure is identified, remedial measures may not be justified if a stable condition has been reached. Remedial work to foundations is usually very expensive and there are risks in providing extra support to one part of the structure in isolation.

With any foundations which are apparently sound but theoretically inadequate, account must be taken of the fact that these have already stood the test of time and coped with extremes of temperature, moisture and ground movement. The most likely causes of future failure of foundations are physical changes such as the removal or planting of trees, adjacent deep excavation for new drains or foundations, increased loadings or altered groundwater movement.

Subsidence may have occurred as a result of underlying or buried features such as fill, peat, leaking drains or mining activities. Differential settlement may be encountered, which can be caused by:
- foundations imposing different loads at different locations
- differences in foundation depth
- shrinkage of clay resulting from drought or local influence of trees

Heave, due to swelling of clay (often associated with tree removal) and direct disruption or damage caused by tree roots, may also be seen.

Correct diagnosis of the cause of foundation failure is critical and should be confirmed, if there is any doubt, by experts. Tapered cracks usually indicate settlement or subsidence and the direction of movement. Crack width monitoring and soils testing may establish whether movement has ceased, or is progressive or cyclical. Old maps, topographic information and neighbours' recollections can provide useful background information on the location of features such as rubbish tips and water courses which might influence foundation performance. A close inspection of adjacent buildings may indicate if a general problem exists.

The problems to look for are:
◊ frost action spalling masonry
◊ sulfate attack
◊ inadequate identification of contaminants
◊ cracking caused by settlement
◊ changes in local vegetation, especially trees

Chapter 2.3　Concrete strips, pads etc

This chapter deals with mass Portland cement concrete used in strip and pad form as foundations for buildings (Figure 2.28). Portland cement had been invented in the 1820s, but its use was by no means widespread in buildings until the later years of the nineteenth century and the early years of the twentieth century. As noted in *Floors and flooring*[66], however, concrete received a big boost after the 1914–18 war, with the Ministry of Health allowing its use for housing with gravel or clinker aggregates.

Characteristic details

Concrete strips

Strip footings are the traditional method of constructing foundations for low-rise buildings in the UK and has been in common usage since the late nineteenth century when concrete became readily available. The Model Byelaws, which were in operation for example during the third quarter of the twentieth century, called for strip foundation widths to be constructed of plain concrete on rock, compacted gravel and sand, and sandy clay in increments of 3 inches from 9 inches to 30 inches, increasing with the loadings on the walls to be carried. These loads ranged from ¼ ton to 2 tons per linear foot. For the poorer soils of loose and soft silts, sands and clays, the limit was 1 ton per linear foot, and the widths were up to 36 inches. The thickness of concrete required was *'not less than its projection from the base of the wall and in no case less than 6 inches'*. Such provisions were deemed to satisfy the Byelaws, and there is no doubt that they provided adequate solutions for the majority of cases.

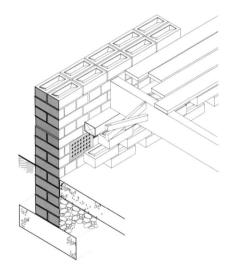

Figure 2.29　Conventional strip footings

The type of footings usually encountered will consist of a 0.3 m thick layer of concrete, cast into the bottom of the trench, which then provides support for the brickwork or masonry walls (Figure 2.29). Strip footings are normally provided centrally under each loadbearing wall.

Traditionally, the depth of the trench was determined solely by the exposed soil and excavation would continue until such time as an apparently acceptable stratum was encountered. As a minimum, it was considered necessary to remove only topsoil or other obviously

Figure 2.28　Concrete strip footings

2.3 Concrete strips, pads etc

compressible material. Consequently, in speculative housing, the underside of the footings might be as little as 0.35 m below ground level, although a depth of about 0.45 m is more typical. This latter figure remains the recommended minimum throughout most of Britain for protecting the foundations against heave caused by frost. However, for building on clay soils, it is now recognised that deeper foundations are required to isolate the building from the swelling and shrinking that occurs in the ground as a result of moisture variations. The minimum depth for building on clay soils recommended in the Code of practice for foundations (BS 8004[68]) is 0.9 m and, where there are trees or other large vegetation, greater depths are likely to be needed. See also Chapter 1.5.

Because of the difficulties of laying bricks or concrete blocks in a very narrow trench, the minimum practical width for strip footings is about 0.6 m. However, in soft soils, it may be necessary to use wider footings in order to spread the load over a greater area. There are also minimum requirements for projection and depth of concrete. In those circumstances it may be necessary to reinforce the footings to prevent cracking.

Trench fill

The trench fill foundation is a variant of strip footings. Rather than placing a minimum thickness of concrete in the bottom of the trench, the trench fill foundation is constructed by filling the trench to within a few tens of centimetres of the ground surface (Figure 2.30). This form of construction has become increasingly economical as labour costs have increased, and the use of mechanical excavation and ready-mix concrete has become more easily available. It has the added advantage of improving safety since the need for working below ground level is minimised, a consideration which tends to be compelling where the ground conditions require a foundation depth of more than 1.2 m.

As man access is no longer required, the width of trench fill foundations is often dictated by the practicalities of setting out and the width of the bucket used by the back-hoe mechanical excavator normally found on a building site. In good ground conditions, a typical width for trench fill foundations is 400 mm.

Although they become increasingly uneconomic and even technically undesirable as the required foundation depth is increased, trench fill foundations are often used to depths of 3.5 m below ground level. Their main advantage over piled foundations is that they enable the ground conditions to be inspected visually and do not require specialised drilling machines.

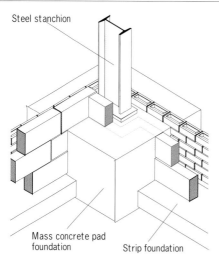

Figure 2.31 A typical mass concrete pad foundation for a rolled steel stanchion. Wall panels may be carried on conventional concrete footings

Pads, piers and caissons

Where the building loads are carried by isolated columns rather than by loadbearing walls (eg in a portal framed structure), it may be preferable to excavate and cast individual pads rather than continuous footings (Figure 2.31). As a rule-of-thumb this form of construction tends to become more economic whenever the distance between adjacent pads exceeds the dimensions of each pad. Panels of brick or block infill between the columns can then be supported either on ground beams spanning between pads or on separate footings, which may be at the same depth as the pads or shallower.

Pads may be circular, square or rectangular in plan and are usually formed using mass concrete of uniform thickness. However, where it is necessary to distribute the load from a heavily loaded column, the pad may be haunched or stepped as shown in Figure 2.32 on page 114, and may require reinforcement.

Where a deeper foundation is required to transfer the foundation loads to a stiffer or more stable stratum, the pad, together with the buried column that it supports, is often referred to as a pier foundation. The use by Sir Christopher Wren of massive masonry piers in the rebuilding of St Paul's has already been referred

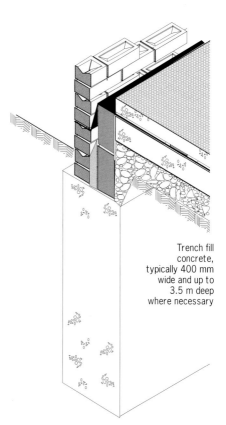

Figure 2.30 Typical trench fill footings which avoid the need for manual working below ground level where deep foundations are required, as in shrinkable clay soils

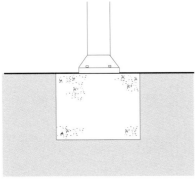
Mass concrete for steel column

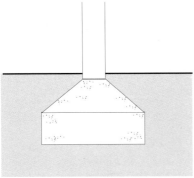
Reinforced concrete with sloping upper surface

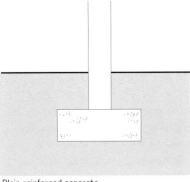
Plain reinforced concrete

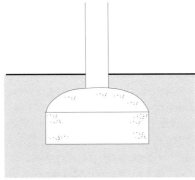
Haunched reinforced concrete

Figure 2.32 Types of pad foundation

to in Chapter 0. Although the distinction is not a clear one, the term pier is commonly used to describe the larger foundations that are required to support a bridge or other large civil engineered structure. An alternative way of constructing foundations of this type, especially in marine conditions, is to sink a prefabricated concrete shell or caisson (a hollow cylinder of steel or concrete 2 m or more in diameter) to the required depth by excavating material either internally or from round the outside, the caisson falling under its own weight, perhaps being assisted by loading with kentledge. The caisson then remains in place and any necessary loadbearing structures are built inside it.

Ground beams
Ground beams usually consist of reinforced concrete sections, which may be cast-in-situ into excavated trenches or be precast with reinforcement loops at the ends for connecting to other beams in the foundations (Figure 2.33).

Rafts and box foundations
A raft foundation consists of a reinforced concrete slab, whose thickness and stiffness are designed to spread the applied wall and column loads over a large area (Figure 2.34). For domestic applications, rafts are often built with thickened perimeters to provide protection against frost heave, in which case they are effectively trench fill foundations with integral groundbearing floor slabs. Downstand edge beams also serve to stiffen the whole of the foundation structure.

Rafts are used where it is necessary to limit the load applied to the underlying soil or to reduce the effects of differential foundation movements as a result of variable soil conditions or variations in loading. Further information on rafts is found in *Floors and flooring*.

Design options for building on soft clay or organic soils (eg peat) are similar to those for building on fill, except that special consideration of the variation of strength with depth is needed. Most soft clay has a desiccated crust which is much harder and stiffer than the underlying soil. The depth of the pressure bulb produced by a foundation tends to be dependent on the foundation width. Therefore, while a raft or wide footing will reduce the applied bearing pressure, the load will be applied to softer soil at greater depth. Consequently there will be a trade-off between reducing bearing pressure and increasing strains due to loading a soil with a greater compressibility. Similarly, where a site is underlain by a thin layer of peat, there may be no advantage in using a raft if the layer is deep enough to be outside the pressure bulb generated by standard strip footings.

Where large amounts of differential settlement are possible, as a result of a variation in thickness of a soft layer across the site for example, rafts can be provided with jacking points. Once most of the settlement is complete, the raft can be levelled by jacking against piers or piles and the gap between the two components then grouted.

Figure 2.33 Ground beams being prepared for incorporation into pad foundations on a steeply sloping site

Where a raft is founded well below ground level to form a basement, it is termed a buoyancy raft or box foundation. For some high rise applications, this form of foundation offers the additional benefit of reducing the ground loading because the weight of the foundation is considerably less than the soil it replaces.

Concrete mixes

As noted in *Floors and flooring*, concrete mix design can be complex, with many permutations of cement type, fine and coarse aggregates, and other additives such as waterproofers. It is nearly impossible to determine exactly what mixes have been used in existing foundations, though some routine, if not very simple, laboratory tests may reveal such details as cement:aggregate ratios.

Strength of mixes

A very great variety of concrete mixes will have been used in the past. The constituent materials may have been of poor quality, and the concrete have been subjected to attack by sulfates migrating upwards from below; inappropriate aggregates may have been used, and many strips will have been laid without adequate compaction.

The Model Byelaw requirements were for *'a concrete composed of cement and well-graded aggregate in the proportion of one-hundred-and-twelve pounds of cement to not more than twelve-and-a-half cubic feet of well-graded aggregate'*. By and large, if it was properly mixed, this was expected to give adequate performance.

The concrete for foundations is now specified by strength. Grades based on characteristic strength of either 30 N/mm² or 35 N/mm² are recommended, although higher strengths may be required for structural reasons. The Building Regulations (England and Wales) Approved Document C4 requires the minimum quality of the concrete to be at least mix ST2 of BS 5328-1[141], or, if there is any embedded steel, mix ST4.

Dimensional stability, deflections etc

It is usual for concrete in foundations, being buried in most cases well beneath the surface, to remain at fairly consistent thermal and moisture conditions. There may be circumstances, though, where it is necessary to investigate possible movements (eg in rafts, cellars or basements).

Reinforcement

Steel reinforcing fabric should comply with BS 4483[142]. Although other forms of reinforcement may occasionally be found in concrete mixes used above ground – for instance, steel fibres or polypropylene fibres incorporated into the wet mix – they will rarely have been used in foundation work.

Additives

Although pulverised fuel ash (PFA) has largely been used in civil engineering structures, it has been also used to some extent to replace part of the cement content in the foundations of large buildings.

BRE investigators have studied the performance of PFA in a number of structures. In addition to fly ash concrete, comparable Portland cement (PC) concrete was available for sampling at all the locations chosen. Measurements of a range of properties were made on concrete cores taken from these structures. Generally, the fly ash concretes showed improved properties when compared with similar PC concretes. These improvements include increased strength, reduced permeability, increased resistance to the ingress of chlorides and suppression of alkali silica reaction. Two-year-old concrete containing fly ash had carbonated to a greater extent than PC concrete from the same structure, but differences were considerably less marked in older structures. Overall, these results indicate that the performance of the concrete in these structures was not adversely affected by the partial replacement of PC by fly ash and, in many cases, significant improvements were observed[143].

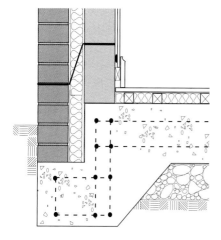

Figure 2.34 The typical edge detail of a reinforced concrete raft, commonly used where loads imposed on the soil must be limited

Main performance requirements and defects

Strength and stability

In the Building Regulations, Section 1E: Strip foundations in plain concrete, gives guidance on the design of traditional strip foundations. This concerns exclusively the provision of adequate **width** and **strength** to limit, to acceptable levels, the settlement of foundations on ground that is compressible and laterally variable in its compressibility.

The quality and thickness of the concrete used in foundations will vary greatly. Traditionally, so-called plain concrete was used. This was site mixed, and the mix proportions often were 1:2:4 cement:fine aggregate:coarse aggregate by volume. The strength of a concrete depended on the aggregate grading, quantity of mixing water, efficiency of mixing, degree of compaction etc. From the late twentieth century the quality control of concretes improved considerably; on level sites, failures due to inadequate specification should be rare. Nevertheless, sloping sites may give problems, particularly where it has been necessary to provide steps in the foundations and adequate overlap is needed.

In new construction, for normal strip foundations of plain concrete (Figure 2.35), the thickness must be at least equal to the amount of its projection from the face of the wall or 150 mm, whichever is the greater. The width of the strip should be that required by Table 12 of Approved Document A.

If a service trench is to be excavated deeper than adjoining building foundations, the Building Regulations (England and Wales) provide for certain minimum dimensions.

For a trench which is less than 1 m from the nearest part of the building, the trench must be filled with concrete up to the level of the base of the foundations (Figure 2.36). Where the trench is 1 m or more from the building, it must be filled with concrete up to a level (below the bottom of the foundations) which is equal to the distance from the building to the trench, less 150 mm.

Whatever the form of construction adopted for concrete foundations, in all cases the adequacy of long term performance does depend in large degree on the condition of the soil underneath them and the proximity of vegetation.

Changes of level
Where a step occurs in normal strip foundations, the two levels of concrete must overlap by a distance at least equivalent to twice the height of the step or 300 mm, whichever is the greater (Figure 2.37). The height of the step must not be greater than the thickness of the concrete strip. There should be no made ground and no wide variation of soil type within the loaded area.

In deep strip foundations 500 mm or more thick, BS 8004 requires any step to be not greater than the concrete thickness and the lap to be at least 1 m or twice the step height, whichever is the greater (Figure 2.38).

Mining subsidence
Foundations for buildings where large ground movements are expected (eg in former mining areas) may be designed to be flexible, with the buildings they support being able to undergo distortion and to return undamaged to their final forms at new levels after the subsidence wave has passed. As already noted, flexible school buildings have been designed and built in former mining areas using the CLASP steel framed system, following advice from the then BRS. The types of ground movement that occur with mining

> **Case study**
>
> **Long term settlement of a pair of semi-detached houses built on trench fill on stiff clay**
>
> Settlement observations from 1951 to 1991 of a pair of semi-detached two-storey houses built on trench fill foundations on stiff clay indicated that settlement was still continuing. After 36 years, deformation of one gable end wall caused minor cracking of the exterior brickwork. It was in this area that the trench fill foundations of mass concrete were cast in a softer, more muddy part of the trench. The settlement of the last 20 years has been influenced by the growth of a nearby ash tree and droughts in 1976, 1989 and 1991.

subsidence were considered, and it was concluded that a flexible building, designed in 10-foot bays, would have to be able to tolerate a differential settlement of 3 inches between adjacent columns and a change of slope of the ground between adjacent bays of 1 inch in 10 feet. The horizontal stretch of the ground surface could be as much as 4 inches in 100 feet. To minimise horizontal forces, the foundations for these buildings were laid so as to prevent any key into the ground; each consisted of a thin concrete slab, with central reinforcement sufficient to take all the tensions that would develop, separated from a smooth surface of compacted graded sand by building paper so that the slab could slide.

The weight of the whole structure was kept to a minimum to facilitate slip under the slab; the three-storey and single-storey buildings were said to impose bearing stresses on the ground of only 162 and 114 lb/sq ft respectively. Because of their light weight and flexible structuring, and the fact that the buildings could tolerate large differential movements, it seems that they would also be suitable for use on poorly compacted fills as well as on ground liable to mining subsidence.

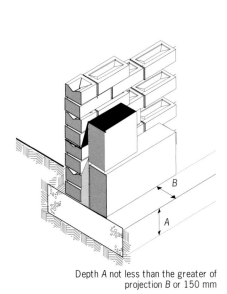

Depth A not less than the greater of projection B or 150 mm

Figure 2.35 Concrete thickness in normal strip foundations

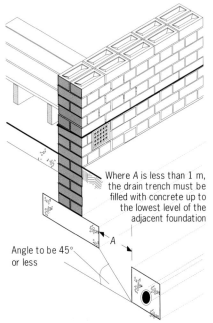

Figure 2.36 Required distance of drains (and concrete protection) from foundations

2.3 Concrete strips, pads etc

Loose sand and gravel

It may be necessary to employ wide strip or raft foundations to reduce the settlement of buildings founded on loose sand and gravel to tolerable levels. There is a possibility that vibrations from machinery, pile driving or traffic may cause long term compaction of the soil, which would in turn lead to subsidence. In very loose sand there is a risk of spontaneous liquefaction under saturated conditions; liquefaction can also be triggered by a rapid rise in the water table, as a result of severe flooding for example. Consequently, very loose sand should be treated using a vibrator prior to construction, or alternatively, where treatment would be uneconomic, piers or piles should be used to transfer the foundation loads to a deeper stratum.

For these special conditions, a suitably qualified geotechnical engineer should be consulted.

Case study

Testing for settlement
Precise levelling surveys of the settlements during construction and early occupancy of four blocks of local authority housing built on soft clays adjacent to the River Clyde were carried out in 1984. Piled foundations had previously been used for four and five storey buildings in this area but strip footings on vibro-replacement treated ground and basements were authorised for this site on condition that certain performance criteria were satisfied. The levelling observations showed that the overall and differential settlements were approximately half the permitted values. Measured soil pressures beneath the foundations of two of the blocks were close to the values assumed in design[144].

Commissioning and performance testing

Where installations are specified by performance, the appropriateness of the specification must be confirmed by testing. One such scheme was examined by BRE investigators in 1984 (see the case study below).

Durability

A few influences on most sites are known to reduce the normally indefinite life of concrete recommended for construction of foundations; the most significant of these being sulfate attack and frost. The care with which the materials are mixed and laid often governs their resistance to degrading influences. Concrete durability depends on the mix proportions and degree of compaction, both being dependent on the quality of workmanship during construction. Washing out fines from ground around foundations can result from leaking water mains or drains. Frost normally affects only uncured concrete, although foundations may be affected if precautions were not taken during the original construction. The relevant codes give extensive guidance on the preferred specifications and precautions for given circumstances.

Breakdown of concrete by chemicals which may be present in the soil in its natural state are due mainly to sulfate attack and be hastened by using unsuitable, poor quality or incorrectly prepared materials.

On the other hand, many chemicals, present as a result of contamination of the site with industrial wastes, can attack concrete. Their presence will be revealed by analysis of samples of the soil taken from boreholes at various depths, or, preferably, of the groundwater that seeps into the boreholes.

BS 8103-1[37] recommends that foundations are taken to a depth of 1 m in clays subject to seasonal movements, and to a minimum depth of 450 mm in sands, chalk and other frost-susceptible soils. In the case of frost-susceptible soils, it advises consultation with the approving authority since these soils are widely variable; an increased depth is also advised in upland areas and in areas known to experience long periods of frost.

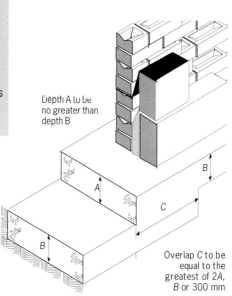

Figure 2.37 Typical requirements for overlap at a step in concrete strip foundations

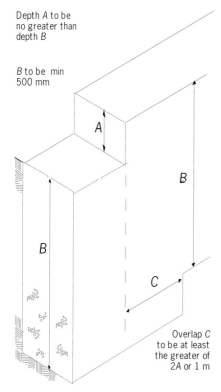

Figure 2.38 Overlap at a step in deep concrete trench fill foundations

Sulfate attack

In most cases, deterioration of concrete foundations is caused by the action of soluble sulfates, which are widely distributed in clays belonging to the London, Oxford, Kimmeridge, Lower Lias, and Keuper Marl formations. The sulfates in these clay foundations may be present as calcium, sodium, or magnesium compounds. These become dissolved in the groundwater, the resultant solution being aggressive to concrete. The amount of sulfates in the soil or groundwater is usually expressed as the concentration of sulfur trioxide (SO_3).

Of the sulfates commonly associated with contaminated land sites (for example sulfuric acid, magnesium sulfate, ammonium sulfate, sodium sulfate and calcium sulfate), the first three are the more aggressive towards concrete. After 15 years in BRE site trials, all the concrete samples exposed to strong magnesium sulfate solution (1.5% SO_3) had lost over 50% of their strength.

The effect of sulfates on concrete depends not only on the concentration and accessibility of the solutions that are formed in the groundwater but also on the quality of the concrete – particularly its impermeability – and on the kind of cement used. Concrete subjected to water pressure on one side (eg in retaining walls) is particularly liable to damage because fresh sulfates are carried into the concrete by the flowing water. Usually, where analysis shows less than 0.2% concentration in the soil, no special precautions are needed. For sulfate contents of more than 0.2% in the soil, sulfate resisting Portland cement should be used.

Sulfate attack due to thaumasite formation is potentially very serious. As already explained in this chapter, thaumasite is a calcium silicate carbonate sulfate hydrate that can occur in certain types of concrete and mortar, either alongside ettringite and gypsum or alone. It is formed in the reaction of calcite, gypsum and calcium silicate hydrate. Unlike ettringite, the amount of thaumasite is little restricted since it is formed from the calcium silicate hydrate fraction of the hardened cement paste.

The breakdown of the calcium silicate binder during thaumasite formation results in a reduction in the strength of the concrete until it eventually deteriorates into a mush. Concrete that has suffered deterioration caused by thaumasite formation typically shows a white pulpy mass between the aggregate grains.

Under cold, wet conditions and in the presence of readily available sulfate and carbonate ions, all Portland cements, even sulfate resistant Portland cements, are vulnerable to the thaumasite form of sulfate attack. The rate of thaumasite formation is generally slow, unless all the following conditions occur, when thaumasite can form rapidly:
- constant high humidity
- low temperature (–4°C)
- initial reactive alumina (Al_2O_3) content between 0.4% and 1.0%
- a supply of available sulfate and carbonate anions

Like blastfurnace slag, PFA is now widely recognised as improving the sulfate resistance of concrete[77].

See also *Avoiding the thaumasite form of sulfate attack: two year report*[145].

There is further information on sulfate attack on concretes in *Floors and flooring*.

Salt crystallisation

If concrete is subjected to wetting and drying cycles caused by groundwater fluctuations, salt crystallisation can occur within the concrete pores, particularly in the capillary zone. If the pressure produced by crystal growth is greater than the tensile strength of the concrete, the concrete will crack and eventually disintegrate.

Chlorides react with the free lime, $Ca(OH)_2$, from the hardened cement paste to form soluble calcium salts. Eventual loss of these salts reduces the strength of the concrete and increases its porosity, making it more vulnerable to the ingress of other chemicals[77].

Acids

Concrete, being an alkaline material, is vulnerable to attack by acids. Prolonged exposure of concrete structures to both organic and inorganic acidic solutions can result in their complete disintegration.

Acids attack the cement hydrates, so degrading the cement binder. The concrete then becomes more porous and vulnerable to further attack. Sulfuric acid and, to a lesser extent, hydrochloric acid are regarded as being particularly aggressive because, in addition to degrading the cement binder, they also produce an expansive reaction.

Studies have shown that, although concrete is theoretically vulnerable to attack by acids, little evidence has been found of the structural weakening of mass concrete exposed to the more common types of naturally occurring acid waters. It seems, therefore, that while concrete, because of its alkalinity, is potentially vulnerable to attack by acid, the rate of attack depends on the type of acid, its concentration, the composition of the concrete, the soil permeability and groundwater movement[77].

2.3 Concrete strips, pads etc

Hydrocarbons

Hydrocarbons such as petrol, petroleum distillates in general, and lubricating oils that are entirely of mineral origin, do not attack concrete. Lubricating oils that contain vegetable oils can cause gradual surface degradation; vegetable oils can readily oxidise to produce acids, and it is these acids that cause concrete degradations.

Despite not readily attacking concrete, the majority of hydrocarbons, because of their low viscosity and surface tension, are capable of seeping through even the densest concrete. If these hydrocarbons contain compounds deleterious towards concrete, the resistance of concrete will depend on the concentration of the aggressive compounds and the rate of ingress of the hydrocarbons[77].

Microbial corrosion

As well as being corroded by chemical agents, concrete can be attacked by micro-organisms; for example, bacteria, fungi and lichens. The actions of certain bacteria, both aerobic and anaerobic, can degrade concrete placed below ground level. Soils are thought to contain both types of bacteria, so when conditions change from anaerobic to aerobic or vice versa the dormant bacterial strain takes over. The combined action of these bacterial strains can result in the formation of sulfuric acid. Also, sites are likely to contain both anaerobic and aerobic areas, and so migration of hydrogen sulfide across the site can lead to the formation of an anaerobic–aerobic cycle, and the formation of sulfuric acid[77].

Other potentially aggressive products

Industrial effluents and waste products may contain many chemicals that will attack concrete: mineral acids, and ammonium and other sulfates, are common examples. Such cases have to be considered individually[77].

Corrosion of reinforcement

There may be visible signs of corrosion of steel reinforcement in ground beams if concrete cover is too thin.

In practice, foundation failure caused by inadequate durability will often be visible only indirectly, as damage to the superstructure: by this stage stability will be affected. See the section on strength and stability earlier in this chapter.

Local building control departments should have information on the presence of aggressive soil or groundwater. Where any doubts exist, samples should be taken for analysis.

Chemical reactions in concrete which involve the cement and aggregate, primarily alkali silica reaction (ASR), are known to cause deleterious changes in the concrete. These changes affect the performance and durability of the concrete and in some cases may have significance for its structural performance (BRE Information Paper IP 16/93[146]).

The following additional considerations apply particularly to the deterioration of concrete in contaminated land:
- industrial effluents, sewage and pollution need special treatment and must be assessed by a specialist, as indeed should problems involving nitrates and other unusual agents (organic acid, fats, oils etc)
- the age of the concrete before the first contact with aggressive agents should be taken into account
- design aspects, including shape, are important for concrete performance
- gases, in the absence of water, are not considered aggressive
- for buried concretes, the permeability of the soil should be considered in relation to the renewal of water or other solution in contact with the concrete. Soil permeability greater than 10^{-5} m/s may require special treatment of the soil[77].

Frost heave

The mechanism involved in frost heave was discussed in Chapter 1.2.

Raft type foundations may crack as a result of incorrect positioning of reinforcement. In clay soils, trench fill footings have been known to fail if compressible material is omitted from the trench sides.

Case study

Examination of concrete removed from the foundations of a public building

The building had been built in 1936 on strip foundations in a sulfate bearing clay soil. After 15 years, movements in the structure led to the underpinning of the south wing followed, 5 years later, by the underpinning of the north wing. Some 30 years after the building was erected considerable movements were still occurring and partial demolition was contemplated. It was reported that the site had become considerably wetter since the underpinning work was carried out and this was attributed to the excavations made at that time breaking into an established water course. Measurements of movements recorded in the structure appear to indicate expansion of the foundations due to sulfate attack, and this was confirmed by BRE laboratory analysis of samples removed from the site.

Case study

Frost heave under a cold store

In a cold store which had been operating for about 4 years at –10°C, damage to the ammonia pipes and distortions of the structure was discovered. It was found that the floor near the centre of the cold store had risen about 75 mm., and samples from boreholes made through the floor contained ice up to 25 mm thick. The boreholes were taken down to unfrozen soil, and the total thickness of ice found by continuous sampling was about 75 mm, showing that all the heave was caused by the formation of numerous ice lenses. Temperature measurements showed that the ground was frozen to a depth of nearly 2 m. The temperature of the cold store was raised very slowly, so that in the thaw the water liberated from the ice lenses could dissipate through the soil, and the structure was lowered safely. Rapid thawing would have led to a foundation failure.

The remedial measure was to supply heat under the insulation to prevent the soil from freezing. This was done by a low-temperature electrical heating grid controlled by a thermostat embedded in the soil.

Work on site

Inspection

The problems to look for are:
◊ footings constructed in poor ground conditions
◊ cracking in supported masonry walls
◊ wash-outs under footings caused by overflowing local watercourses or leaking drains
◊ breakdown of concrete from acid or sulfate attack

Chapter 2.4   Piles

Piles are used to transfer loads from buildings to the supporting ground, and are used in a wide range of situations where conventional strip footings are inappropriate. They are particularly used where soft or loose soils overlay strong soils or rocks at depths that can be reached conveniently by driving or boring. They are often the most economic type of foundation when very heavy loads must be supported or uplift forces need to be resisted. Large piles are extremely useful for limiting the settlements of large structures on deep stiff clays; smaller versions can provide appropriate foundations for houses and other small buildings on stiff clays liable to shrinkage and swelling. The technique has been in use for many years (Figure 2.39).

Piles of many different types and methods of installation have been developed to suit the wide variety of soils described in Chapter 1. However, they are not a universal solution to every foundation difficulty and specialist advice will normally be necessary before they are specified, both in new buildings and in remedial work in existing buildings.

This chapter includes all types of piled foundations. They fall into two main types:
- bored and dug, including short bored and secant
- driven and jacked piles, steel, concrete and timber

The construction of large basements is now normally carried out by constructing a stable perimeter wall within the ground and excavating the soil from the enclosure. Piles of various forms can be used for this kind of construction, and the techniques are described in Chapter 3.

The use of mini-piles in remedial work is described in Chapter 2.1.

Characteristic details

Types of pile
There are basically two types of pile:
- end-bearing
- friction

End-bearing piles derive the greater part of their support from bearing forces at the base. They act largely as columns transferring loads through soft deposits, usually to dense granular soil or rock at the foot of the pile. Friction piles, on the other hand, develop most of their support from friction between the shaft and the soil, usually firm clay.

Both end-bearing and friction piles can be formed in a number of ways. The two main ones are:
- displacement piles
- non-displacement or replacement

Displacement piles are installed by forcing the soil out of the way and they can be either preformed or cast-in-place. The main types include:
- all preformed piles of timber and reinforced concrete, continuous or jointed
- tubes and shells of steel or concrete, closed at the lower end, which are left in the ground after driving and then filled with concrete
- piles formed by driving steel tubes fitted with detachable shoes or temporarily closed at the bottom with plugs of concrete or gravel – the shoe is left behind or the plug hammered out as the tube is extracted while concrete poured through the tube runs out at the bottom to fill the void

Figure 2.39 A lorry-mounted boring rig in operation in 1970

Non-displacement or replacement piles are formed by casting concrete or cement grout in a bored or excavated hole. H-section rolled steel piles are frequently referred to as small displacement piles.

Non-displacement or replacement piles can be formed by:
- driving an open tube, in sections if necessary, and extracting the soil core as the tube is sunk. When the bearing stratum is reached the tube is cleaned out. It is withdrawn as the concrete is poured or left as a permanent lining
- augering a hole and filling it with concrete. A range of techniques is available for preventing collapse of the borehole and for placing the concrete at the bottom of the hole beneath a stabilising liquid

Casings, coatings, linings and sleeves

These may be needed for cast-in-place piles in certain ground conditions, either to protect the concrete from flowing into voids round the shaft or to protect the fresh mix from aggressive soil conditions.

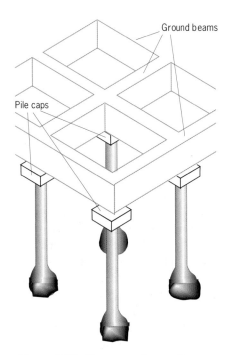

Figure 2.40 Pile caps and ground beams

Caps

The upper end of a pile often needs to be adjusted on site so that an appropriate connection can be made to the remainder of the foundations. This adjustment normally takes place by means of trimming or by fitting an additional item of construction called a cap. Piles are often capped in groups; that is to say, connected in groups so that they act together rather than independently (Figure 2.40).

Short bored piles

More than half a century ago, the general problem of providing satisfactory foundations for houses and other light buildings on shrinkable clay became urgent. Although at this time the practice of stringent control of the vegetation in proximity to the building could reduce appreciably the incidence of damage to existing and future structures, such control was not easy to ensure, and alternative ways of tackling the problem were developed, of which the centre-piece was the short bored pile.

Rainfall in the south east of the UK had been below average in the first years of the 1939–45 war, with correspondingly larger than normal shrinkages in clay soils. There were large numbers of reports of cracks in houses (precursors of a similar spate 30 years later) prompting BRS to develop as early as 1942 a foundation system based on short bored piles for houses built on clay[147].

For houses and light industrial buildings where strip or pad footings are unsuitable (eg on clays liable to substantial shrinkage or swelling) (BRE Digest 240[43]), short bored piles are eminently suitable. The piles support ground beams in loadbearing wall construction and should be located at the sides of openings and beneath load concentrations. It seems likely that, had this very practical solution to the problem of foundations on clay been more widely adopted from that time onward, virtually all of the many cases of damage that occurred in houses built in subsequent years would have been avoided. Instances of damage, whether due to seasonal swelling and shrinkage of clay soils or due to the growth or removal of trees, would certainly at the least have been greatly reduced.

BS 8004[68] recommends that short bored piles be considered as an alternative for low-rise dwellings on soft ground, made ground or clays more than 2 m deep, particularly if the water table is high.

Driven small-diameter shell piles can be used for similar structures on other difficult soils and where there is a high water table, but the costs may be greater.

Mini-piles

Since the 1970s, piles less than 300 mm diameter (and frequently less than 150 mm diameter), known as mini-piles, micro-piles or pin-piles, have been increasingly used for remedial operations and work in confined spaces. Many of these piles have been installed through the substructures of historic buildings found to be deteriorating or settling unduly. Remedial work of this type needs extensive site investigation and careful design. The works are normally undertaken by specialist civil engineering contractors. However in many cases, mini-piles have been used to underpin walls and stabilise and repair cast-in-situ floor slabs of low-rise buildings. They have also been used in the construction of new walls and floors on weak soils, and fills and shrinkable clays. These piles are often installed by small specialist contractors, some of whom may not be familiar with the principles of foundation behaviour or the particular problems of pile design and construction[136].

Chapter 2.1 describes several examples of the use of mini-piles in remedial underpinning.

Equipment used for constructing or placing piles

Bored pile machines

These machines consist of a vertical guide containing a kelly bar (rod) which holds a cutter. The kelly bar is rotated by hydraulic power from a truck or crane which supports the bar and cutter in operation, and moves them to new points of work. Cutters are often in the form of augers.

Though a fairly rarely used technique, in suitable soils, special under-reaming cutters can enlarge the pile base to more than twice the shaft diameter. The augers are either track-mounted or suspended from a crane. (The motor produces considerable noise and vibration.) If the soil conditions are suitable and temporary casings are not required, short, small-diameter piles can also be bored by tractor or lorry-mounted augers.

The holes drilled may form the basis of in situ concrete piles or contain a concrete base and a steel column. The bored pile machine can be used to install column bases and columns at below basement level leaving the main excavation of the basement until a later stage of construction. Holes of up to 2 or 3 m in diameter and up to 25 m deep can be drilled with normal equipment, and up to 60 m deep with special fittings.

Firm and stiff clays, capable of standing unsupported long enough for pile boreholes to be drilled and concreted without the need for temporary casings, are frequently overlain by fill, loose granular soils or other unstable soils which could easily run into the bore when the auger is withdrawn. There are three main methods of reducing this risk:
- using a short lead-in liner for the auger where the depth of unstable ground is shallow
- keeping the borehole full of water or drilling mud (bentonite suspension) so that the liquid pressure at any level in the borehole is greater than the natural water pressure tending to disturb the borehole wall. The liquid level must be maintained and, when bentonite mud is used, its properties must be continually checked for soil contamination; this leads to a slower rate of boring than for an open hole. More importantly, when either water or bentonite mud is used, the concrete cannot be poured through a funnel at the top of the borehole. It must be placed below the stabilising liquid through a tremie pipe, and the liquid then floated off to storage
- using hollow stem continuous-flight augers. Very little spoil is extracted while the flights are penetrating the soil. When the required depth is reached a plug at the bottom of the hollow stem is expelled as cement grout or concrete is pumped down the stem. The pile forms from the bottom as the auger, which continues to rotate in the direction of boring, is withdrawn to remove the spoil. Care is required to ensure that spoil does not remain in the borehole to contaminate and weaken the concrete or grout

Piles excavated by continuous-flight augers in diameters from 300–750 mm can be formed in waterbearing sands and gravels, stiff clays and soft chalk to depths of about 30 m. Where necessary, reinforcing cages up to 12 m long can be lowered into the wet concrete after the auger has been removed.

Pile driving

Pile driving may take one of two forms, depending on site conditions:
- drop hammer
- vibrator

Drop hammers are powered by compressed air, hydraulic pressure or diesel engine. Control of the movement of the hammer and the alignment of the pile is provided by leaders forming part of the driving rig or hung from the crane that carries the hammer. The piles may or may not be driven with a plastics or wood cushion between the hammer and the pile, depending on the circumstances.

Where the soil conditions are suitable, piles and casing tubes can be installed by vibration. The pile or casing is vibrated along its axis at a frequency and amplitude depending on the system used, the weight and dimensions of the pile, and the soil properties. This causes the resistance to penetration of the soil to be broken down, permitting the pile to sink under its own weight and the weight of the vibrator. The method is mainly used in loose sands and silts.

Large rigs impose heavy loads on the soil, and the temporary works needed to move and operate them over weak ground can be extensive.

Materials used for pile construction

Timber

Timber piles are not used now for buildings, although being used on a considerable scale at one time. They can usually be installed by general civil engineering contractors, and are mainly used in temporary works and marine applications where resilience to impact loading is important. They are generally made in the round log form – although for certain applications, can be squared – and are treated to preserve integrity above water table levels.

Timber piles tend to be cheaper than steel or concrete piles but are unsuitable for heavy loads. They should be prevented from splitting during driving by hoops fixed at the tops and steel points at the bottoms.

Steel

Steel piles of H, box and heavy tube sections are available in a range of sizes and can be lengthened or shortened if required during driving. One form of pile uses light corrugated steel shells to provide a permanent casing for concrete. Depending on the length required, a number of shells are assembled on a mandrel for driving. After driving, the mandrel is withdrawn and the shells are filled with concrete. Another permanently cased pile is constructed of steel strip edge-welded to form a continuous helix with a flat shoe welded to the bottom. The casings can be shortened by flame cutting during driving or lengthened by butt-welding. Steel piles are normally bottom-driven on a plug of dry concrete and concreted with only short bonding bars connecting the shaft to the pile cap.

Where soil conditions are problematical, the steel may be specified in one or more of the following ways:
- to be high yield steel
- to have protective coatings
- to be of heavier section than would otherwise be needed, when the outer surface is considered to be sacrificial, depending on durability requirements.

Corrosion conditions in a marine environment can be severe, and steel piles are normally grit blasted before coating (eg with epoxy formulations) to achieve something like 175 μm thickness.

Concrete
Reinforced concrete piles are usually made of segments of relatively small cross section, in various lengths (say 4–10 m) joined by special couplings, and fitted with detachable points to suit different driving conditions. Reinforcement is normally of high tensile steel strand. Where conditions are suitable, lengths of up to 20 m may be used. The segments are best cast in the factory from high quality concrete capable of withstanding hard driving, but sometime are precast on site. Other forms of reinforced concrete driven pile use precast hollow shells or cylinders threaded on to a mandrel and installed by a combination of top and bottom driving through the mandrel. Pile lengths can be altered during installation by changing mandrels and adding or removing shells. When the required penetration or resistance is reached the mandrel is withdrawn, reinforcement placed and the hole concreted to form an integral pile.

Withdrawable-tube piles are driven only to the depth required by the ground conditions. In favourable situations and where the piling rig can extract the tube, they can generally be installed more cheaply than most other piles. Withdrawable tube piles have the advantage that the concrete is cast in direct contact with the soil. In some proprietary systems a plug of concrete is driven out of the bottom of the tube to form an enlarged bulb before the tube is withdrawn and a reinforcing cage is lowered into place before the bulk of the concrete is poured (Figure 2.41).

Great care must be taken to ensure that the shaft concrete is not lifted as the tube is withdrawn and that squeezing ('waisting') of the wet concrete does not occur in soft soils. Various proprietary systems have different methods of compacting the concrete as the tube is extracted; these include:
- tamping
- low-frequency vibration of the tube
- driving of precast units into the wet concrete

Full-length reinforced concrete piles are usually installed by general civil engineering contractors, but segmental and shell piles and cast-in-place piles are installed by specialist contractors.

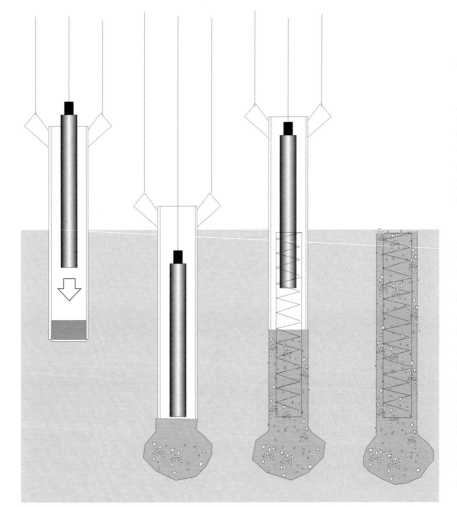

Figure 2.41 Stages in the construction of a temporarily cased pile

Main performance requirements and defects

Strength and stability
As already described above, a pile may be a slender shaft of timber, steel or reinforced concrete driven into the ground, or a column of concrete or cement grout cast in a hole bored from the surface of the ground. It can have a flat, pointed or bulbous contact area with the soil at its base but normally has a vertical or near vertical interface between the shaft and the surrounding soil.

Pile spacing

Piles are normally specified to meet requirements for magnitude of loading or to minimise settlements. The piling scheme therefore must relate closely to the distribution of loading and take account of shear forces and bending moments in pile caps, ground beams and rafts. Shear and bending forces are generally least when numbers of small piles are used in preference to a few large diameter piles. On the other hand, a single large pile beneath a column may eliminate the need for a pile cap, but lateral stability then may be a problem. In cohesive soils the settlements of piles under working loads are smallest when the piles are long and slender, provided they can be placed far enough apart not to influence each other unduly. The centre-to-centre spacing should normally be not less than three pile diameters[148].

Pile bearing forces

Whatever the actual form of the base, vertical bearing forces come into play when the pile settles under load (Figure 2.42). Where the pile penetrates to rock or firm support, heavy reinforcement cages can be installed in larger diameter bored piles; very large compressive and tensile loads can then be carried.

It is possible for the base to be embedded in soil capable of sustaining much higher loads than could be carried by a similar contact area near the ground surface. Loose, granular soils are compacted by pile driving and their bearing capacity thereby increases. On the other hand, driving piles into dense sand or gravel beds may cause disturbance and a consequential reduction in bearing capacity as well as damage to the piles. Except when piles bear on rock, significant movements are necessary for the bearing capacity at the base to be developed (up to 10% of the base breadth or diameter); under working conditions only a small proportion of this extra bearing capacity may actually be used.

When a pile moves under compressive or tensile loading, vertical frictional or shear forces are developed between the shaft surface and the surrounding soil. Quite small shaft movements cause these forces to increase rapidly and in many soils they may be substantially larger than the bearing forces at the base. Frictional forces frequently reach maximum values at settlements (or uplift movements if the pile is loaded in tension) of less than 2–3% of the shaft diameter, but in stiff clays further movement can cause a reduction in the frictional resistance.

Relative movement between a pile and the surrounding soil must occur for frictional and bearing forces to be developed. In common with other types of foundation, it is not possible to form a piled foundation which will completely eliminate settlement. There is always a small component of settlement due to the elastic compression of the pile shaft. Clay soils consolidate under load for long periods as water is slowly forced from between the soil particles. Structures on piles in clay can, therefore, continue to settle for some years after completion.

Piles are frequently used to transfer loads through soft silts and clays to firm ground beneath them (Figure 2.43). If the soft strata are disturbed during installation of the piles or are later subjected to loading from, say, the floor of a building, these strata may settle more than the piles. The relative movements between the shafts and the soil then become negative and the piles are subjected to additional loading which must be carried by the bases.

Shaft resistance

Considerable research has been devoted to deriving resistance values from loading tests of various types of pile in differing ground conditions. This research indicates a great deal of scatter (ie variability in loads carried), with measured values ranging widely. There are a number of possible complications in establishing a correlation between the various factors.

Wherever possible, final designs should be based on values derived from load tests of prototype piles at the specific site.

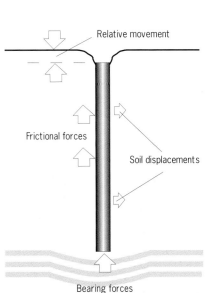

Figure 2.42 Vertical bearing forces come into play when the pile settles under load

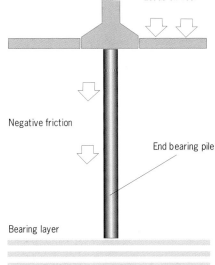

Figure 2.43 Piles are frequently used to transfer loads through soft silts and clays to firm ground beneath them

Skin friction
Unless due account is taken in the design, negative skin friction may produce large settlements or even over-stressing of piles. Negative skin friction can be reduced by using coatings such as bitumen or by sleeving the upper part of the pile with a low friction material. However, coatings of this type are expensive and it is normally preferable to increase the penetration depth in order to increase the load carrying capacity of the pile.

Noise and other unwanted side effects
Noise and vibration
In urban areas, long periods of noise and vibrations are unlikely to be acceptable so some form of bored pile, particularly the continuous-flight augered pile, would probably be more environmentally suitable. Piling with hollow stem continuous-flight augers is probably the quietest technique in wide use in the UK.

Loud noise and some vibration often accompany hammer driven piling but driving under controlled conditions does not necessarily produce harmful vibrations. Noise levels can be reduced by enclosing the hammer and leaders inside a heavily insulated box. Some cast-in-place piles are installed using an internal drop hammer acting on a plug of concrete at the bottom of the tube. These are quieter and cause less vibration than top driven piles. Heavy rigs with leaders can often be replaced by drop hammers hung from cranes with the casing tubes temporarily supported by trestles.

With vibration driving of piles, the chief source of noise is from the motors driving the vibrator.

Ground heave
Where piles are to be installed close to buildings, heave associated with driving large displacement piles must be minimised even if the vibration caused by driving them can be tolerated. Open steel sections cause less ground movement than other driven piles and heave can be reduced by pre-boring holes for larger displacement piles. If heave is to be prevented altogether, bored piling is usually necessary.

Heave of the ground surface can cause problems when closely spaced piles are driven on open sites. Piles already installed may be lifted during the driving of additional piles and the point resistance reduced. Tensile forces are set up in nearby piles and damage can result if the concrete of driven shell, cast-in-place piles is still relatively green.

Commissioning, performance testing and monitoring
Concrete pile integrity can be tested relatively simply by sonic or vibration methods which can identify imperfections.

Various kinds of tests may be used on one or more piles at the installation stage to determine their likely future performance (eg for design verification purposes):
- maintained load tests
- constant rate of penetration tests
- constant rate of uplift tests

Tests may reveal, *inter alia*:
- ground conditions in which piles continue to settle without any increase, or with a disproportionate increase, in load
- residual settlement on removal of load

Typical settlement limits which were used by the former Property Services Agency in the UK Government's building programme were 10 mm at design verification loading, and 8 mm residual settlement. Tests may establish whether or not the acceptance criteria have been met, and records of these tests and driving resistance measurements should be recorded and made available for subsequent examination.

The establishment of ultimate bearing capacity of driven piles in more recent years has become a sophisticated operation; for example, using wave analysis, when specialist knowledge is required for its interpretation.

Techniques for instrumentation to monitor the performance of piles following their installation include:
- slope measuring devices
- load cells

One slope measuring system uses electrolevels to measure the very small changes in slope in a pile when it is subjected to load. These are precise tilt measuring devices that can resolve changes in angle to fractions of seconds of arc. The changes in slope measured can be used to calculate deflections, bending moments, shear forces and soil reactions. The use of electrolevels has been shown to be cheaper, more reliable and more sensitive than strain gauge systems[149].

Monitoring pile performance
Instrumentation systems for monitoring the performance of piled and piled raft foundations have been available for a number of years. Systems have been developed to suit most types of piles, including tubular and H-section steel piles, and bored cast-in-situ concrete and precast concrete segmental piles. Measuring instruments can be fitted after the piles are formed or driven so that they do not have to withstand the pile installation forces, and can be replaced if they become inoperative[150].

Health and safety
The relevant British Standard is BS 5573[151].

Durability

Steel
Steel piles are easily handled and capable of hard driving to carry heavy loads; they can be driven in long lengths. H-sections and open ended box and tube sections cause less ground movement than other displacement piles. Steel piles are liable to corrode above the soil line and in disturbed ground, but can be protected by coatings and by cathodic techniques.

The corrosion of steel requires the presence of both oxygen and water (except in the special case of anaerobic bacterial corrosion). A number of studies have found that in undisturbed natural soils the type and amount of corrosion of driven steel piles is so small that it is negligible, irrespective of the soil resistivity and soil pH. There is, however, a difference in corrosion rates between piles driven into disturbed and undisturbed soils, which is thought to be due entirely to the differences in oxygen concentration. Corrosion of steel piling in UK soils has been found to range from nothing to 0.03 mm/year with a mean of 0.01 mm/year.

There is no evidence that driven steel piles are subject to corrosion by the action of sulfur reducing bacteria except where they project above ground or mud levels in a marine situation[77].

Concrete
Cast-in-place piles need to be constructed from high quality concrete mixes. There is, however, a danger that high strength mixes may lead to shrinkage cracking, which can have a detrimental effect on durability.

Precast piles and piles formed of segments or shells, cast under controlled conditions from dense well compacted concrete, can withstand hard driving and are generally resistant to attack by chemicals in the soil, groundwater or seawater. For maximum durability of piles in aggressive soils and groundwaters, dense concretes and smooth shaft surfaces are therefore required. These are best achieved by using precast segments or shells. Cast-in-place piles, though, can have high resistance to attack if the correct mixes are used, and great care is taken in compacting the concrete and extracting casings. For particularly adverse conditions, permanent casings may be advisable; protective coatings can also be applied to the steel[148].

Sulfate resisting cements have been used for many years in severe conditions (BRE Digest 250[152]) although risk of attack on concrete is difficult to predict with certainty.

Recent site and laboratory investigations carried out by BRE on problems of sulfate attack have shown that thaumasite has been responsible for the deterioration of concretes and mortars specifically designed to give good sulfate resistance. There are concrete mix designs to minimise deterioration due to the thaumasite form of sulfate attack[33,145].

On the other hand, examination of several hollow, precast concrete piles made with Portland cement after they had stood for many years in ironstone heavily contaminated by sulfuric acid showed no more than superficial signs of deterioration when cores were taken and analysed. This particular examination indicated that the poor state of some of the concrete was more likely to be due to poor initial compaction than to subsequent chemical attack.

Timber
The Romans used timber piles, a favoured technique being sessile oak piles some 450 mm in diameter with pointed wrought iron shoes driven into soft ground. Provided the piles remained below the water table, they survived quite well, as recent excavations of the foundations of Roman structures have shown.

Unless below the permanent water table, where conditions will prevent wood rot, timber piles are at risk from attack by wood-rotting fungi. Either heartwood of a durable or very durable species must be used, or one of lower natural durability that readily accepts wood preservative treatments (BRE Digest 296[154]); recommended treatments are given in BS 5268-5[155] and BS 5589[156]. If timber is used in the round or sapwood is present, preservative treatment is essential irrespective of heartwood durability. For freshwater piling, where the decay hazard is exceptionally high, only certain timbers should be used; and, with marine piling, at risk of attack by marine borers, again only certain timbers should be used. (Section 4 of BS 5589.)

Case study

The behaviour of mini-piles in sulfate contaminated soils

An investigation was carried out on embedded mini-piles in sulfate contaminated (magnesium or calcium sulfate) soils for periods of up to two years. It was shown that the presence of sulfate and magnesium ions increased the shear strength of the clay at all ages of test and the strength of the pile–clay bond. After one year all piles in contact with sulfate had been attacked. The most marked deterioration was shown by those embedded in the clays with increased calcium sulfate concentration and with the lowest magnesium sulfate concentration. The clay–concrete bond increased between six and twelve months for all sets of piles except those embedded in the clay with the highest magnesium sulfate concentration which showed a loss of over 50%. The biggest gain in strength (50%) was shown by the piles embedded in the clay with increased calcium sulfate[153].

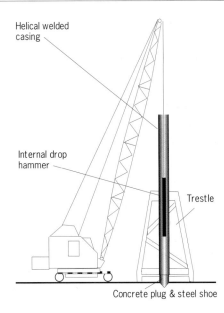

Figure 2.44 Driving a cased pile

Work on site

Site investigation
All the soils affecting installation of piles and behaviour of the foundations under working conditions should have been identified before any decision was taken to use piling. The size and type of structure, the depth of fill (or weak soils) above the supporting strata, and the strengths and compressibilities of the main supporting strata should have been also determined.

Any subsequent variations in seasonal groundwater levels should be recorded, and water that is flowing or under artesian pressure analysed for chemicals likely to damage pile materials[148].

Workmanship
During driving of precast reinforced concrete piles, the pile heads should be protected from damage by impact forces using helmets and packings; special steel points are required when driving to rock. Reinforcement must be incorporated in precast piles to resist tensile forces set up during handling and driving, and resulting from lateral or uplift forces in service. Unless specified otherwise for particular proprietary systems, concretes for driven cast-in-place piles (Figure 2.44) and bored piles in which the casings are withdrawn should be highly workable so that arching does not cause waisting as the casings are lifted. Highly workable mixes are also necessary for large bored piles without casings to enable the concrete to flow freely between the bars of reinforcement cages. Concrete must be placed carefully in casings to ensure that a dense homogeneous pile is formed; but where temporary casings have to be removed vibrators must not be used because of the risk of lifting the compacted concrete and reinforcement with the casing. Plasticisers are normally added to concretes and grouts used in piles formed with hollow stem continuous-flight augers[148].

Supervision of critical features
All testing and monitoring of piles and their installation should be carried out by competent and experienced engineers.

Inspection
The problems to look for are:
◊ settlement of ground beams connecting piles
◊ groundwater containing chemicals likely to damage piles
◊ original records of pile driving resistance not available
◊ breakdown of concrete from acid or sulfate attack
◊ rot in sections of old timber piles above the water table

Chapter 3

Basements, cellars and underground buildings

This third chapter deals with parts of buildings other than foundations lying below ground level. For the purposes of this book, a basement is taken to be that part of a building which is below ground level and which is used as habitable accommodation; whereas a cellar, also below ground level, is not used as habitable accommodation. Basements are generally reasonably well finished and heated (Figure 3.1); cellars are usually unoccupied areas probably built to store coal (Figure 3.2), and for other long term storage requirements. Cellars are typically poorly finished, and may be damp and lack adequate ventilation.

Existing cellars about to be converted into habitable accommodation, or basements about to be upgraded, may be inadequate in terms of thermal insulation, ventilation or provision of heating.

Cellars which have been used for fuel storage are likely to be contaminated by salts. Original finishes may have deteriorated and subsequent use of lightweight plasters in a relatively damp environment may have accentuated dampness and condensation problems.

Figure 3.1 A basement in the Victoria and Albert Museum, probably built for storage but now used for recreational purposes

Figure 3.2 A coal cellar (with accumulated leaves) from the nineteenth century which has remained derelict for many years

3 Basements, cellars and underground buildings

Basements tend to be difficult to ventilate properly and may often require special measures to avoid condensation. Bathrooms and other utility rooms, which require only limited natural lighting, are often located in basements and invariably generate high moisture vapour levels; the vapour will need to be extracted or diluted by ventilation. There can be risks of surface and interstitial condensation with balconies and access ways above basements; these problems are analogous to those sometimes encountered with flat roofs. Using basements as habitable accommodation will usually require particular attention to be paid to the likely performance of the building fabric and the servicing systems if satisfactory living conditions are to be provided.

Many cellars and basements, because they are at low levels and often near to water tables or tree root systems, are at risk of structural damage. This topic is dealt with separately in Chapter 3.1. Furthermore, most old cellars have no dampproofing and penetrating dampness can be a significant problem. This topic is dealt with in Chapter 3.2. Other matters which need consideration include ventilation, thermal performance, condensation, lighting, fire protection, and means of access and escape in case of fire, all of which may be more difficult to achieve in basements than in rooms above ground; these topics are covered in Chapter 3.3. With underground buildings, also known as earth sheltered buildings, many of the performance requirements are similar to those of cellars and basements, but, because the whole building is underground rather than just a part underground, there may be additional considerations relating to means of access, lighting and other services.

Basements in the housing stock

Around 550,000 dwellings out of a stock of around 20 million dwellings in England have all or part of their habitable accommodation in basements[4]. Of these, over three quarters date from before 1919. Dampness is the most common defect encountered, though it is not quite as common as one might expect, with just over 1 in 20 of all basements showing some signs of penetrating or rising damp. The occasional structural problem was observed in basements, but instances were too few to draw any statistically valid conclusions. Information for the other countries of the UK was not available for this book but it is probably the case that the vast majority of basements nationwide date from before the 1914–18 war.

In urban areas, cellars were commonly used for the storage of solid fuel. The circular cast iron manhole covers set into pavements in front of Victorian and Edwardian dwellings are still a familiar sight (Figure 3.3) although the use of gas for heating now means that little coal is now stored. The other main use of

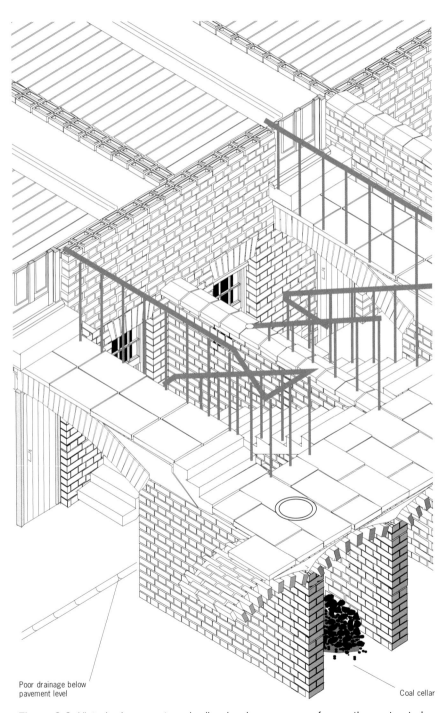

Poor drainage below pavement level

Coal cellar

Figure 3.3 Victorian basements and cellars in urban areas were frequently constructed below the level of adjoining streets. Pavements were carried on brick arches over the coal cellars

3 Basements, cellars and underground buildings

cellars was for general storage, or, in a few dwellings, and taking advantage of more stable temperatures at low level, for the storage of wine.

A current estimate of the number of houses built with some form of basement construction, mainly to accommodate sloping ground, is in the region of 10,000 per annum.

Changing attitudes to cellars and basements in housing

As part of a study of utilisation of space in housing, an examination was made in the late 1970s by BRE investigators of the feasibility of providing habitable basements in newly constructed housing[157]. In the study, alternative methods of constructing basements were examined, basement house plans were developed and a comparison was made, on a cost and amenity basis, of basement houses and comparable non-basement houses. It was shown that, in general, basement houses were more expensive on both flat and sloping sites than the nearest comparable non-basement houses. When measured as cost per unit of useful shell area (ie the floor area excluding circulation areas and garage), the basement house was 22–28% more expensive on flat sites and 8–11% more on sloping sites. However, the notional plot area required for basement houses is less, giving scope for higher densities of development and reduced land costs.

A further move to encourage the incorporation of basements into new housing came with the formation by the British Cement Association of the Basement Development Group in 1991. There was a realisation that there was scant information on structural design and that the information on waterproofing needed reviewing. Cost estimates for various site layouts were prepared, and builders and owners encouraged to consider incorporating a basement.

The estimates showed that a house wholly below ground might cost 11.5% more and a house with a basement partly below ground might cost only 4.1% more than a house without a basement. Where ground conditions would require deeper (2 m) foundations then the percentage increase reduces to 9.7% and 2.0% respectively. However for the same size house (typically 129 m^2) the site can be 3 m narrower, with a potential saving of 21% of land costs. Energy costs will also show a possible saving of between 4.4–5.6% and 6.1–9.5% for semi-detached and detached properties respectively[158].

A number of potential problems need to be faced when considering the conversion of a cellar into habitable accommodation, and these are considered in turn in the following chapters.

Basements in the non-housing building stock

There is no generally available information of the incidence of basements in the non-housing sector. However, it is a matter of common observation that many buildings built in areas where land values are very high do exploit the additional potential for accommodation and car parking provided by one or more basements. As will be seen in the following chapters, very deep basements of several storeys below ground level do have considerable implications for fire precautions and means of escape.

3 Basements, cellars and underground buildings

Chapter 3.1 Structure

Cellars and basements by their nature form the lowest parts of buildings, and may be required to provide whole or partial support to the remainder of the carcass. The assessment of their structural condition, of whatever materials they are made, is therefore of paramount importance in assessing the structural condition of the building as a whole. The older the building, the more crucial it will be to determine the structural condition of the cellars and basements.

Figure 3.4 The undercroft of St Paul's Cathedral

Characteristic details

Crypts
As was noted in *Floors and flooring*[66], large arched or barrel vaulted basements of major public buildings such as castles date from Roman times. Later medieval church crypts had more complex intersecting vaults with piers at relatively close centres. In parallel with the later barrels came the rather more slender ribs of the intersecting voussoir arches, infilled with stone slabs to carry the ground floor above. In Renaissance times, vault spans tended to become larger (Figure 3.4).

Domestic cellars and basements
Floor construction
Cellar floors are often poorly constructed and in a poor state of repair. Some comprise compacted earth or rock. Concrete floors in older properties may be only a thin layer of weak concrete. Many cellars have stone or slate flag floors, in which the flags have either been laid directly onto the soil below or have been bedded in a thin layer of weak concrete. Cellars with suspended timber floors are rare.

Occasionally the cellar floor will have been replaced or upgraded with a modern concrete floor which includes a dampproof membrane. In many cases, where the cellar lies beneath only part of the building footprint, the ground floor above and alongside is of mixed construction (ie some rooms may have solid concrete floors and others may have suspended timber).

Wall construction
Where a cellar is located in soil rather than rock, certain walls within the cellar will act as retaining walls; they will vary considerably from building to building. In a few cases the cellar may simply have been dug into the bedrock and the rock then forms the walls. In soil, more usually, the walls will be of stone or brick construction, typically poorly constructed without any vertical barrier to dampness. Sometimes they are finished internally with a coat of sand and cement render or, occasionally, plaster, but this is often poorly applied.

Ceiling construction
In most cases the ceiling will comprise nothing more than the suspended timber floor of the room above. At best this may be covered with some kind of wooden or plaster boarding to provide a ceiling. It is rare for the cellar ceiling to be in good condition. As a consequence the ceiling is likely to be very draughty. A more enlightened owner will have taken the opportunity to improve the thermal performance by placing sheet insulation between the joists or by means of quilts suspended on netting. Sometimes the cellar ceiling is formed in vaulted stone or brick, or occasionally in suspended concrete. In these cases the ceiling is likely to provide reasonable resistance to airflow from the cellar into the rooms above[159].

3.1 Structure

Non-domestic cellars and basements

Most city centre commercial properties and many industrial buildings include the provision of basements. These basements may be of considerable depth where they are used to provide underground parking facilities. In commercial developments, in congested city areas, basements usually occupy the whole site area to maximise the use of valuable land space below ground. As such they will provide the walls that are used to support the site boundary and which should at the same time avoid settlement or damage to adjacent roads or property.

The conventional methods for constructing these retaining walls are time consuming and expensive. That this is so is largely due to the extensive nature of the temporary works needed for their construction.

However, new developments in techniques have made it possible to reconsider the whole concept of retaining wall construction so that construction time and money can be significantly reduced. The methods now available make it possible to install the main wall structure (permanent works) before any excavation of the basement is started.

Once the wall has been established, excavation can proceed. As most retaining walls of this type require support from the permanent structure built within them, temporary support is needed as an interim measure as the excavation takes place. The most favoured method of support, ground anchors, enables the working area of the site to be free of any obstruction to the continuity of the works. Internal raking shoring, by contrast, leads to a great deal of obstruction which inevitably causes inefficient sequencing of construction operations and to increases in cost.

Three basic methods for providing perimeter construction are in use today:
- diaphragm walling
- contiguous bored piles
- secant piling

All these methods involve operations that require the services of a specialist subcontractor.

Diaphragm walls

Diaphragm walls are commonly used in clay and sand with gravel areas. The resulting wall is substantially watertight. The method of construction is illustrated in Figure 3.5 and works as follows.

Guide walls, spaced to be the final wall thickness apart are first constructed by the main contractor (stage 1 in Figure 3.5). Between these walls a trench is excavated to the required depth by the specialist contractor. As excavation proceeds, the sides are prevented from falling in by keeping the excavated area filled with bentonite slurry. This is a 3–6% mixture of a form of diatomaceous earth in water. The reasons why this mixture is so effective in preventing a trench of considerable depth in sand, sandy clay or gravel from collapsing involves a complex chemical process and will not be considered in detail here. Suffice to say, it works extremely well.

The methods of excavation vary with the specialist contractor in question but usually involve hydraulic grabs, rotating wheel cutters or other means. As excavation proceeds, so it is kept full with the bentonite liquid and the excavation carried out in bays in the manner illustrated. Alternate configuration for the bays is usually adopted unless the engineer requires a wider spacing to maintain the stability of the adjoining boundaries.

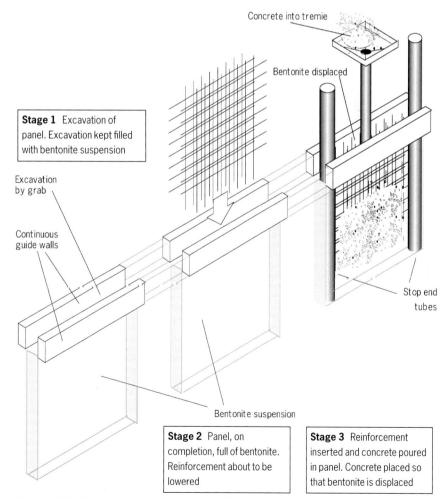

Figure 3.5 Stages in the construction of a diaphragm basement wall using a bentonite slurry to maintain the walls of the excavation until the reinforcement and concrete are placed

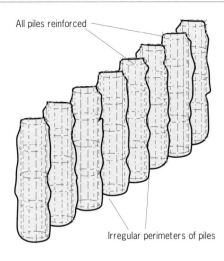

Figure 3.6 Cast-in-situ piles conform to the final shape of the excavation, leaving a rough surface when exposed

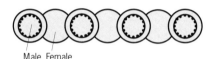

Light construction
Male piles reinforced with mild steel bars

Heavy construction
Both piles reinforced with rolled steel sections

Heavy construction
Male piles reinforced with helical binders

Figure 3.7 Light and heavy forms of Libore secant piling

Once a bay has been excavated, prefabricated reinforcement is lowered into the excavated area, displacing some of the bentonite as it does so (stage 2). The surplus liquid is pumped back to storage tanks, where it is cleaned ready for further use. A special quality of the bentonite is that it does not adhere to the reinforcement.

Concrete is now poured into the trench by means of a tremie (stage 3). Bentonite is displaced as the pour proceeds and is fed back to the storage tanks. On completion, all the bentonite is displaced. As the reinforcement does not cling to the bentonite, the normal bond between the concrete and the reinforcement is obtained. The bentonite may become contaminated with sand or soil fines during the process, but it is easily cleaned before reuse.

The particular advantages of this method are that:
- installation is free from vibration and excessive noise
- walls are constructed with minimum disruption to adjacent areas
- these walls form a dual purpose: they avoid the need for any temporary sheeting to the excavation and they become the final structural wall of the permanent works. (Some form of internal facing is usually needed for cosmetic purposes)
- diaphragm walls are substantially watertight
- diaphragm walls need support and this may be provided by the permanent structure within the basement or by ground anchors acting outside the walls

With the last point above, either temporary support will be needed until the permanent structure within the basement is complete, or ground anchors can be installed at an appropriate moment in the excavation sequence and no temporary support will be required.

Contiguous bored piles
As the name suggests, bored piles are installed as close together as possible to form the perimeter wall before any excavation takes place. The accuracy of placing the piles depends upon the type of pile and the method of placing, but piles cast-in-situ will conform to the final shape of the excavation. Figure 3.6. Contiguous piles are normally restricted to ground conditions where naturally dry soil conditions exist. However, a method of sealing the gaps between piles to provide a watertight structure is under development.

As with diaphragm walling, temporary support is usually provided by ground anchors or raking shores within the basement area. The advantages are similar to those for diaphragm walls, with the exception of watertightness. Against, is the need for much more effort to provide an acceptable internal face finish than needed with the diaphragm wall.

In cost terms, the differences between diaphragm walls and contiguous piles are likely to be small. Choice is a matter of comparing costs, together with any special requirements that the supervising engineer may feel favours one or the other method for structural reasons.

Secant piling
Secant piling is a development of the bored pile principle. The name secant is used since adjacent piles cut into each other, forming a cut out in the shape of a secant. In this configuration, a watertight wall can be achieved.

Until relatively recently, secant piling was mainly used in heavy civil engineering for dock walls, major retaining wall structures, and in cut-and-cover railway construction and similar situations. There is now a lighter weight system on the market which claims to be competitive with the standard contiguous bored pile approach. As the methods used are different, they are described separately.

Libore secant piling

This system is named after the boring rig used. This is of very heavy construction and capable of boring through almost any type of strata, whether dry or waterbearing, and with ancillary equipment which can break up boulders. In spite of its robustness, the equipment is capable of forming retaining walls in restricted areas, with minimum disturbance to adjoining property, services and people.

The first stage of installation involves boring alternate piles (female) at centres less than twice the diameter of the selected pile size. As a second stage, the gap is closed by cutting the male piles into the female piles (Figure 3.7). In this way the surfaces are bonded together and a watertight joint results. As will be seen from Figure 3.7, the female piles are not reinforced unless a very heavy wall is needed. For heavy situations, the female piles are reinforced with steel 'T' sections or square box arrangements of reinforcement (allowing room for the secant cut-out). Alternatively, all piles can be reinforced with steel 'T' sections.

As might be expected, with the heavy equipment involved the result is more expensive than the methods previously described. Its value is in situations requiring watertightness and strength characteristics which coincide with difficult boring conditions with which the other systems cannot cope.

Stent wall secant piling

This approach operates in quite a different way from the Libore method. Here the retaining wall is formed of alternate piles of bentonite/cement/PFA (B/C/PFA) and reinforced concrete piles with secant interlocks. The installation sequence is shown in Figure 3.8. The B/C/PFA piles are constructed first, along a line 100 mm outside the line of the following concrete piles. Once they have set to the right strength, they are followed by the concrete piles which, on installation, cut secants out of the B/C/PFA piles. The result is a watertight wall which is claimed to be no more expensive than a diaphragm wall or contiguous pile wall.

Again the construction is silent, with the advantages similar to the diaphragm wall, but claimed to be cheaper.

Main performance requirements and defects

Walls forming a cellar or basement usually act as retaining walls; and since they are completely or almost completely located below ground level, they often need to resist quite large lateral loads.

Where a building is piled and also has a basement, probably in an urban area, the pile heads must be finished below basement floor level. Driven piles may be installed from the ground surface and cut down as excavation proceeds; alternatively, piles may be installed from the base of the excavation. However, the ground conditions are likely to be inferior to those at the surface, especially in wet weather, leading to problems with heavy piling rigs.

Bored piles can be formed from the ground surface prior to excavation and concreted only up to the required level, the remainder of the bore being temporarily filled with sand. These piles help to reduce the heave when the basement is excavated in clay soils liable to swelling as the load is removed (BRE Digest 315[148]).

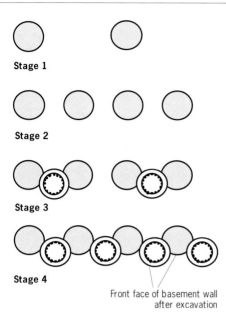

Figure 3.8 Stages in the formation of Stent secant piling

Case study

Secant piles
The interaction of a secant pile retaining wall with the over-consolidated London Clay in which it was constructed has been monitored by BRE investigators. The wall was 20 m deep and formed by 1.18 m diameter bored piles at 1.08 m centres. The depth of embedment was 11 m. The slab bridging the wall with the adjacent wall acted as a prop, but there was some compressible packing between the slab and the wall. Lateral earth pressures and porewater behind and in front of the wall were monitored during wall construction and subsequent excavation. Vibrating-wire strain gauges were used to estimate the bending moments in the wall and the prop load. Observations gave some insights into the complex interaction between the retaining wall and the stiff clay, particularly where it relates to the use of retaining structures in over-consolidated clay soil where the initial in situ horizontal stresses may be much larger than the vertical stresses[160].

Figure 3.9 In spite of the mixture of materials in this domestic cellar, no structural distress is apparent

Vertical and horizontal loads

Basements have to resist horizontal loads from the retained ground. Also, hydrostatic pressure can contribute uplift forces on the structure. For new construction, simple design rules for structure such as those used for walls in the Building Regulations, Approved Document A1/2[39], are not appropriate and specific calculations are required.

So far as an existing building is concerned, some reliance can be placed on the fact that the building has already stood the test of time, but any changes in use and circumstances that might disturb a structural condition that has existed for many years still must be considered. In particular, restraint provided by internal walls and floor structures to the outer basement walls may be significant and should not be underestimated. They should not be removed without adequate assessment of the consequences. Also, it is important to take into account the effects of changes which might influence the moisture level in the ground around the building.

Horizontal loads can be imposed on walls by nearby foundations (surcharging). The likelihood and magnitude of surcharging should be evaluated when considering the placement of additional foundations.

To assess theoretical load capabilities, actual wall, floor and foundation constructions will need to be identified. Site investigations are advisable to assess loadings from retained ground and groundwater.

Specific structural guidance has been produced by the Basement Development Group. British Standards Codes of practice on foundations (BS 8004[68]), masonry (BS 5628-3[161]) and loads on walls in water retaining structures (BS 8007[162]) all have some relevance.

The effect of basements on foundation design

The additional load on the ground caused by a building may be reduced by providing a basement or a light hollow box foundation, the weight of which is less than the weight of ground excavated. No additional load is applied to the ground until the total weight of the building and the subsurface structure is greater than the total weight of ground excavated; until these conditions are achieved the building actually tends to float (see the case study below).

Excavations for the construction of large basements may lead to heave in certain circumstances. For example, when the new Shell Centre was being built on the south bank of the River Thames, it was estimated that excavations in the London Clay for the large basements would create upward displacements of the underlying Bakerloo Line tunnels of 20–30 mm. See also Chapter 1.1.

3.1 Structure

Distress in cellar and basement structures

Settlement of foundations or horizontal displacement of walls may result from ground movements, clay heave or excessive loading. Bulging or cracking of enclosing walls would suggest that the walls have inadequate strength.

A visual survey should identify the more obvious signs of distress such as bulging walls, uneven floors, cracking and misalignment. Panelled finishes on walls may need to be removed to allow inspection of the condition of the structure behind (Figure 3.9).

> **Case study**
>
> **Measuring distress in basements at the Mansion House**
> Automatic instrumentation systems were designed and installed to monitor the movements of the Mansion House during the construction of the Docklands Light Railway (DLR) extension to Bank Station. Ground movements below foundation level were measured by 'strings' of BRE electrolevels installed in tubes grouted into boreholes. Electrolevels on beams and wall plates were used to determine vertical and horizontal movements of the structural elements of the basements. An independent check on basement movements was obtained by a water levelling system. Attention was focused on the ballroom area of the building which has had a long history of structural damage from tunnelling operations in the vicinity of Bank underground station. Here, as well as the wall plate electrolevels, crack monitoring was carried out by displacement transducers; wall movements were monitored at ceiling level by invar wire extensometers. BRE load cells were also mounted on pre-tensioned bars installed to tie this part of the building together prior to construction of the DLR overrun tunnel at Bank. The instrumentation provided comprehensive data on the pattern, rate and extent of building and ground movement, enabling informed decisions to be made on the timing of construction phases of the DLR overrun tunnel and subsequent Mansion House refurbishment.

Work on site

A basement constructed well before the superstructure of a building can present problems and, in waterlogged soil, may require continuous dewatering until much of the building has been completed. If tanked, upward displacement caused by flotation is a real possibility.

Workmanship

Work on construction and waterproofing of basements is normally carried out by specialist contractors with experience of the various methods and practices.

Supervision of critical features

Before any work is specified, it is important to check which walls are earth retaining and which are freestanding. If part of a partition, floor or built-in structural timber is to be removed during rehabilitation work, it will also be important to check the structural implications.

> **Inspection**
> The problems to look for are:
> ◊ cracking of masonry walls
> ◊ cracking or movement of vaults
> ◊ leaking surface water drains causing washout of foundations
> ◊ sulfate attack in walls
> ◊ rot in built-in structural timber

Chapter 3.2 Waterproofing

Dampness and mould growth are often indications of condensation but can also be a result of penetration of moisture from the ground. The diagnosis of the cause of dampness in basements tends to be complex; however, dampness in unoccupied existing buildings is rarely due to condensation. This chapter deals with penetrating moisture from the ground. Condensation, which is linked to ventilation and heating regimes, is dealt with in Chapter 3.3.

Penetration of groundwater may be due to the water table being high or the excavation for the basement was in impermeable subsoil. Any history of flooding in the basement, and whether the water table level ever exceeds that of the basement floor level is useful information in these circumstances. Flooding can lead to lifting of waterproofing treatments if they are inadequately loaded. A detailed investigation of the ground may be warranted in order to establish the water table level and the soil type. In some situations it is possible to lower the water table by improving the drainage system around the outside of the building.

If there is an external light well, it will be important to check that it is not bridged by debris, that drainage is still functioning and not silted up, and that air bricks are not obstructed by soil or vegetation. The floor will need to be examined for signs of moisture penetration. The survey should include any constructions such as garden walls and arches under steps which abut the main structure as these are a potential source of dampness. Plumbing leaks may also be in evidence.

Floors and walls of basements are often impregnated with hygroscopic salts (Figure 3.10), particularly if the basements have been used to store solid fuel. If lightweight plaster has been applied to walls, this will accentuate any dampness problems.

Remedial methods for dampness occurring in basements are given in this chapter. In respect of walls and ground floors, dampproofing arrangements were also discussed in *Walls, windows and doors*[101], and *Floors and flooring*[66].

Since it is the cellar or basement which bears the brunt of any inundation of flood water, it is dealt with in this chapter rather than in Chapter 1.3 (Figure 3.11).

Characteristic details

Basic description
Cellars are often damp, suffering from moisture passing through poorly constructed walls and floors (Figure 3.12). Where dampness problems are not too severe, simply increasing the level of ventilation in the cellar can help to dry them out. Excessive moisture should be avoided because it could lead to rot in the timber floor above the cellar.

BS 8102[163] provides guidance on the level of protection against water

Figure 3.10 Salts from dampness in the structure is one of the problems to be faced in converting a basement to living accommodation

3.2 Waterproofing

Figure 3.11 Flooding in a basement. The lower courses of brickwork are also saturated

from the ground required for new construction. This may be adapted for use in improving the resistance of existing construction.

For new construction it is important that the client has a clear idea of what present and future use will be made of the basement. Exclusion or control of moisture and, in some circumstances, water vapour is the chief consideration, and BS 8102 gives four grades of basement usage; these are listed in Table 3.1 together with acceptable forms of construction. The grades can also be used for assessing existing properties when refurbishment work is being considered. The previous use is likely to be for some other purpose than that being proposed and dampness was probably then of less concern.

Existing tanking systems are usually based on asphalt and have been in general use for many years. Since asphalt is capable of accommodating a degree of movement in the structure, and provided the workmanship is satisfactory, it is comparatively rare to encounter a failure.

Materials used in liquid applied membranes for applying dampproofing fall into three categories:
- bitumen emulsions and solutions, some containing rubber latex
- polyurethane compounds
- epoxy resins.

All these materials can provide protection, although they depend on good quality workmanship for complete integrity.

Waterstops

Waterstops are needed in in-situ reinforced concrete construction where day joints are to be formed during the casting process, and where the structure is intended to be resistant to water penetration. Several types are available and they prevent the ingress of water in different ways.

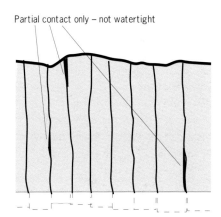

Figure 3.12 Poorly constructed piled walls may not be waterproof

Table 3.1 Level of protection to suit basement use[163]

Grade	Basement use	Performance level	Construction	Comment
1	Car parks, plant rooms etc (not electrical)	Some seepage and damp patches tolerable	Reinforced concrete to BS 8110	Groundwater check for chemicals
2	As a for Grade 1 above but need for drier environment (eg retail storage)	No water penetration but vapour penetration tolerable	Tanked or as above or reinforced concrete to BS 8007	Careful supervision. Membranes well lapped
3	Housing, offices, restaurants, leisure centres etc	Dry environment required	Tanked or as above or drained cavity and DPMs	As for Grade 2 above
4	Archives and controlled environment areas	Totally dry environment	Tanked or as above plus vapour control or ventilated wall cavity with vapour control and floor cavity and DPM	As for Grade 2 above. Check for chemicals in groundwater

3 Basements, cellars and underground buildings

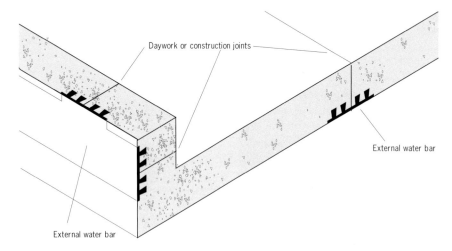

Figure 3.13 Dovetail-section rubber or plastics strips (water stops) cast into the shuttered face of the pour

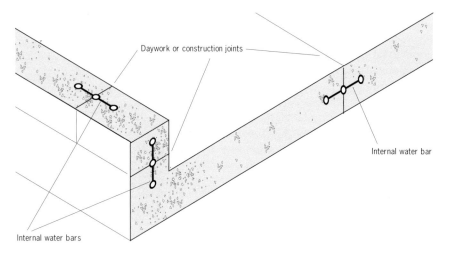

Figure 3.14 Dumb-bell section rubber or plastics strip cast into the open face of the shuttering

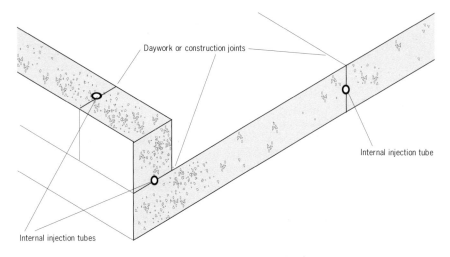

Figure 3.15 Perforated tube cast into the open face of the pour for later injection of resins

Rubber or flexible waterstops

The most common forms include extruded sections designed to provide a continuous barrier to the ingress of water through joints in the concrete structure. Strips of rubber or plastics, dovetailed on one side, are fixed to the face of the shuttering and cast into the wet concrete. These are external to the structure (Figure 3.13). Where used horizontally they must be cleared of debris before placing the concrete. They will only resist the passage of water from the face on which they are fixed.

Waterstops designed to function in the middle of the wall are difficult to install as successful placing of concrete cannot be guaranteed. Strips of rubber or plastics of dumb-bell shape are cast into the open face of the pour. These are internal to the structure (Figure 3.14).

Water-swellable waterstops

These function by the sealing pressure being developed when the hydrophilic material absorbs water. In strip form they are placed against the concrete joint before the next pour. They can be attached to rubber or PVC waterstops to provide a combined system.

Cementitious crystallisation waterstopping

The product comprises cement, fillers and chemicals mixed on site as a slurry and applied to the face of the concrete before the next pour. Salt crystallisation within the pores and capillaries of the concrete provides the waterstopping.

Post-injected waterstopping

A perforated or permeable tube is fixed to the first pour of concrete in the joint, leaving the open ends accessible. The second pour is made and when the concrete is hardened a polyurethane or proprietary fluid is injected into the tube to seal any cracks or fissures in the construction joint (Figure 3.15).

3.2 Waterproofing

Main performance requirements and defects

Dampness

During housing rehabilitation, basements which have been in use only as utility rooms are often considered for conversion to living areas. This change of use requires very careful assessment. If basements are to be converted to provide living space they must be protected from penetrating dampness, associated mould growth and moisture related deterioration of components (Figure 3.16). The practicability and cost of conversion should be evaluated carefully, particularly bearing in mind structural implications, risk of flooding and the presence of positive water pressure in the ground related to a high water table. Also, basements are difficult to ventilate and an assessment of condensation risk in the finished basement is advisable before work starts[164].

There is always a risk that alteration works will exacerbate problems of dampness. For example, lowering floors to improve headroom, and extensions to increase plan area, present considerable structural and dampproofing problems. Investigations should be made of groundwater levels and the causes of any previous dampness be established.

Few masonry basement structures would have comprehensive external tanking incorporated when built and retrospective installation is often discounted on cost grounds. Designers have to decide whether asphalt tanking is required or whether one of the cheaper proprietary waterproofing systems would be satisfactory. Proprietary systems available include waterproof renders containing special additives, paint-on high-build coatings and moisture resistant lathing materials that allow plastering of damp walls. When considering the use of any of these products, it should be borne in mind that discontinuities in the waterproofing are difficult to avoid and are a common cause of dampness later.

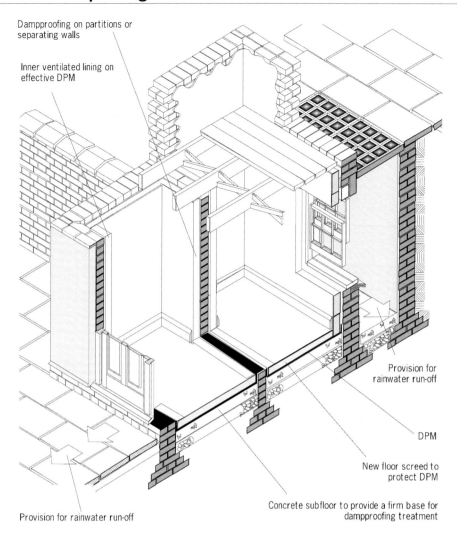

Figure 3.16 A basement on a sloping site which is to be converted to a living area must incorporate the necessary components and features that will provide effective dampproofing

An alternative approach, which is more of a palliative than a solution, is to improve drainage to lower the local water table around the structure. This technique is only practicable in some soils.

Penetration of moisture from the ground or flooding commonly results in surface dampness, salting and timber decay. Structural movement can cause failure of any dampproof membrane (DPM) applied internally or externally. Holes for pipes or removal of internal waterproofing for electrical fixings may cause localised leaks. There may be difficulty in providing adequate integrity at junctions between internal and external walls and in achieving correct overlapping between vertical and horizontal DPM materials. Internal treatments may lift if not restrained. High water table and high wall salt content will increase the difficulties of providing waterproofing.

Obvious signs of a dampness problem are visible salting or tide marks on walls or floors and decay of timber components. Estimates of the moisture content of masonry is obtained using a calcium carbide meter or by weighing and drying samples of wall material. Moisture content of timber can be checked with an electrical resistance meter, but readings will be inaccurate if salt content is high. To assist in the specification of suitable remedial measures, a detailed ground investigation may be needed to determine the soil type and the position of the water table.

3 Basements, cellars and underground buildings

Case study

Results of a comprehensive survey of moisture content of basement walls

Cardiff Bay covers the estuaries of the Rivers Taff and Ely and had one of the highest tidal ranges in the world of 14m. Construction of a barrage across the mouth of the bay between Penarth Marina and Queen Alexandra Head has created a large area of freshwater. Locks and sluices are incorporated into the barrage to allow small craft to pass through and to release the flows from the two rivers (Figure 3.17).

Many of the properties in the city near to the impounded water have basements which range in size from those under the whole area of the building to a small coal storage bunker usually located under the front entrance. Some basements have already been converted to habitable accommodation incorporating dampproofing, some are used only for storage, and a few have insufficient headroom to be of any practical use.

With dampness already being visible in the walls of many of the properties, concern was expressed that the effect of impounding a large area of water, even though the resulting water level was well below that experienced during high tides, would be to increase the levels of dampness in the walls. Accordingly Cardiff Bay Development Corporation wished to establish and record moisture levels in the walls both before and after impoundment.

BRE was contacted to prepare a methodology for sampling the moisture in the walls and to undertake the sampling. The need to measure moisture in masonry accurately had arisen on many occasions at BRE and for this reason the Gravimetric method was developed. (A full description of the method is contained in BRE Digest 245[137]). Briefly, samples are drilled from the mortar or masonry and weighed as found (ie damp). The sample is then oven-dried, reweighed and the percentage moisture content is calculated. A further test is also carried out to check for the influence of salts in the sample which establishes the hygroscopic moisture content.

A method statement was prepared covering procedures for locating suitable walls for sampling, carrying out the laboratory tests and presenting the values obtained. The preferred locations for sampling for each property were
- a party wall
- an external earth retaining wall
- a freestanding internal wall

Heights of samples above floor level were to be 300 mm, 900 mm and 1.5 m, and measurements were to be taken to locate where the sampling was actually carried out. Where necessary, holes were filled after sampling. In total, nine drillings were taken in most of the 94 properties examined, though in a few of them some of the locations were inaccessible. Results were presented as a single sheet for each property giving a plan of the basement including the location of the lines of drillings with dimensions from fixed points. Moisture and hygroscopic moisture content values were recorded for each drilling together with a note of the material sample – brick, stone or mortar.

Results given in the Figures alongside are only the first round of measurements reported for a height of 300 mm above floor levels. They show a fairly wide range of values, probably typical of moisture content levels in basements in older properties in tidal and riverine areas generally. Three pie charts present the measurements recorded at each of the three locations. Figure 3.18 shows the values obtained for party walls, Figure 3.19 for earth retaining walls, and Figure 3.20 for internal walls.

The second round of sampling will be taken as close to the first set as possible and should enable any differences caused by the impounding to be apparent.

Figure 3.17 The main locks of the Cardiff Bay barrage under construction

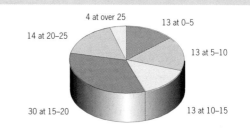

Figure 3.18 Party walls: numbers of cases at moisture content (%) 300 mm above floor levels

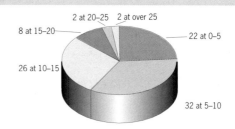

Figure 3.19 Earth retaining walls: numbers of cases at moisture content (%) 300 mm above floor levels

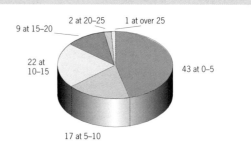

Figure 3.20 Internal walls: numbers of cases at moisture content (%) 300 mm above floor levels

3.2 Waterproofing

Remedial work to cure dampness problems

All methods for waterproofing existing buildings aim to provide a moisture resistant envelope to walls and floors. There are several types of envelope system in common use, ranging from asphalt tanking to ventilated dry lining. In rehab work, a more traditional alternative is to provide a drained cavity by building a new inner leaf. The remedies in this section are essentially those given in BRE Good Repair Guide 23[165].

In the treatments that follow, several result in loss of space because of the additional wall thicknesses required. In each case, there must be a sound concrete subfloor, minimum 150 mm thick, as a stable base.

Drained cavity

This is a tried-and-tested solution, but there is a space penalty, and it relies on effective gravity drainage or installation of a sump and pump (Figure 3.21). The method is not suitable in conditions of high water pressure or high water tables. It can be completed by non-specialist contractors.

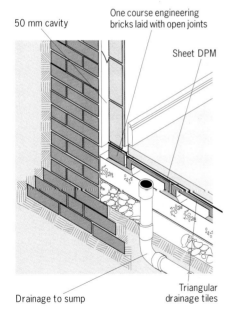

Figure 3.21 Dampproofing a basement by means of interconnecting cavities on walls and floors. This method can only be used where drainage is possible. The drained cavity wall and floor construction provides a high level of safeguard. Providing a ventilated cavity and horizontal DPM prevents moisture ingress

Case study

A defective basement DPM

The BRE Advisory Service was asked to investigate the causes of puddles of water in the basement areas of a 10 year old house. The external walls of the basement had been built directly off a concrete floor slab and footings. An external land drain had been provided. The 150 mm thick slab had a layer of thermal insulation, covered by a polyethylene DPM and a 63 mm screed. PVC tiles or textile carpets were then laid on the screed.

Early dampness problems attributed to condensation and insufficient time being allowed for the screed to dry before PVC tiles were laid, had all been rectified before BRE were called in. Substantial puddles of water were seen on two occasions during the eighth year of occupation. The first occasion coincided with a period of heavy rain, while the second did not. Just prior to the BRE examination, there had been a period of exceptionally heavy rain, although only a small area of dampness was seen.

Observations of the internal walls of the basement showed that many of these had tide marks of salts to a height of 200 mm or so. Removing some of the plaster showed that these walls had been built off the slab with a DPC laid generally, but not exclusively, on the top of the first block course.

The external walls appeared to have a brushed-on DPM on the surface of the inner brick leaf, and then covered with render and plaster. There appeared to be little attempt to provide continuity between the vertical and horizontal membranes. However, measurements with an electrical resistance meter showed readings which could be considered satisfactory.

The possibility that moisture could result from condensation was investigated by measuring the surface temperature of the screed and adjoining walls with a surface thermometer, and the air temperature, relative humidity, and dewpoint temperature with an aspirated hygrometer. Surface temperatures were measured at 16–18 °C; with a dewpoint temperature of 10 °C, condensation was not occurring.

Although construction water had most probably been the cause of earlier problems, this had now dried out. No leakage from the heating system was apparent.

So far as ingress of water from the exterior is concerned, there appeared to be a weakness at the junction of the wall and slab DPCs. For a tanking to be successful, the horizontal portion needs to be laid beneath the slab, and well lapped with the vertical. Although there was now no direct evidence to be seen on site, it was concluded by the BRE Advisory Officer that it was possible that some movement in the building had occurred which opened up a crack between slab and wall, and which subsequently closed. Otherwise, water penetration during recent wet weather could have been expected.

It was pointed out to the building owner that painting a coat of bitumen onto brickwork is not normally a reliable method of preventing water from outside from penetrating into a basement.

However, as no water appeared to be entering the basement at the time of the inspection (the already mentioned small patch of damp was drying out), perhaps all that would be needed would be to make good the existing screeds and plasterwork and redecorate. A more robust solution would be that afforded by a cementitious tanking system, with a proprietary render on the inside leaf of the external walls at the junction with the floor. The owner and his agent were recommended to read BRE Good Building Guide 3[164].

Case study

Failure of newly constructed tanking

The waterproofing membrane for a basement was a bituminous treatment applied in liquid form to the concrete walls and floor. The membrane was then supported and protected by a blockwork inner leaf with mortar filling the narrow cavity. When the basement was inspected, water was lying on the floor and bitumen was running out from mortar bed joints. Removing areas of blockwork revealed that the bitumen had not set (Figure 3.22). Site records showed that application of the bitumen was during cold weather and had been quickly followed by the blockwork and mortar infill, preventing evaporation of the solvent and curing of the bitumen. As the prime requirement was rapid completion of the project, the recommended solution was to remove the existing blockwork, mortar and uncured bitumen, and rebuild using the drained cavity system.

Figure 3.22 Bitumen running from bed joints indicated that the waterproof membrane had not set. This was confirmed by removal of the blockwork

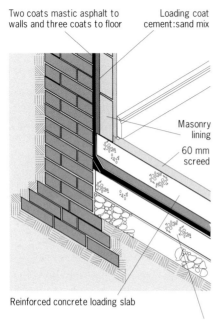

Figure 3.23 Dampproofing a basement by mastic asphalt tanking. The structure itself does not prevent water ingress, and protection therefore depends on a total barrier system applied externally or internally. The system may also need a vapour control layer, depending on the use to which the basement is put

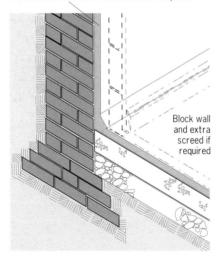

Figure 3.24 Dampproofing a basement by cementitious render coats

Wall and floor finishes, together with existing flooring materials and screed need to be removed. A course of engineering bricks is then laid on the slab, creating a 50 mm cavity with the existing wall and a drainage channel, and leaving perpends open at intervals. Above the brick course a physical DPC is then laid and a new blockwork wall constructed, tied into the original wall with stainless steel wall ties at standard spacings. A drain or sump is installed, a self-draining underfloor layer of triangular drainage tiles and a sheet DPM (which is lapped up the side of the engineering bricks) are then laid. As an alternative to tiles, a heavy-grade plastics dimpled sheet, which will also act as a DPM, may be used. Finally, a new screed, at least 50 mm thick, is laid and the walls replastered.

Mastic asphalt tanking
This is a costly but durable solution (Figure 3.23). There is a significant space penalty, in terms of area and usually height as well, and the system needs to be installed by specialist contractors.

The existing flooring, back to subfloor slab, is first removed ensuring that all surfaces are left with an adequate key. Horizontal brickwork joints are then raked out to a depth of 25 mm, and coated with proprietary high-bond primer. Glazed brickwork needs to be hacked or bush hammered. On smooth concrete, a wire brushing with either the addition of a proprietary cement:sand slurry with plasticiser or a light application of a proprietary high-bond primer may be necessary. External angles of masonry and concrete must be rounded.

A two-coat asphalt angle fillet is then built up at wall-to-floor and wall-to-wall junctions, and three coats of asphalt to a total of 30 mm on floor slab and 20 mm on walls are then completed (BS 8102). Joints between successive coats should be staggered by at least 150 mm on floors and 75 mm on walls. Finally, at least 50 mm protective cement:sand floor screed is laid and a brick or blockwork lining wall is built, backfilling the cavity progressively against the asphalt with a cement:sand mix.

Cementitious render or compound
Correctly mixed render or compound, properly applied to a stable background, should last for many years (Figure 3.24). Specific allowance must be made for services if these need to be run in the wall. The cementitious layer is vulnerable to accidental puncturing unless special wall fixings are used, or an inner blockwork wall is added. The method can be used in areas having high water tables, provided particular care is taken, and in this respect a specialist contractor is more likely to achieve a satisfactory solution.

First, a cement corner fillet is built at wall-to-floor and wall-to-wall junctions. Then, three coats of proprietary mix (thinned with clean water) are applied by trowel. Successive coats must be lapped in strict accordance with the render manufacturer's instructions, and curing completed. If using a cementitious compound, the substrate must be dampened and two coats applied to manufacturer's recommendations, followed by a loading coat and floor screed. Walls may be skim coated.

Self-adhesive membrane
This provides a durable solution and a suitable surface for services and fixings, but with a space penalty. The method can be used in areas having high water tables, but specialist contractors are not usually necessary.

First, the brickwork is cleaned, and all flooring removed down to the subfloor slab. All brick surfaces are flush pointed or rendered if the masonry is uneven (see *Walls, windows and doors* for suitable mixes), and the concrete slab cleaned and dried. Wall-to-floor and wall-to-wall fillets are then constructed, and the membrane applied to dry wall and floor surfaces following the manufacturer's instructions, allowing at least 150 mm overlap at the joints. The floor membrane is then protected and a blockwork lining wall built, progressively backfilling with cement:sand mortar. Finally, a new floor screed of at least 50 mm thickness is added and the walls replastered. Some contractors have suggested that when applied internally, water pressure can cause detachment. Careful workmanship and attention to detail will result in a satisfactory job.

Liquid applied membrane
This too provides a durable solution and a suitable surface for services and fixings, but with a space penalty (Figure 3.25). The method can be used in areas having high water tables, and it is not usually necessary to employ specialist contractors.

All flooring is first removed down to the subfloor slab. The brickwork is then cleaned and flush pointed, and a final clean given to all surfaces to be coated.

All products should be used strictly in accordance with the manufacturer's instructions, particularly any ventilation requirements during and after application. The procedure is usually as follows:
- wall-to-floor and wall-to-wall fillets are constructed
- one or more liquid coats are applied and allowed to cure
- a new floor screed is laid
- a new inner leaf is constructed (normally having the cavity backfilled with cement:sand mortar) with the floor membrane being protected from damage during building operations

Ventilated dry lining
This has the advantage of only marginally decreasing space in a room (Figure 3.26). However, it is only suitable if the dampness is slight, and it does not protect against groundwater under pressure. The dimpled plastics sheeting is vulnerable to accidental puncturing unless special fixings are used. Specialist contractors are not usually necessary, though. The linings have a life of 20 or more years.

First a high density dimpled polyethylene sheet is laid on the floor, turning up the wall by at least 150 mm and overlapping to manufacturer's recommendations. Next, a new floor screed at least 50 mm thick is laid. Proprietary dimpled plastics sheeting is fixed to the wall surface by nailing, screwing or special plastics plugs, leaving a gap top and bottom for ventilation. Finally the wall surface is plastered, or covered with plasterboard, while retaining the ventilation gaps.

Partition walls
Partitions must be dampproofed, with careful attention to detail. There are three approaches.
- The partition is completely removed, dampproofing and screed are installed by one of the methods already described, and the partition then rebuilt. This is normally only advisable for partitions which are not loadbearing and not contributing to the stability of the building
- A DPC is inserted at the base of the partition, overlapped with the floor and wall dampproofing. This is usually only appropriate when the partition is not connected to an external wall
- The external wall dampproofing is continued along the partition. This does not prevent moisture entering the wall masonry but stops dampness penetrating to wall finishes. Full protection may be needed, or a ventilated dry lining may be adequate. Timber door frames in partitions must be protected by continuous dampproofing around the sides and base of the opening. This may be the only solution for situations where the partition is connected to an external wall and where the partition cannot be removed because it has a structural role

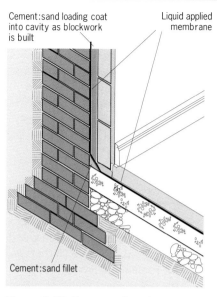

Figure 3.25 Dampproofing a basement by liquid applied membrane

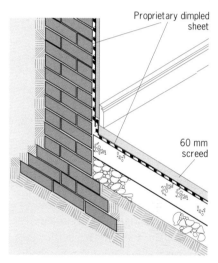

Figure 3.26 Dampproofing a basement by dimpled sheet dry linings

Existing DPCs at ground level in external walls
If an existing DPC in an external wall is defective, it must be repaired or replaced before a new dampproofing system is installed in a basement. There should be effective lapping between the horizontal DPC and the vertical dampproofing.

Installing lining walls and new ceilings to basements may reduce ventilation around ground floor joists. It may be necessary to improve ventilation to joists, and to isolate joist ends from walls, where possible, with DPC material or by using joist hangers.

Door and window frames in basements

The dampproofing layer in a basement must be taken into the reveal to abut the frame. Lining to walls is usually stopped at the edge of the reveal and plaster, or adhesive fixed plasterboard taken round to complete the reveal. Interior sills, of durable or preservative treated timber, should be fixed to the wall linings.

Door and window frames of durable or preservative treated wood can be retained in position if there is no indication of rot and if they are likely to remain dry. If there is a risk of wetting, frames should be isolated from damp masonry by a physical DPC which laps the wall dampproofing.

Replacement timber frames should be durable or treated as specified in BS 5589, and isolated from damp masonry. If excessive wetting is likely, aluminium or PVC-U frames may be fitted.

Fixing services

If services are run behind ventilated dry linings, moisture resistant fittings and a waterproof seal at outlets should be provided. With cementitious dampproofing, services should be run in recesses in the walls. Services can be run on dry internal partitions or inside hollow skirting systems. Basement floors cannot normally accommodate heating and water pipes.

> **Case study**
>
> **Dampness in a hotel basement**
> Dampness and efflorescence was showing on plaster at the base of the walls in a hotel and the BRE Advisory Service was called in to advise. It was thought that water was leaking inwards through the kicker/wall joint and it was recommended that the joint be sealed with a resin. Dampness resulting from a flood in a basement during construction of the hotel may have masked the wall joint leakage. These separate incidents emphasise the importance, when dealing with any problem, of obtaining holistic appreciation of the planning, design, construction and subsequent performance of a building if the causes of problems are to be correctly identified and remedies found.

Chimney breasts

If a fireplace is to remain in use, dampproofing can be taken to the chimney breast reveal and the heating relied on to maintain dryness round the breast and hearth; otherwise an envelope treatment will be required around the breast. Alternatively, the chimney breast can be removed, new support provided for the stack and the whole basement wall dampproofed.

See also Chapter 2.1 in *Building services*[20].

Built-in timbers

Built-in, non-structural timbers which would be vulnerable to rot if sealed behind dampproofing should be removed. With embedded structural timbers, the safest solution is to replace them with materials less sensitive to moisture. Structural timbers can only be retained if they are sound and of durable species, and if dampness can be minimised in the supporting structure. If there is any timber rot, the cause should be investigated and preventative measures taken.

See also Chapter 2.1 in *Floors and flooring* and Chapter 2.1 in *Building services*.

Durability

In basements, because of the greater risk of contact with penetrating moisture or exposure to high humidity levels, materials and components are more prone to premature failure. Special care should be taken to isolate timber from dampness.

Few old masonry basements have a fully effective tanking system and some form of dampproofing will be needed to protect against groundwater.

Adequate access must be allowed for cleaning underground drains and drainage gullies, sunken areas and light wells adjacent to basements. Drainage should be adequate to eliminate risk of flooding from water run-off in periods of very intense rainfall.

Timber decay caused by wet or dry rot, and salting and spalling of plasterwork, caused by migration of salts, are common problems. The lack of effective tanking can lead to dampness and even flooding. Corrosion of metal components such as plaster angle beads, and electrical boxes and conduit, may be hastened in a basement environment. French drains and similar drainage structures may silt up over time. Mechanical equipment such as drainage pumps may become obsolete and uneconomic to repair.

Obvious indications of long term problems are decaying timbers, spalling brickwork, salting, crumbling plaster, mould and dampness.

Drainage systems under and around basements should be carefully examined and replaced if suspect. Where drains are to be covered over by new construction they should not only be checked for efficient discharge but tested by air or water methods to detect any leakage.

Investigations will need to be carried out to discover whether sulfate or frost damage to mortar has weakened walls.

Maintenance

Most membranes will have been placed beneath loading coats, so will be inaccessible. Of great importance is that they are not punctured by subsequent alterations. Cementitious systems within a basement may crack if there is any movement in the structure, but it will be obvious where repairs are needed.

Specific advice may need to be given to occupiers on how to attach shelves and other fixings to basement walls.

3.2 Waterproofing

Work on site

Treatment of areas which have been flooded

Immediate action after flood waters have subsided is important in reducing reoccupation times and in minimising repairs and replacement. Immediate action in a building should include:
- checking for structural damage
- switching off electricity supplies
- shutting off the gas supply
- checking drainage systems are clear
- removing soft furnishings that are wet
- draining structural cavities
- beginning the drying process

Close attention should be paid to personal hygiene in the cleaning process because of risk of contamination of flood waters by sewage[166].

Structural damage

Houses that have been buffeted severely by flood waters, floating baulks of timber or other debris may suffer structural damage by:
- undermining foundations on sandy subsoils
- compaction of certain soils not previously flooded
- in clay areas, ground heave and, later, shrinkage as the subterranean water dries out, leading to cracking of masonry
- floors weakened to the extent that they affect the stability of adjacent walls

Cracks exceeding 5 mm in width, or several narrower cracks occurring together will need further investigation.

Draining

The building must be drained thoroughly. Although most of the flood waters may have subsided or been pumped away, some might remain:
- under basement floors. Where the level of the bottom of the floor – this is usually in concrete but may be earth in older buildings – is below the levels of the adjoining drains, water is likely to be trapped
- within sumps, pits or access to drains. These must be drained and cleared
- radon ventilation systems
- between the outer and inner leaves of basement walls. Holes may need to be drilled in the vertical brickwork joints in the inner leaf between every fourth brick to allow water to drain away to the interior of the basement

Cleaning

Mud and silt must be cleared scrupulously and the areas sprayed with disinfectant. Local authority environmental health officers may be able to give advice.

The building should be inspected for signs of trapped mud, particularly:
- inside wall cavities. Bricks should be removed at intervals, carefully, and the mud raked or flushed away. If this requires the removal of insulation layers, these should be reinstated
- in air bricks and vents which provide essential air flow to boilers. Balanced flues of boilers and gas water heaters should also be checked

Drying

Drying out may have to continue for many months before the building can be reinstated completely. The walls – both bricks and plaster – absorb large quantities of water during flooding. A solid wall, one brick thick, for example, may absorb as much as 55 l/m^2 and then take over a year to dry to its former state. As the walls dry, efflorescence may appear on the surface. Efflorescence should be removed by brushing when the wall has dried completely, taking proper precautions against inhaling dust. In some cases, however, salts may have been introduced into the walls by sea water; over the years, the walls can attract more moisture and give similar symptoms to those of rising damp. In these cases plaster must be removed and be replaced with new plaster capable of withstanding the actions of the salts. See also Chapter 10.2 in *Walls, windows and doors*.

Thin walls dry more quickly than thick ones; other things being equal, a thick wall of twice the thickness of the same material as a thin one will take four times as long to dry.

Stone walls are likely to dry out more rapidly than those of brick, since they are often less porous. However, if the walls have rubble cores these may need to be drained just like cavities.

The following points should be observed:
- warm air should be kept flowing through the building by both heating and ventilation
- windows and doors kept open to give good ventilation, even when the heating is on. Measures might have to be taken to prevent housebreaking
- loose floor coverings and carpets should be removed for disposal
- impervious wall coverings stripped to help the walls dry out
- pictures and furniture removed or kept away from damp walls and other areas of dampness
- cupboard doors kept open

If the heating system is operable, it should be run with the thermostat set to 22 °C or above and as much ventilation as practicable.

Reinstatement

During cold weather, wet walls may be damaged by frost causing the surface of the brickwork to crack and powder away. Some walls may expand because of the dampness and contract on drying, producing fine cracks which usually can be dealt with simply by repainting or skimming over with a plaster filler. Some types of wall plaster soften readily when wet and crumble when they dry out again. Others may expand and contract to such a degree that replacement is needed.

Probably the greatest danger caused by flood waters, even months after inundation, is rot in timber. The longer timbers remain wet, the more likely outbreaks of rot will occur. Splitting in timber can be minimised by ensuring the timbers dry on both faces, which is helped by removing panelling and skirtings.

All timbers, including door frames and skirting boards, attached to or embedded in damp walls are vulnerable and should be moved away or cut back from the walls.

Case study

A flooded basement
The basement of a health building was flooded to a depth of 450–600 mm. A specialist firm of contractors were commissioned to insert a chemical injection dampproof course and to replaster affected walls. Subsequently problems arose with the plaster bulging and becoming detached.

Drilled samples were taken from the affected walls by the BRE Advisory Service and, after laboratory testing, moisture contents in excess of 22.5 % wet weight were established for the walls. Further tests carried out on render samples demonstrated that sulfate attack on the render had occurred and this was the reason for the bulging of the plaster. Remedial measures were discussed with the building owners: the Advisory Officer considered that the walls were probably too wet for a chemical system to be successfully injected, but insertion of a physical dampproof course might prove successful although there were practical limitations on the use of this method. Other approaches for consideration included a proprietary render system and a dry lining based on dimpled plastics sheeting.

Salt contamination from sea water or other sources will affect the readings obtained when using an electrical moisture meter. Where it is possible to accurately measure moisture content of timber components, target values should be 24% or lower when measured during the months of October to May and below 22% for the remainder of the year[167]. It is recommended that timbers are inspected six months after they appear to have completely dried out and again after another 12 months. There are several types of rot likely to affect the timbers; rots will show up as brown or white strands, small orange or white blotches, cracking or splitting of the wood, soft areas which offer no resistance to a sharp instrument or penknife, and, in extreme cases, fungal growth. See *Recognising wood rot and insect damage in buildings*[168] and *Remedial treatment of wood rot and insect attack in buildings*[169].

Wood blocks and other coverings like vinyl or linoleum which have been stuck to concrete screeds may have lifted because the adhesive has weakened on wetting and require relaying. Some materials, particularly wood block and strip, may have swollen and become damaged.

Impervious floor coverings should not be laid until drying out is acceptably complete. Testing should be carried out using a hygrometer, and readings should be in the range 75–80% (see Chapter 1.4 of *Floors and flooring*). Valuable timber panelling must be dried thoroughly and not be replaced until the backing walls are completely dry. In timber framed walls (eg partitions) some plasterboard panels should be taken away to expose the timber framework, any insulation removed as necessary, and the wall linings not replaced until drying is complete.

Panelled doors are unlikely to be affected seriously unless the panels are made of the type of plywood which expands because the adhesive is sensitive to water. Modern flush doors are often more severely damaged and require replacement. Other doors and windows may stick but should not be eased by planing the edges until drying (and associated shrinkage) is complete.

Metals are likely to have escaped serious damage unless flooding is by sea water. However, steel reinforcement embedded in concrete may corrode and expand causing long term damage. In other cases the drying process should leave the metals unscathed, although locks and hinges should be oiled to prevent rusting and seizing up.

Redecoration should be delayed until walls have dried thoroughly, and new coverings should be confined to porous coatings like emulsion paint rather than wallpaper. Walls should be treated with a fungicide if there are signs of mould growth.

It is essential that all electrical installations and appliances that have been immersed in water or found to be in a damp condition are disconnected, examined, thoroughly dried out and tested. Particular care must be taken with any electrical equipment. Cables in good condition should not be affected by immersion but junction boxes etc will[170].

Water may be trapped in ducts or conduits containing cables and should be opened up to assist drainage. Once the cleaning and drying is completed, the installation should be tested for earth continuity and insulation resistance as laid down in the most up-to-date IEE Regulations, and an inspection certificate issued. The electricity installation should be inspected every month for the first six months after the initial test, and at least twice again in the following six months.

3.2 Waterproofing

Workmanship
The water table may have to be lowered by pumping while work on basements is carried out.

Inspection
The problems to look for in damp basements are: ◊ ineffective existing tanking ◊ water table higher than basement floors ◊ external wall cavities bridged by debris ◊ blocked land drains ◊ blocked air bricks ◊ plumbing leaks ◊ hygroscopic salts in walls (from old solid fuel storage) ◊ former use of lightweight plasters The problems to look for after flooding are: ◊ defective electrical wiring and apparatus ◊ defective gas supplies ◊ contaminated water supplies ◊ drains blocked with silt

Chapter 3.3 Other aspects of performance

Where basements or semi-basements are used for habitable accommodation, or cellars have been converted to living areas, similar standards of performance to above ground accommodation are expected, especially in relation to noise, thermal insulation, ventilation and condensation, daylighting, fire and means of escape and precautions against radon. Practical considerations often mean, though, that these ideals cannot be fully met.

Cellars in older properties are still frequently used for accommodating service entries and meters (Figure 3.27).

Characteristic details

Basic description of types of below-ground constructions

Topics discussed in this chapter are influenced to a considerable degree on the shape and size of the cellar or basement in relation to the building, to ground levels, and to means of access. In the last case, whether there is independent access from the outside (Figure 3.28) or whether access is provided only from within the building (Figure 3.29).

Figure 3.27 Services in the cellars of older properties are frequently maintained in hazardous and damaging conditions

Figure 3.28 Access to this basement is provided from street level independently of the dwelling above

Figure 3.29 No access from street level is provided to these basements from the fronts of the dwellings

3.3 Other aspects of performance

Five types of below-ground construction need to be considered (Figure 3.30):
- a building which is entirely below ground (Figure 3.31). It may or may not be stepped
- a full cellar or basement, which is completely located below ground level, to an above-ground building
- a partial cellar or basement which is completely located below ground level
- a semi-basement or cellar
- a stepped construction

A cellar or basement may extend under the whole area of a building or only part of a building. Occasionally a cellar or basement may extend beyond the floor plan of the building: for example, where part of a cellar once formed a coal hole under an adjacent pavement. Semi-basements or cellars are similar to basements or cellars, but have at least one or more walls exposed as an external wall instead of acting as a retaining wall. Stepped constructions are buildings located on a sloping site and in which the lower level rooms are dug into the sloping ground to form a partial or full semi-basement.

Main performance requirements and defects

Noise and other unwanted side effects

Where basement walls separate dwellings, they are required to resist the transmission of sound as provided by Part E of the Regulations. Care should be taken to detail the junctions of separating walls with other elements such as perimeter walls, floors and partitions[171].

Thermal performance

Thermal performance of ground floor slabs is dealt with in Chapter 1.3 of *Floors and flooring*[66]. The thermal performance of basement floors is also briefly mentioned in the same chapter of *Floors and flooring*.

As *Principles of modern building*[9] pointed out, construction laid directly in or on the ground does not lose heat in the same way as walls and roofs. The ground acts to some extent as an accumulator of heat, and in fact most of the heat loss occurs within a short distance of the edges adjacent to external walls. Consequently the average heat flow is dependent on the size and shape of the construction.

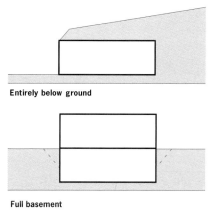
Entirely below ground

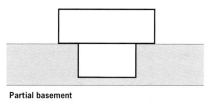
Full basement

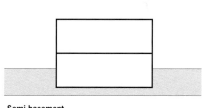
Partial basement

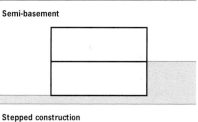
Semi-basement

Stepped construction

Figure 3.30 Types of basements and cellars

Figure 3.31 An underground house near the Pembrokeshire coast

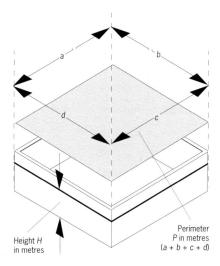

Figure 3.32 Criteria for calculating the perimeter-to-area ratio and depth (H) of a basement for establishing the U-value

The heat flow pattern in basements is rather complex and time dependent. The U-values can be calculated using the steady-state component averaged over the basement, which provides an approximation of the heat losses which is adequate for most purposes. The procedure is explained in BRE Information Paper IP 14/94[172]. The deeper the floor level below the ground, the better the U-value.

Alternatively, Table 3.2 below, taken from IP 14/94, gives U-values of uninsulated basement floors in terms of the perimeter-to-area (P/A) ratio and the basement depth (Figure 3.32 on page 151). Linear interpolation is appropriate.

In theory, of course, a basement is well insulated, being in contact with the ground which constitutes an infinite thickness of insulation. However, domestic basements are seldom entirely below ground and areas of the enclosing walls are often exposed to exterior conditions. In addition, a basement wall just below ground level will be exposed to temperature fluctuations similar to those for an exposed wall. For this reason, wall insulation should extend well below ground level.

While the ground around a basement has insulating properties, it also creates a structure with a high thermal capacity which will respond slowly to heating. To provide a more rapid response, it will be necessary to consider insulating the external walls and basement floor.

If internal or external insulation has already been added, it will be important to find out if the insulation is of a type which is suitable for use in potentially damp conditions; exterior insulation will need to be frost resistant.

In summer, basements tend to provide good thermal comfort, provided cross-ventilation is available and heat gains (eg from cooking) are controlled.

Inadequate insulation of basement walls in contact with the ground and above ground level is the most common problem. Water saturation of insulation on external walls below ground level can occur where water permeable insulation is used. Thermal bridging by other structural members such as beams and columns may occur. The insulation value of solid construction may be reduced if the structure is saturated.

Ventilation and condensation

Many domestic cellars are poorly ventilated. This is due to insufficient air vents having been provided at the time of construction, and those that have been provided are often blocked up. As a consequence, the air quality is likely to be poor which results in musty smells. Any increase in ventilation of the cellar using unheated air will have an impact on heating costs for the house.

As well as considering ventilation of the cellar it may be worth considering ventilation in the rest of the house. Open flue heating appliances in the cellar or basement must be adequately ventilated.

Dampness and mould growth are often indications of condensation, but can also be a result of penetration of moisture from the ground. The diagnosis of the cause of dampness in basements tends to be complex; dampness in unoccupied buildings is rarely due to condensation. However, condensation is common in occupied basements because ventilation and heating are usually poor.

Basements tend to be difficult to ventilate properly and may often require special measures to avoid condensation. Bathrooms and other utility rooms, which require only limited natural lighting, are often located in basements but can generate high moisture vapour levels. This vapour will need to be extracted or diluted by ventilation. There can be risks of surface and interstitial condensation with balconies and access ways above basements; these problems are analogous to those sometimes encountered with flat roofs. See also Chapter 1.11 in *Roofs and roofing*.

Basements which have been used for fuel storage are likely to be contaminated by hygroscopic salts. Previous use of lightweight plasters in a relatively damp environment can accentuate dampness and condensation problems.

The type of construction and wall (and possibly floor) structure will need to be identified to determine the extent of thermal bridging. Thermographic techniques, while expensive, may be valuable in identifying the exact location of thermal bridging. Consideration should also be given to the adequacy of cross-ventilation in hot summer conditions.

BS 5250[173] refers to condensation risk relevant to basements. BRE Digests 163[174] and 245[175] also may be helpful. The BRE audiovisual package *Remedies for condensation and mould in traditional housing*[176] offers simple, practical advice.

Table 3.2 U-values for uninsulated basement floors

Perimeter-to-area ratio (m^{-1})	Basement depth (m)				
	0.5	1.0	1.5	2.0	2.5
0.1	0.21	0.19	0.18	0.17	0.17
0.2	0.34	0.32	0.29	0.28	0.26
0.3	0.45	0.41	0.38	0.35	0.33
0.4	0.55	0.50	0.45	0.42	0.39
0.5	0.63	0.56	0.51	0.47	0.43
0.6	0.70	0.62	0.56	0.51	0.47
0.7	0.76	0.67	0.60	0.54	0.49
0.8	0.82	0.71	0.63	0.57	0.52
0.9	0.87	0.75	0.67	0.59	0.54
1.0	0.91	0.79	0.69	0.61	0.55

3.3 Other aspects of performance

BRE Information Paper IP 14/94 gives data for determining the insulation needed to achieve U-values of 0.45 W/m²K for basements, as currently required by the Elemental Method of satisfying the Building Regulations. However, new U-values are proposed for new housing, which will affect the above calculations. IP 14/94 also gives information on basement U-values which can be used either for the Target U-value Method in the Regulations, or for calculating heat losses in more general terms. The paper supported the 1995 edition of the Building Regulations.

Precautions against radon
The various options for reducing the levels of radon entry into cellars and basements are discussed in the feature panel on pages 154 and 155.

Poor ventilation practice can increase radon entry. A combination of well sealed or rarely opened windows downstairs, and poorly sealed or regularly opened windows upstairs should be avoided, as this can increase the stack (or chimney) effect within the house which results in radon being drawn into the house from the ground. Similarly, unused chimneys or unsealed loft hatches can increase the stack effect. Vents should not be cut through timber floors to provide ventilation to combustion appliances such as open fires as these can become major radon entry routes. Existing vents should be sealed and alternative sources of ventilation, such as through-the-wall vents, provided[159]. See also Chapter 3.1 in *Building services*[20].

Case study

Dampness and condensation in a converted basement used for storing household goods

This basement was in a Grade II listed building, the main part of which was built around 1840 with a further extension in the 1890s. The house was in very poor condition. It had been converted to provide five flats for elderly people together with a storage room for each flat in the cellar. The structure was of brick with a slated roof; the chimneys had been removed above roof level. The cellar external walls were earth retaining with the outside ground level approximately 300 mm below the new ceiling level. Ventilation had been provided by means of air bricks inserted into the outside walls above ground level and below ceiling level. No dampproofing had been provided and the walls were painted with emulsion paint.

Dampness and fungal growth on stored items (Figure 3.33) had been a continuing problem. An extract fan had been installed which operated intermittently, switched by a humidistat. Occupants complained about the noise produced by the fan when operating. The BRE Advisory Service was called in at this stage.

Measurements were taken of internal conditions using chart recorders. Values of temperature at 16–17 °C and relative humidity of around 87% were recorded for a long period of time. With the extract fan in operation, the humidity reduced to 65–70%. Rusting of metal components and mould growth were inevitable with these high levels of humidity. Discoloration of the emulsion paint and some visible dampness highlighted the inadequacies of these conditions.

The BRE report proposed that air louvres were inserted in the fire doors for each store room. These had to be of a type that would self-seal in the case of a fire. It was also noted that external air ventilation was being impeded by growth of vegetation and a regular maintenance regime was recommended.

It was considered that the main extract fan should be retained, but fitted with a time clock to prevent operation during the night hours. It would have been ideal to have ventilated these store rooms by passive stacks which could have been routed through the old chimneys. The removal of these stacks had removed this option.

One of the store rooms was in excellent condition due to there being a small freezer. Just enough heat was emitted by the freezer to raise the ambient conditions to allow storage of books and other papers without deterioration (Figure 3.34).

Figure 3.33 Fungal growth on items stored in a basement

Figure 3.34 Items stored in a basement adjacent to that illustrated in Figure 3.33. Sufficient heat was emitted from the small freezer to keep the moisture levels below dewpoint

Options to reduce radon levels

Sealing walls and floors
Where a house with a cellar or basement has low-to-moderate radon levels (up to say 400–500 Bq/m^3), it may be sufficient to carry out simple sealing of the walls and floors – all cracks and gaps found in the walls and floors will need to be sealed if this is to be effective. Cracks and gaps are likely to be reasonably easy to locate within the basement or cellar, as there are unlikely to be any floor or wall coverings to hide them from view. However, owing to poor construction and less demanding standards of finishing there are likely to be many cracks and gaps needing sealing. Consequently effective sealing is likely to prove difficult and time consuming.

In most cases the floor above the basement or cellar will be of timber construction and major gaps and cracks, particularly large openings around services, will need to be sealed. Less common are vaulted cellars, with concrete floors above. With these there will probably be fewer opportunities for sealing. If the cellar is rarely used, or perhaps never used, one possible option might be to seal it off from the rest of the building while still retaining good ventilation.

Sealing to ground floors
In cases where the basement or cellar is only located under part of the building, sealing of adjacent solid ground floors can be considered. Sealing floors to the occupied part of a dwelling may be more complicated owing to the need to lift carpets, move furniture and possibly strip skirting boards. This solution is only likely to prove successful with low-to-medium radon levels (up to 400–500 Bq/m^3), and where it is possible to seal major gaps and cracks. Large gaps or holes in suspended timber floors can, and probably should, be sealed. However, completely covering a timber floor with an impervious material such as polyethylene sheet is not recommended as this could lead to timber rot.

Replacement floor
If the cellar floor is in a very poor condition – perhaps being just compacted earth or cut directly from the bedrock – laying a new floor may be an option. The new floor should have a radon-proof membrane included to help to prevent radon passing through the floor slab. This will also double up as protection from damp penetration. It is doubtful whether replacing a single floor will provide a significant reduction in radon levels, and installing a sump within the soil or fill below the new floor may be a useful precaution. The sump can then be activated if radon levels remain high after the new floor has been laid[159].

Whole-house positive pressurisation
Positive pressurisation involves blowing air from the loft space down into the house to reduce the pressure difference slightly between the dwelling and the underlying soil, and to increase ventilation and therefore dilution of radon within the dwelling. The system works best in dwellings that are reasonably airtight. It is usually only recommended for use with indoor radon levels up to about 750 Bq/m^3, although if the house is very airtight it might be effective with higher radon levels. These measures may prove effective in reducing radon levels in the occupied part of the house, but the radon level might then remain high in the cellar. If this is the case, separate provision for the cellar may be needed, or the cellar sealed from the house.

An important advantage of using positive pressurisation as a solution is that it is simple to install.

See also Chapter 3.3 in *Building services*.

Increased underfloor ventilation
Suspended timber floors at ground floor level are often poorly ventilated. Increasing underfloor ventilation can help to lower radon levels and will help to reduce the risk of timber rot. For low to moderate radon levels (up to say 400–500 Bq/m^3), it is worth considering increased natural underfloor ventilation. This can usually be achieved relatively easily by upgrading existing airbricks or by providing additional ones. In some cases the underfloor void is open to the cellar. If it is, additional underfloor ventilation may also help to ventilate the cellar space. For higher radon levels (above 400–500 Bq/m^3), or where it is difficult to increase the natural underfloor ventilation, mechanical underfloor ventilation may have to be considered. See also Chapter 1.5 in *Floors and flooring*.

Increased natural ventilation to a cellar
For low-to-moderate radon levels increased natural ventilation of the cellar may be worth considering. This works by diluting radon within the cellar, introducing natural ventilation through airbricks, wall vents, disused coal holes and, in semi-cellars, vents in windows. If the cellar is completely below ground, air will need to be ducted into it. As the cellar is not used as living accommodation, the localised draughts and reduction in temperature within it caused by increased ventilation are probably acceptable. However, increased ventilation is likely to result in a need for increased heating in rooms above. To minimise problems, and to prevent draughts entering the living accommodation above, any large gaps in the floor between the cellar and the rest of the house should be sealed.

Fire and means of escape
The Building Regulations include special requirements for the structural fire protection of basements that are more onerous than for superstructures, and are set out in Approved Document B1[177]. The Regulations would generally have to be complied with if conversion or material alterations are contemplated.

The Regulations also make provision for venting of heat and smoke from all basements except the very smallest and shallowest. Smoke outlets or vents ideally need to be provided for each separate space.

In many domestic cellars access is through a door inside the house, and it is quite common to find the cellar staircase located immediately beneath the staircase giving access from the ground floor to the first floor. Walling concealing the staircase is often constructed of timber framing clad in timber boarding. Therefore fire and smoke can move directly from the cellar to the ground and first floors.

The Building Regulations make provision in Section 17 of the Approved Document for firefighting shafts approached through firefighting lobbies to be provided in buildings with basements 10 m or more below ground or access level, although not all firefighting shafts need to be provided with firefighting lifts. The fire resistance of walls and floors of firefighting shafts in basements need to comply with the requirements for these shafts in the remainder of the building; for example, all construction separating the shaft from the remainder of the building needs two hours, and from the firefighting lobby, one hour.

Options to reduce radon levels (cont)

Increased natural ventilation to the rest of the house
While improvements to the way in which a house is ventilated can help to reduce indoor radon levels, increased ventilation can affect indoor comfort so this may not be the best solution. Nevertheless, for low levels of radon, it may be worth considering. Any changes to ventilation must be permanent; simply changing window opening patterns is unlikely to be sustainable in the long term.

Mechanical ventilation of a cellar
Where the cellar is large, or the radon level is moderate to high (above 400–500 Bq/m^3), natural ventilation may not be sufficient to lower the radon level adequately. In these cases a mechanical system may be more appropriate.

A fan can be fitted to blow fresh air into the cellar. This will have two effects:
- dilution of radon in the cellar
- slight increase in air pressure within the cellar so as to counter the natural flow of radon from the ground into the cellar

If the cellar is not well sealed from the dwelling above, the increased airflow may cause cold draughts within the home. Therefore it is best to seal cellar doors and any larger gaps in the floor above the cellar.

The increased air movement in the cellar will cool the air, so this is generally not an acceptable solution if the cellar is to be used for long periods as, say, a workshop or games room. Care must also be taken that any water pipes are lagged and that goods such as wines stored in the cellar are kept away from the direct fan draught. Increased air movement, however, can help to dry out dampness within the cellar, and may help to avoid fungal growth and timber rot. This solution may actually improve the general condition of the cellar, making it a more usable space.

Mechanical ventilation can also be fitted into the cellar to extract radon laden air. This can be achieved by:
- fitting a fan to extract air from the cellar, in which case the cellar then acts as a large radon sump under the house
- providing of additional fresh air inlets on the opposite side of the house to the fan to provide cross-ventilation and dilution of radon levels

Extracting radon laden air can increase the flow of radon into the cellar through the cracks and gaps in the walls and floor, so radon levels in the cellar will equalise with levels in the ground beneath and may be higher than before. Therefore, in the absence of precise measurements, extract ventilation should only be considered when the cellar is used for storage and rarely accessed, or when it is sealed from the dwelling above.

With both supply and extract ventilation systems there is a risk of some fan noise. Fans should not be located under noise-sensitive areas. If noise proves to be a problem a silencer may need to be fitted.

The increased air movement in a partial cellar may also benefit floor voids in any adjacent suspended timber floors. This can be assisted by providing ventilation holes from the cellar. If the radon level is very high or the size of dwelling is such that ventilation to the cellar is insufficient to influence the radon level in the adjacent floor void, or both of these, a second supply fan to ventilate the floor void might be needed.

Changes to the ventilation of a cellar can influence airflow through the earth under adjacent solid ground floors. For example, mechanical extract ventilation from a cellar under one part of the house could result in radon being drawn from the soil beneath the rest of the house, so that the cellar acts as a large sump, reducing the need for action elsewhere.

Sump system to a cellar
If the radon level is moderate-to-high (above 400–500 Bq/m^3), and the floor of the cellar is of reasonable quality concrete or is a good condition stone flag floor, a sump system is likely to be the most suitable option.

It is doubtful that a sump system would be appropriate if the floor comprises severely cracked concrete, is made up of poorly jointed flagstones, has been cut directly from the rock, or comprises exposed earth. For the sump system to work effectively there will need to be some permeable soil or fill beneath the floor slab. Where the water table is above or at the same level as the floor of the cellar, radon is likely to enter through the walls rather than the floor, so installing a sump would prove ineffective as it would fill with water and prevent the flow of air through the soil.

If the sump is a viable option it can be located anywhere within the cellar that is convenient. The benefit of this system is that sumps have been shown to be the most effective method of reducing moderate-to-high levels of radon to below the action level. The sump system should not alter any of the indoor conditions within the cellar or dwelling above, although some form of isolation may be needed if a boiler or open flued combustion appliance is operated within the cellar.

The exhaust pipework and the fan unit to the sump can be located internally or externally, depending on preference. Internal fans may need to be insulated to reduce noise, and the routing of pipework be boxed in or carefully routed if taken through the dwelling above. External fans can be boxed in to improve aesthetics, or can be hidden from view with plants.

If the area of ground floor adjacent to the cellar is large, or the sump system fitted to the cellar is not providing adequate radon reduction to the rest of the house, it may be necessary to fit an additional sump beneath the adjacent solid ground floor. In some cases a second sump can be constructed behind a wall of the cellar and under an adjacent solid floor, manifolded to the existing sump system so only requiring the expense of running one fan. This is termed a multiple sump system.

A floor over a basement of a dwelling in single occupancy, with a basement floor area less than 50 m^2, is required to have a full half hour fire resistance rather than the modified half hour criterion applicable to the first floor construction of a two-storey house. More exacting requirements apply to larger basement areas and basements offering multiple dwelling accommodation. In the case of a compartment wall between basement flats, non-combustible construction is required. Only limited fire protection can be attributed to existing lath and plaster.

Overlaying ceilings with fire resisting boards will increase the structural load on the floor.

Means of escape from basements in the event of fire is a major consideration. The provision of an alternative means of escape from basement bedrooms in houses is recommended in Section 2 of BS 5588-1[178]. Where a flat is situated in a basement, and the flat is not provided with its own entrance, an alternative means of escape must be provided.

Timber floors over basements may be found with inadequate or complete absence of fire protection. Where a timber stairway to a basement forms part of the fire separation between flats it should have fire protection. Gas services may be found passing through non-ventilated voids in the basement construction – escapes of gas can build up to dangerous levels in these voids.

3 Basements, cellars and underground buildings

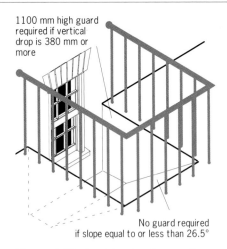

Figure 3.35 Provisions for the guarding of light wells

Windows into light wells may not be large enough to act as an alternative means of personal escape; the minimum dimensions needed are 850 × 500 mm, with sill height at 600–1100 mm.

A careful evaluation of the suitability of the existing layout and construction should be made to try and ensure that the level of fire protection would meet current standards. This may require exposing the existing construction.

BS 5588-1, Section 1.1, although intended for new-build, provides guidance applicable to rehabilitation.

Soundly constructed masonry walls are usually able to provide fire resistance for periods longer than required by building regulations. This inherent resistance to fire is however easily compromised if any unsealed holes or gaps are left through a wall. See also Chapter 6.1 in *Walls, windows and doors*[101].

Daylighting

Adequate natural light is generally more difficult to achieve in basements than in rooms above ground level, and needs careful consideration when assessing the rehabilitation potential of basements. This applies particularly in schemes involving the conversion of older properties into flats. See also Chapter 1.6 in *Walls, windows and doors*.

In accordance with Section 18 of the Housing Act 1957[179], local authorities could issue their own regulations controlling the ventilation, lighting and protection against dampness of basement rooms used for human habitation. The Act specified an average minimum floor to ceiling height of 7 ft 0 in. In the absence of local requirements made under this Act, there was no mandatory prohibition on windowless habitable basement rooms, provided they had adequate mechanical ventilation and artificial light. Building Regulations no longer control the heights of rooms.

Where windows are provided in light wells extending below ground level, the wells need to be provided with guarding (Figure 3.35).

Durability

Durability of the materials from which basements and cellars are constructed have been dealt with in detail in other volumes of this series. However, basements in areas having high water tables inevitably will be at risk from salts being transported from adjacent ground. In particular the risk of sulfate damage to mortar should be investigated before conversion to habitable accommodation is attempted.

Work on site

Inspection

Existing lighting levels can be evaluated subjectively or measured with a light meter. BRE Digests 309 and 310 offer a technique for estimating how future changes in window area will affect natural lighting level.
The problems to look for are:
◊ blocked air bricks
◊ condensation and mould growth
◊ penetrating damp
◊ sulfate attack
◊ means of escape inadequate
◊ no guard rails round light wells
◊ radon in particular areas
 (see Chapter 1.2)

Chapter 4 Public and other utilities

This chapter deals with water, electricity and gas installations outside the building, but within the boundaries of the site. Installations within the building footprint are dealt with in *Building services*[20].

A building site can become a cat's cradle of criss-crossing buried and overhead service wires, and buried pipes and tubes. Often, for old buildings, no complete record exists of what is present. For buildings built since the 1950s full records should exist, although experience warns that the services are not always exactly where indicated. Stories abound of the catastrophic penetration of a pile driven into the roof of a tunnel, or an excavator bucket lifting or breaking a water main (Figure 4.1) or, worse still, a hospital's power cable.

Moves were afoot in the 1970s to rationalise the then multiplicity of service connections to buildings, and Government departments with responsibility for large building programmes took the lead in the development of the common trench[180]. In essence, the common trench consisted of a profiled excavation with three different stages of backfilling. At the lowest level was the water main, with the level above occupied by gas on one side of the trench and high voltage electricity on the other. The final level then accommodated low voltage electricity on one side of the trench with telecommunications and TV cables on the other. Examples of the common trench will be found in buildings built for the Government estate, though its use has not been widespread.

Figure 4.1 A lead water main caught by an excavator bucket while digging a soakaway. The water main did not leak – perhaps because one of the few advantages of lead being its ductility

Nowadays there is at least a measure of agreement on the colour coding of the various pipes and cables used for services:
- low voltage electricity cable – black
- high voltage electricity cable – red
- water pipe – blue, though older pipes will be in black
- cable TV – green
- telecommunications cable – grey
- gas pipe – yellow

See also BS 1710[181].

Chapter 4.1 Water supply

As stated in *Building services*[20], by 1991, 99% of dwellings in England were supplied with cold water from public mains. However, there was still a small residue of dwellings at that time (around one in a thousand) which had no piped water whatsoever[182], with most of the remainder being supplied from wells. The situation is understood to have changed little. Dwellings without piped supplies tend to be in rural areas above the 200 m contour level, especially those also remote from other communities.

Figure 4.2 A domestic well dating back to the seventeenth century. Before the 1939–45 war, in rural areas supplies of drinking water were often obtained from shallow underground aquifers

Water is supplied in England and Wales by statutory water undertakers who have been controlled by the provisions of the Water Act 1945[183] and in Scotland under the Water (Scotland) Act 1980[184] by local water authorities. Installations in buildings were required to comply with relevant byelaws made under the Water Acts. These byelaws, based on the Model Water Byelaws[185] primarily covered prevention of waste, undue consumption, and misuse or contamination of water. Byelaws were replaced by the Water Regulations 1998[186] which came into force on 1 July 1999.

Conservation of drinking water was dealt with comprehensively in *Building services*. This book deals primarily with the supply of mains water up to the building, whereas *Building services* deals with its distribution within the building.

So far as local supplies are concerned, the composition of natural waters is governed by the nature of the ground from which they originate. Surface waters from lakes, rivers or reservoirs are maintained by rainfall in which the main dissolved constituent is carbon dioxide – up to 5 ppm. If the ground within the catchment area is relatively insoluble (eg mainly granite), the water will be soft and its acidity (high acidity indicated by low pH) will be controlled by the amount of decayed vegetation or peat within the area. Alternatively, if the water drains over or through chalk, it is likely to be very hard.

The composition of deep-well waters depends on the ground in which they are located but the composition and temperature are much more constant than for surface supplies (Figure 4.2).

Characteristic details

Water mains were commonly made of wood until the beginning of the nineteenth century, with lead pipes serving individual buildings. Although lead was used widely for the supply of drinking water until the middle years of the twentieth century, belated recognition of its harmful effects means that it is not now used. Lead as a material is quite durable if fatigue and creep failure are avoided, and some dwellings are still served by lead piping. Water supply pipes to individual dwellings commonly consisted of lead until around the 1939–45 war when iron was introduced, to be followed in the 1950s by copper.

Leakage problems were very common until the first introduction of cast iron mains in London in the second half of the eighteenth century. At first, the caulking of joints was of lead, and the external coating was coal tar solution. Although concrete mains were used in small numbers from the early twentieth century, their use did not become widespread until prestressing techniques were developed in the 1920s[19]. Until the 1970s, most mains were laid in cast iron; since then the use of polyethylene has gradually increased.

Cast iron is a ferrous metal alloy containing more than 1.7% carbon. It has been extensively used in the manufacture of pipes and is by far the most common material used for buried pipelines. The main types of cast iron are grey cast iron, in which most of the carbon is in the form of carbon flakes, and ductile iron, where most of the carbon is of graphite nodules[77].

For supplying fresh water, bare iron or steel is unacceptable owing to the severe rusting that occurs. Galvanizing gives effective protection against almost all hard waters but not against acid waters or soft waters with a free carbon dioxide content of 30 mg/cm^3 or more. The ability of the water to put down a protective, inert calcareous scale is of prime importance in considering the suitability of galvanized steel.

Water conditioners will undoubtedly remove scale, but this may not be entirely beneficial, depending on the quality of the water to be treated.

Main performance requirements and defects

Requirements for water supplies
Requirements for water supplies are covered in Water Regulations[186] and the related guidance documents. Further guidance can be obtained from BS EN 805[187].

Frost
Freezing water mains were at one time a common experience. Mains now tend to be buried at appropriate depths, a minimum of 750 mm in most parts of the UK, although depths up to 1350 mm may be needed where the risk of freezing is greater. Pipes should not be brought closer to ground level simply to provide easier access to underground stop valves.

Commissioning and performance testing
BS 6700[188] refers to the need to pressure test underground pipelines and to inspect for conformity with the specifications for valves and hydrants.

Durability
Corrosion of metals in soil is predominantly an aqueous electrochemical process. However, conditions in the soil can vary from being atmospheric (dry) to complete immersion; as a result corrosion can range from the negligible to the rapid. The prevailing ground conditions and the corrosiveness of the soil depend on a number of interrelated factors, the most important of which are:

Soil type, structure and texture
- clay
- chalk
- sand
- humus (organic content)

Soil moisture
- type of soil moisture (eg free groundwater, gravitational water (rainfall), capillary water)
- water content and dry matter content
- water holding capacity
- water table fluctuations

Micro-organisms

Chemical and physical–chemical factors
- pH (acidity or alkalinity)
- oxidation reduction (redox) potential
- percentage carbon content
- resistivity of the soil
- presence of aggressive ions

The heterogeneous nature of soil and the variety of factors that contribute to the operation of the corrosion cell make soil corrosivity difficult to predict. However, examination of the electrochemical mechanism shows that the conductivity of soil can be a major factor in corrosion and, as a result, soil resistivity measurements have become one of the most widely used measurements of soil corrosivity, with values of resistivity ranging from mildly or non-corrosive soils at more than 5000 ohm cm to very corrosive soils at less than 700 ohm cm[77].

Cast iron
Cast iron is susceptible to corrosion by sulfate reducing bacteria. As well as showing localised pitting corrosion, it undergoes a form of selective leaching known as graphitization which reduces the mechanical strength of the material. Graphitization results in the loss of the metal, leaving behind the network of carbon flakes. The surface of the cast iron takes on the appearance of graphite and can easily be cut with a knife. Failure of cast iron pipes 6 mm thick within a year and perforation within four years as a result of microbial corrosion are quite common. The intrinsic corrosivity of the sulfide films which can develop on ferrous metals is high.

Cast iron possesses no useful resistance to mineral acids. Hydrochloric acid will cause corrosion at any concentration and temperature. Dilute sulfuric, nitric and phosphoric acids are also aggressive. Similarly corrosive are well aerated organic acids. Guidelines drawn up by the Water Research Centre (WRc) on the use of ductile iron pipes state that highly acid soils (pH less than 5) are corrosive towards cast iron pipes even when they are protected by a system of zinc coating and loose polyethylene sleeving[77].

Lead

Although lead is inherently durable, it has some ability to dissolve in water – to a greater degree in soft water than in hard water. In areas served by soft water supplies there is considerable concern about the risks to health and some replacement has been carried out.

Steel

Corrosion of the outside of the pipe is the main concern. Although some galvanised pipe may be seen, the use of coal tar solution treated pipes was also common. Both types will by now almost certainly have been replaced.

Copper

Copper is essentially immune to corrosion in most underground environments due, mainly, to the formation of a protective oxide film over the entire surface of the metal. It is, however, attacked by oxidising acids such as nitric acid and concentrated sulfuric acid, and corrosion has also occurred when buried in certain soils, but there is no direct relationship between any single feature of soil composition and the rate of corrosion[77].

Plastics

Most common polymers are susceptible to oxidative degradation; this can be promoted either thermally or photo-catalytically. For some, prior exposure to sunlight can increase their susceptibility to thermal degradation, and should therefore be stored in appropriate conditions before use. When buried in the ground, they are relatively protected, especially in anaerobic conditions.

The specification for plastics pipes to be used in contaminated ground should receive careful consideration. Diffusion of solvent molecules into a polymer results in swelling of the plastics by interfering with the normal intermolecular attraction in the plastics. If the extent of swelling is slight then the only effect is an increase in the permeability of the plastics. However, in extreme cases the disruption of the normal intermolecular forces can result in a loss of mechanical properties such as tensile strength, elongation at break, tear resistance and puncture resistance, and can eventually cause ductile failure. The extent of swelling depends on the nature of the solvent and the chemical characteristics of the polymer.

Environmental stress cracking (ESC) is craze or crack growth brought about by the simultaneous action of mechanical stress and surfactants (eg alcohol, soap and detergents). The mechanism of ESC is not well understood. The fracture surface appears brittle although under the microscope it shows a degree of ductility. Surfactants do not solvate the polymer, nor do they attack it chemically, nor produce any discernible effect on the polymer in the absence of stress. Similarly, in the absence of the surfactants, similar stress levels do not cause the level of cracking associated with ESC.

Microbial degradation is essentially a chemical process involving enzymes produced by micro-organisms (eg bacteria and fungi). These enzymes react with specific sites on the polymer causing hydrolytic and oxidative reactions, and resulting in chain scission. The molecular weight of the polymer is reduced and there is deterioration in the physical properties of the material. For biodegradation to occur, the following general conditions must exist in addition to the presence of micro-organisms:

- presence of oxygen, moisture and mineral nutrients
- temperatures ranging from 20–60 °C, depending on the type of micro-organism
- a pH between 5 and 8

Fungi can also cause cracking and crazing of some plastics, for example polyurethane, by tunnelling into them. Stressing the plastics can increase the rate of degradation.

The biodegradability of plastics is influenced also by the plasticisers, lubricants and stabilisers that are often added to the polymer. These low molecular weight compounds are usually much more sensitive to microbial attack than the polymer molecules. The loss of these additives from degradation can result in changes in the physical properties and the performance of the plastics. They can also act as a source of nutrient for the micro-organisms, increasing the extent of biodegradation on a polymer that is otherwise bio-resistant.

Permeation of the plastics by other materials can also occur and is determined not only by diffusion of the permeant through the plastics but also by the degree to which the permeant is soluble in the plastics. The mechanism by which an organic chemical permeates a plastics materials differs with the type of polymer. In the case of polyethylene, and polybutylene for example, the structural integrity of the polymer is not affected by the majority of organic chemicals.

A survey by the Water Research Centre on the effect of contaminated land on permeation of the pipeline carrying a water distribution system found that the large majority of incidents involved contamination resulting from spillage of petrol or other petroleum products such as heating oil.

In the case of PVC-U, if levels of contaminant in the surrounding groundwater are low, permeation of contaminants through the PVC-U is not a problem. On the other hand, certain organic contaminants may cause PVC-U to soften and swell until eventually it loses its structural integrity; at this point permeation becomes significant.

Polyethylene has poor corrosion resistance to oxidising acids; nitric acid and sulfuric acid at high concentrations are highly corrosive. Hydrochloric acid (HCl) does not attack polyethylene chemically but can have a detrimental effect on the mechanical properties of the polymer. HCl can easily diffuse into the polymer, where it disrupts the intermolecular forces between the polymer chains.

High molecular weight polyethylene is resistant to microbial degradation. A number of soil burial tests have shown that the maximum weight loss that can be expected for a high molecular weight polyethylene due to microbial degradation is 1–3%[77].

Soils
Light sandy soils and chalk are not generally aggressive. On the other hand, acid peaty soils, and ground which contains large amounts of cinder and builders' rubble, can be very corrosive to steel, copper and aluminium; these therefore require highly acid resistant and protective coatings. Heavy, anaerobic clays may be corrosive to ferrous metals since these provide a favourable environment for sulfate-reducing bacteria but aluminium, lead and copper are resistant to this type of attack. Saline environments cause severe corrosion of aluminium and galvanized steel but have little effect on copper or lead; lead, though, can be severely corroded if it is connected to copper. In aggressive soils, backfills of chalk or sand may prove beneficial.

Work on site

Inspection
The problems to look for are: ◊ lead mains ◊ leakage of pipes ◊ seized stop valves ◊ hidden stop valves (covered by soil or tarmac) ◊ aggressive ground conditions

Chapter 4.2 Wastewater drainage

This chapter deals with wastewater drains and private sewers up to the point where they discharge to public sewers, even though the public sewer may lie in land forming part of the building site and therefore, nominally, within the ownership of the building owner. Requirements for public sewers are outside the terms of reference of this book.

Three different types of drainage systems have been used in the UK, depending on the age of the development and local circumstances:
- combined wastewater and surface water
- partially combined wastewater and surface water
- separate wastewater and surface water

Excessive load on wastewater treatment plant at times of heavy rainfall has led to an increase in the use of separate systems, and this form of arrangement is now common.

So far as technical requirements in legislation and Standards are concerned, the Building Regulations Part H[189], the Building Standards (Scotland) Regulations Technical Standards Part M[190], and the Building Regulations (Northern Ireland) Booklet N[191] deal with wastewater drainage both within the building, and externally to the building. British Standards separate these provisions into two distinct series: within the building BS 8301[26] and BS EN 12506-1 to 12506-5[192], and externally to the building BS EN 752-0 to 752-7[193].

The Building Elements series of books follows the British Standards practice, and accordingly above ground drainage within the footprint of a building is described in *Building services*[20].

Within existing legislation in England and Wales, the priority for wastewater drainage remains, in order of priority:
- public sewer
- private sewer
- septic tank
- cesspool or cesspit

Characteristic details

Details of systems

A wastewater drainage system normally consists of a network of pipes laid from a building to fall to a local authority sewer, a private septic tank, a treatment plant or, occasionally, to a cesspit (Figure 4.3).

The aim should be to design a system with the minimum of:
- excavation of ground
- length of drain runs
- number of manholes
- number of sewer connections
- risk of blockages

Drains should be laid in straight runs between access points at an approximately constant gradient. With long pipe runs, this is achieved by using sight rails set up at each manhole or inspection chamber at a fixed height above the required invert level. Using a 'boning' rod whose top edge is aligned with the sight rails, a series of pegs or pins are then driven to invert level at regular intervals along the trench. The pegs or pins are removed as the pipes are laid. The longitudinal alignment of the pipes is controlled by maintaining a fixed gap between the pipe sockets and a side line set up at half pipe level. This side line is normally stretched between steel pins located at each manhole or inspection chamber, although intermediate pins may be needed on very long lengths.

Access to drains is required for testing, inspection, maintenance and removal of debris. The traditional means of access is the inspection chamber, which becomes a manhole when it provides working space at drain level. However, manholes are expensive and rodding eyes may therefore be substituted, particularly at the head of shallow branch drains. A rodding eye consists of a vertical or inclined length of pipe with a suitable cover or cap at ground level, while at its base it makes a curved junction with the drain.

The design of the drainage system must ensure that waterborne waste is carried away efficiently with minimal risk of blockage or leakage of effluent into the ground. There are two main aspects of design:
- hydraulic
- the selection of suitable pipes, and procedures for burying the pipes to ensure that they are adequately protected and do not cause unnecessary subsidence of nearby foundations

Most buildings in the UK are connected to mains drainage systems. In England, only around 3.5% of dwellings are not on mains drainage; most of these are in rural

4.2 Wastewater drainage

areas, indeed it is only the most isolated of dwellings that do not have mains drainage[4].

Certain basic features such as cesspools, and in some areas even septic tanks, may be considered to be undesirable and the possibility of connecting to mains drainage perhaps ought to be investigated. In many areas, local authorities encourage separating rainwater and waste systems to eliminate any surcharging and surface flooding from drains during heavy rainfall.

In a survey[194] carried out for one water undertaking in 1981, of all the sewers surveyed, 84% of pipes were of 300 mm or less in diameter; 87% were laid at depths of less than 3 m, with an average depth of 2.1 m; and 82% of the sewers consisted of clay pipes.

Common domestic drainage layouts

Before the widespread introduction of mains drainage in suburban areas, it was common for individual dwellings to be served by septic tanks. On the subsequent connection of dwellings to mains drainage, these tanks became redundant, though many will have survived, buried beneath pavings or gardens.

Drain lengths, of course, depend on the age of the property, the particular type of layout adopted for the drainage system, whether the drainage is from the front or the back of the dwelling, and the position of the sewer. For drain length estimation it is necessary, therefore, to consider the most common layouts for each type of dwelling*.

* Some data on drain lengths for domestic buildings were collected during BRE studies on drain function and designs but the sites in this programme did not cover all dwelling types. The bulk of this previously unpublished information was obtained from the data tapes containing the original survey material for the EHCS 1981 for metropolitan and urban conurbations and provincial towns. Front and back garden widths and lengths were taken as the basis for the calculations. It was clear that plot sizes became progressively smaller after the mid-1960s. Apart from some changes following the demolition of houses standing in large gardens and redevelopment of the sites, plot dimensions for most of the dwellings concerned will have remained exactly the same, so the information is still considered to be broadly valid.

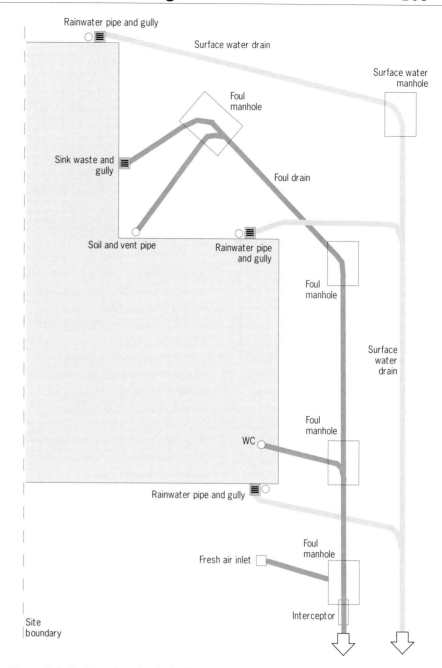

Figure 4.3 Drainage layout typical of interwar suburban housing

Drain lengths in the following paragraphs are given to the front boundary. An additional figure is needed, therefore, to cover the distance from the boundary to the connection to the main sewer in the road.

Estimated foul drain lengths for dwellings are as follows.

Old terraced houses
It is most likely that these dwellings will be served by private sewers. The length of drains would be given, therefore, by the sum of the length of the private sewer per dwelling and the length of the pipework from the house to private sewer. This sewer could run through either the front or back gardens of the terrace but its distance from each house is unlikely to be at the extremities of the plots. BRE measurements on about forty 1920s terraced houses gave an average figure of just under 4 m for this distance; assuming this figure, the overall length of drain run per

dwelling is 9 m. The maximum possible length occurs if the connecting private sewer is positioned at the end of the back garden (ie the sum of back garden length and width, 15 m). This arrangement is possible in very old property with drains previously serving outside toilets, but is now uncommon. The minimum drain length likely with a terraced house occurs where the bathroom and kitchen are at the front of the house, in which case the dimension is given by front garden length – about 4 m[*].

Semi-detached houses
It is likely that most are served by traditional drains, connected directly to public rather than private sewers; that is, the drains run through the length of the front gardens. From experience it is clear that most of these dwellings will have rear-positioned kitchens and bathrooms. Using this assumption, drain lengths are given by the sum of the front garden length and building depths (ie 15 m). The minimum length with traditional arrangements is when the drain only runs to the front of the dwelling (ie 7 m), although this is not thought to be a very common practice. BRE work has shown that an average figure for the drain length from house to private sewer is about 4.5 m (taking an average of 50 dwellings). For a garden 8 m wide, the drain length is therefore about 12.5 m. From the EHCS the median and mean of front garden length is 7 m, for back gardens it is 15 m and 21 m respectively. House depths median is 8 m and mean is 9.5 m.

[*] A detailed explanation of the method of calculation is not appropriate for inclusion in this book, but an outline of it in relation to terraced houses is as follows. Most terraced houses built before 1976 have front gardens between 1 m and 5 m in length; the median and mean values for these houses are 4 m and 4.5 m respectively. Widths: median 5 m and mean 5.5 m. Back garden length: median 10 m and mean 13.5 m. House depths: median 8 m and mean 10.5 m. Mean and medians are relatively close for front garden lengths, and to a lesser extent for front garden widths, indicating an approximately normal distribution. The other principal dimensions show a greater scatter reflected by the wider variation between mean and median. The medians are used for the calculation of drain lengths (to avoid buildings with larger plot sizes unduly weighting the results). A similar approach is used for the remaining house types.

Semi-detached bungalows
The most likely arrangement, as for semi-detached houses, is a drainage system running from the rear of the house through the length of the front garden, a total length of 16 m. The minimum but less likely length, with the drain running only from the front of the house, is 7 m. From the EHCS, the median and mean of front garden length is 9 m and 9.5 m. Back gardens: 13 m and 16.5 m respectively. House depths: median 9 m and mean 9.5 m.

Detached houses
Many dwellings of this type have been built since 1944 so that private sewer systems are probably as likely to be used as direct connections to the public sewer. However, by chance, the drain lengths likely from both arrangements are similar, assuming that kitchens and bathrooms are positioned at the rear of the building. For private sewer arrangements, using data from BRE studies, the estimated length is 17.5 m, while for runs through the front garden direct to the public sewer the figure becomes 17 m. The minimum likely drain length, assuming the kitchen is at the front is 8 m (the front garden length) but this is thought to be a less common arrangement. Likely drain lengths for detached bungalows will be very slightly greater than for detached houses. From the EHCS, the median and mean of front garden lengths: 8 m and 9 m. Back gardens: 19 m and 22.5 m respectively. House depths: median 9 m and mean 10.5 m.

Flats
The nature of the data on plot sizes for flats is such that it is not feasible to give estimates on drain lengths. Modern high rise flats have a wide variation in drain lengths before connection to a sewer.

National and regional differences

Legislation covering drainage has always been framed in fairly general terms with few specific requirements regarding design. It is inevitable, then, that there are discernible, albeit minor, differences in practice throughout the UK. For example, in Sheffield where hilly and rocky ground is extremely common, steep gradient systems have been in use for a long time, contrary to current practice in other areas. Also in the exceedingly flat areas such as those found near Peterborough, very flat gradients have been used for many years for long drains, with success, as BRE studies have shown.

Large cities often show tendencies to resist changes in practice. In London areas, intercepting traps were still specified by some authorities until the 1980s and the use of PVC for underground drainage systems was slow to be permitted on a wide scale. London, of course, had its own Bylaws for many years. Other distinct differences in detailed practice can be found; for example, 150 mm rather than 100 mm drains for dwellings have been required in Manchester since the early 1960s.

With common legislation in England and Wales, Welsh practice is not thought to differ widely from English. Practice in Scotland is governed by its own regulations and other legislation, in particular the Sewage (Scotland) Act[195], so practices common in England, such as private sewers, can differ from those in Scotland. Sealed access covers, positioned within manholes, have been common practice in both legislations, and in more recent years it is known that proprietary plastics drainage systems with rodding eyes have been widely used in new estates; so present Scottish and English practice is probably similar in many respects, although cesspools are not now permitted in new construction in Scotland.

4.2 Wastewater drainage

Effect of terrain
Apart from the obvious effects on choice of gradient and direction of the drain run, the use or not of private sewers, and the method of connection to the public sewer, will depend to some extent on terrain.

Hilly areas
For the greater part of the twentieth century, the choice of drain gradient has been very restricted. Maguire's Rule, mentioned in the introduction, was generally used and consequently very steep or very flat gradient systems were uncommon. In very hilly areas public sewers were not installed following the contours of the land but, for example, made to descend steep hills in a series of steps using backdrop manholes. Where necessary house drains were connected to the sewer via such manholes. Because of their size and cost, these manholes were unlikely to be situated within the curtilage of a private dwelling; and to avoid a multiplicity of them, one backdrop manhole would serve a number of dwellings via a shallow private sewer.

Research has shown that the formerly widely held fears concerning the use of very steep gradients were unfounded, hence many systems which have been installed in hilly areas during recent years will have steeper drain runs and fewer backdrop manholes. The design of these manholes has also tended to change, with 45° backdrop slopes instead of vertical ones. The Mascar Bowl system is frequently used in these situations, employing a hemispherical bowl of around 0.6 m diameter, at or near the surface with a central outlet which connects with the main drain using a suitable length of pipe.

Flat areas
Using Maguire's Rule to design domestic drainage systems in very flat areas generally did not lead to very deep drains because of the relatively short length of pipework involved. Connecting sewers, being of larger diameter could be laid at flatter gradients according to the rule and so, again, even with the longer lengths involved, drains did not become excessively deep. Nevertheless the modern tendency has been to allow flatter gradients and therefore recent developments are likely to have shallower drains than those of 50 years ago. There is also a tendency to use smaller diameter pipes.

Types of pipe
Drain pipes fall into two categories: flexible pipes, which either were or are made from materials such as plastics and pitchfibre, and rigid pipes, which were or are made from concrete, vitrified clay, asbestos cement or cast iron. The behaviour of the two types of pipe in the ground is significantly different: while rigid pipes have an inherent strength, flexible pipes will deform under the application of loads and require support from the surrounding backfill material to prevent excessive deformation.

Most, but not all, pipes are now supplied with flexible joints. In addition to allowing rapid assembly, these joints allow a limited amount of lengthwise movement to take place between adjacent lengths of pipe and still remain watertight. The watertightness of the drain is therefore less likely to be impaired by ground movements caused, for example, by seasonal shrinkage and swelling of clay soils, mining subsidence, or settlement of nearby foundations. In some circumstances, for example when using a cut length of rigid pipe, it may be necessary to form a caulked joint; the flexibility of the drain can, however, be maintained by ensuring the distance between flexible joints does not exceed a standard pipe length. Where drains connect to manholes or other structures, the projecting

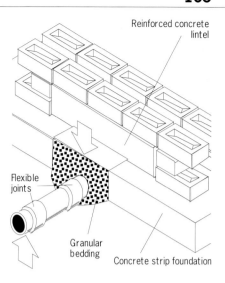

Figure 4.4 A widely used method of protecting a drainage pipe passing through the structure where large amounts of differential movement were expected. Using a separate lintel may not provide sufficient seal against gas entering the building, and is not now recommended

length of pipe should be kept as short as practicable and, ideally, should not exceed two pipe diameters. In cases where large amounts of differential movement between the structure and the pipe are anticipated, it is prudent to provide a short length of pipe (say 0.6 m long) to connect to the pipe projecting from the structure; the flexible joints at the ends of the short length of pipe then act as a pair of hinges which allow the pipe and the structure to move independently (Figure 4.4). One widely used detail in the past was to insert a precast lintel over the pipe, but there could be a risk that the point of entry is not gastight.

Pipe sizes and gradients

The design of drainage pipelines has undergone many changes over the years. The Model Building Byelaws required every drain to be properly supported, protected from damage, laid at a proper inclination, and provided with suitable watertight joints. It was to be of adequate size and if, intended for the conveyance of foul water, to have an internal diameter of not less than 4 inches. There were other byelaws governing the design of manholes, such as the material was to be of brick, concrete or cast iron, to have inverts with proper channels and benching, and to be provided with step irons where the depth of manhole exceeded 5 feet.

The Building Regulations 1965[25] required that the joints of a drain or private sewer 'remain watertight under all working conditions, including any differential movement as between the pipe and the ground or any structure through or under which it passes'.

Table 4.1 Pipe sizes and gradients

Number of dwellings connected to drain	Peak flow (litres)	Pipe dia (mm)	Flattest gradient	Notes
1–5	<2.5	100	1 in 40	1 in 70 for short branches with peak flow >l litre/s (=1 WC)
5–20	>2.5	100	1 in 80	Possibly down to 1 in 130
10–150	>2.5	150	1 in 150	Possibly down to l in 200

It is impossible to say with certainty what range of practices will be encountered on site, depending as it does largely on the age of the layout, and the quantities of effluent and its constituents. However, some indication can be obtained from Table 4.1, taken from BRE Digest 130[196] (now withdrawn), covering the sizes and gradients of small domestic drains.

The flatter gradients were permissible only where satisfactory joints and a high standard of workmanship and supervision could be guaranteed.

High-rise flats, often surrounded by communal grass plots under which the connecting drains ran, were normally equipped with 100 mm (for blocks up to 10 storeys high) or 150 mm internal stacks. The drains connecting these blocks often served from two to four stacks per block and a number of blocks before joining the public sewer. Therefore pipe sizes vary from 150 mm to about 450 mm in diameter. Drain depths are often of 2 m or more, sometimes due to the provision of underground car parks, and also because of the large distances concerned.

After the 1950s, practice in the jointing of pipes underwent a profound change. Clay, concrete, asbestos cement and some types of iron pipes became available with factory prepared flexible joints. These joints were strongly recommended in preference to cement mortar or other rigid forms of jointing for the following reasons:
- a minimum of skill was needed and jointing was more rapid and reliable
- a test could be applied immediately after laying the run of pipes
- rectification of faults became relatively quick and easy
- because of the speed of laying, the time the trench had to be kept open was reduced to a minimum, with a possible saving in pumping and strutting and less risk of the trench bottom becoming adversely affected by weather or of damage to the pipes (eg by vandalism)
- less interruption in pipelaying caused by freezing temperatures

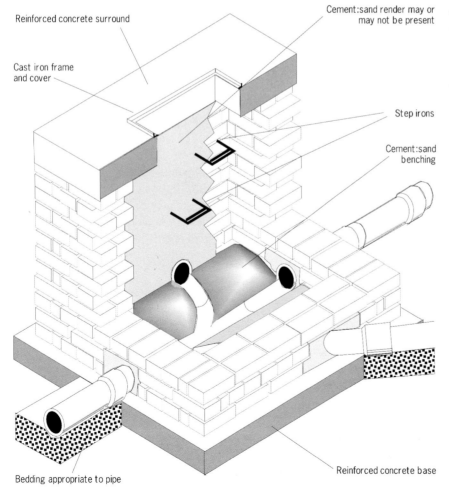

Figure 4.5 A typical manhole constructed in the 1950s

- the flexibility of the joints reduced the risk of breakage of the pipeline through movements due to settlement of the soil or of structures to which the pipeline was connected

Since that time, the use of flexible joints has further increased until the practice is all but universal.

European Standards are now available for the design of drain and sewer systems outside buildings. These consist for the most part in British Standards versions of the various parts of EN 752:
- Generalities and definitions (BS EN 752-1[197])
- Performance requirements (BS EN 752-2[198])
- Planning (BS EN 752-3[199])
- Hydraulic design and environmental considerations (BS EN 752-4[91])
- Rehabilitation (BS EN 752-5[200])
- Pumping installations (BS EN 752-6[201])
- Maintenance and operations (BS EN 752-7[202])
- Construction and testing of drains and sewers (BS EN 1610[203])
- Pressure sewerage systems outside buildings (BS EN 1671[204])

Inspection chambers and manholes
Since Victorian times, inspection chambers and manholes for domestic work have been mostly constructed in one brick English bond upon concrete slab bases. A bond sometimes encountered in manhole construction is water or manhole bond, which was supposed to give better waterproofing and to remove the need for render, though this may have been wishful thinking.

It is commonly understood in the industry that an inspection chamber, while it may be sufficiently large and sufficiently shallow to insert and manipulate rods for cleaning purposes, will not be large enough for a person to enter. With deeper drains therefore, sufficient space is needed to accommodate both the rods and the operator – hence the term manhole.

Where drains are shallow and sites are relatively flat, the most common construction will be found to consist of a brickwork shaft on a concrete slab from invert to ground level, following the practices described in the popular construction textbooks of the time[205]. The deeper the invert, the larger the chamber. It was common to build a chamber 24 × 18 in (say 600 × 450 mm) up to 2 ft (600 mm) deep to the invert, and increase the size to 42 × 30 in (say 1200 × 750 mm) where the invert was at 5 ft (1.5 m) or more. Where a large chamber was formed below ground level, a shaft having not less than 36 inches (900 mm) clear internal space should have been constructed. Step irons of cast iron inserted into the brickwork during construction provided hand and foot holes (Figure 4.5).

The invert in the manhole was normally formed from half-round section pipes complete with the necessary bends; the space between the edge of the channel and the inner wall of the manhole rounded off with a cement mortar haunching, commonly called benching, so that any surcharging could easily drain away. Cement rendering was commonly used for the manhole sides even though it had a tendency to crack and fall off into the channel and pipework, forming blockages.

During the interwar years, the manhole closest to the house collected the branch drain from stacks and gullies, and was generally very shallow, around 0.3–0.6 m deep, though the depth of other manholes depended on the terrain and the distance to the sewer.

For sloping sites, to restrict the amount of excavation and to ensure suitable gradients to maintain self-cleaning velocities of flow for the drainage runs, manholes may be seen with backdrops, ramps or vortexes. In some parts of the UK a backdrop was known as a tumbling bay (Figure 4.6). A backdrop manhole is defined in BS EN 752-1 as a manhole with a connection, by means of a vertical pipe at or just above invert, from a drain or sewer at

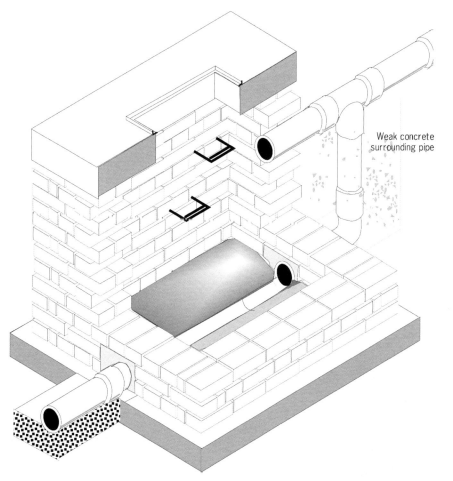

Figure 4.6 A typical backdrop manhole constructed in the 1950s on a steeply sloping housing site

a higher level. Such manholes may also have the backdrop sloping (Figure 4.7); another alternative is for the drop to take place within the manhole, with a rodding eye capped or open at high level (Figure 4.8). The open version is commonly used in situations where a new drain is to be added to an existing manhole.

A vortex manhole is normally circular: the sewage enters at high level and is directed onto the wall surface, flowing round the circumference of the manhole and at the same time dropping under the influence of gravity until it reaches the invert. The Mascar bowl is a form of vortex inspection chamber.

Circular precast concrete chambers became available just before the 1939–45 war in various diameters from 36–54 inches. British Standards, Part 2 of BS 5911[206] and BS 5911-200[207], followed.

In the mid-1960s, manhole construction changed with both PVC and clayware manufacturers offering complete drainage systems including factory made manhole and access fittings in a variety of materials and designs (Figure 4.9). PVC pipework and branch fittings came to be used in both brick and concrete manholes, giving level invert connections between branches and main drain runs, and marking a distinct change in practice for domestic systems. However non-level inverts were still used, even in preformed manholes.

Recent tightening of health and safety legislation governing work in confined spaces, has led to the need to avoid, as much as possible, maintenance engineers entering drainage systems. However, the old recommendations for minimum manhole dimensions to permit rodding and working within the confines of the manhole also depended upon size of the largest pipe entering the manhole, and the depth of the manhole. For manholes less than 1.5 m deep the minimum sizes of rectangular manholes ranged from 750 × 675 mm to 1200 × 750 mm or more, and circular ones from 1000–1200 mm diameter or more depending on pipe sizes. For those greater than 1.5 m deep, the minimum sizes ranged from 1200 × 1000 mm to 1350 × 1225 mm or more for rectangular manholes. For circular manholes greater than 1.5 m deep, the minimum sizes were the same as those less than 1.5 m deep.

Minimum shaft sizes for manholes greater than 3 m deep depended upon the means provided for access. Where steps were built-in, the minimum size was 1050 × 800 mm. Where ladder access was provided, 1200 × 800 mm, and where a winch only was used, 900 × 800 mm.

Inspection chambers made in plastics (BS 7158[208]) (eg polyethylene, polypropylene, PVC-U, and reinforced thermosetting resins) did not become available until the 1960s. They have mainly been used in conjunction with plastics drains and were also governed by BS 8301, though their limited size meant that they were restricted to shallow inverts not exceeding 2 m.

Gullies
European Standards, including prENs 1253-1[209] and 1253-2[210], have been published.

Cesspools, cesspits and settlement tanks
The cesspool traditionally has been a chamber constructed below ground level having a size to contain all the foul and wastewater produced by a dwelling over a period of time. When full, the chamber needed to be evacuated by suction tanker. Construction typically would have been of rendered one brick thick walls on a concrete base, with a reinforced concrete top containing a sealed access. Although these facilities were occasionally constructed up to the 1950s, modern cesspools tend to be factory made plastics or GRP tanks.

The Model Byelaws made under the Public Health Act 1936 required that a cesspool be impervious to liquids both from outside and from inside, be properly covered and ventilated, and be sited so that it does not become a nuisance to occupants of buildings. Slightly revised criteria, including the minimum size of 4,000 gallons, were introduced in England and Wales in Part N 17 of the Building Regulations 1965[25]. The current Approved Document is Part H2; a new draft Approved Document has been published[211].

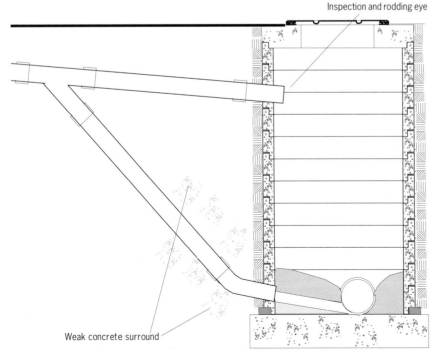

Figure 4.7 A sloping backdrop manhole

4.2 Wastewater drainage

More detailed criteria are also currently to be found in BS 6297[212], although this Standard is now becoming increasingly out-of-date.

The requirement for watertight construction usually meant that the walls of manholes had to be rendered internally. Where permeable ground conditions existed on site, and after the building control officer had left the site after giving final approval, it was not unknown for a header 'accidentally' to be knocked out of the cesspool wall so that the need for emptying became less frequent!

Septic tanks

Small on-site plants for the treatment of sewage from one or more dwellings or other small buildings in areas remote from public sewers have been permitted for many years, and will be found in a vast variety of designs. The principle of operation, solids settlement by gravity, is the same for all. Blackwater flows into one side of one or more tanks. The solids are partially broken down by anaerobic action (under a scum or crust where faecal matter is present), and the partially treated water is drained off to ground through a pipe at the other side of the tank or tanks for final treatment by a suitable soakaway, depending on local ground conditions. The outflow pipe from the tank is prevented from clogging by using a T-shaped pipe or scum board. Solid detritus, including waste matter that cannot be broken down, accumulates as a sediment in the bottom of the chamber.

Design criteria for small modern sewage treatment works can be found in BS 6297. More recent provisions are to be found in BS EN 12566-1[213].

The following parameters are considered in the Standard:
- structural, hydrostatic and backfilling loads
- watertightness under test
- hydraulic efficiency under test
- access
- durability
- installation instructions
- operating and maintenance instructions

Most prefabricated chambers will be found to be made from reinforced concrete, glassfibre reinforced resin or polyethylene. Most constructed in-situ chambers were built of rendered brick or concrete block on reinforced concrete bases fitted with reinforced concrete covers.

Greywater reuse

Since the 1980s there has been a considerable increase in interest both in the UK and in the rest of Europe in the reuse of wastewater, or greywater (eg for flushing toilets). Standards are being developed for small on-site treatment and infiltration systems for processing domestic greywater.

Two main stages are normally involved in this type of treatment:
- settlement to remove solid matter
- biological treatment to remove specific pollutants or to reduce biochemical oxygen demand to levels which are statutorily required for consent to discharge

There are many ways in which treatment may be carried out, but one of the most common methods is to use a conventional septic to remove gross solids followed by processing through one or more beds of reeds (*Phraunites communis*). The common reed has an ability to transfer oxygen from its leaves via its roots to the surrounding soil. Bacteria thrive in this oxygen-rich environment, and so aerobic treatment of wastewater becomes possible.

There are two basic types of reed beds, one of which has the surface permanently flooded with the wastewater, and the other in which the level of the water is below the surface of the bed. It is the latter kind that is of most interest in the UK climate. The method of construction of the bed takes one of two general forms:
- horizontal
- vertical

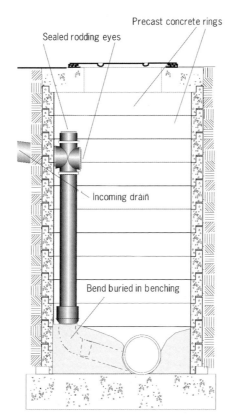

Figure 4.8 A backdrop manhole with the drop contained within the manhole

Figure 4.9 A typical plastics manhole or inspection chamber

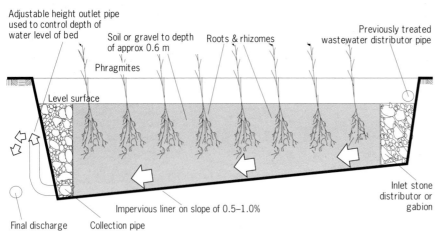

Figure 4.10 A section through a typical horizontal flow reed bed

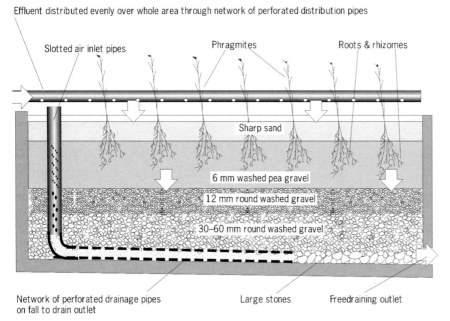

Figure 4.11 A section through a typical vertical flow reed bed

In the **horizontal** type the wastewater flows from a discharge pipe at one end of the bed construction, through a filter trench filled with gravel and a 0.6 m deep bed planted with reeds, to a discharge point at the other end of the bed (Figure 4.10).

Primary beds are normally sized at around 5 m² per person equivalent, and 600 mm deep – the maximum depth of the reed rhizomes. The filter medium is normally washed and graded gravel placed in lined or unlined basins.

Horizontal flow beds require less maintenance than vertical flow beds to be described next. Weed growth may be controlled sufficiently by flooding the surface.

In the **vertical** type the water flow is fed intermittently downwards from a perforated discharge pipe at the level of the reeds through several layers of sand and gravel of various sizes (Figure 4.11). More than one bed is needed in order to allow resting and recovery of the sand layer since choking of the filter material is one of the most common modes of failure of this type of bed.

Vertical flow beds are normally 1–1.5 m in depth, and are sized according to likely demand. Typically the size will have been 0.8 m² per person equivalent for the first stage and 0.25–0.5 m² for a second stage, although more recent information suggests 1.0 m² to accommodate biological oxygen demand, and 2.0 m² for nitrification are more appropriate. Small beds need a greater area per person equivalent than do larger installations, ranging from 2.0 m² per person for a 4-person installation to 0.8 m² for a 100-person installation. Beds should not be oversized as detrimental weed growth might ensue[214,215].

See also *Reed beds for the treatment of domestic wastewater*[216].

Reed beds need annual maintenance, including cutting of the reeds. They also need to be fenced off from grazing animals.

Small wastewater treatment plants are covered in BS 7781[217]. See also BS EN 12255[218].

Pumps and lifts

Under certain circumstances – for example, where there are very deep basements containing sanitary facilities – it will be necessary to provide pumping facilities to raise effluent to the level of sewers. Pumps may be driven by electricity or by diesel engines. Installations are governed by the provisions of BS 8005-2[219] and BS 752-6[201].

There are also three European Standards relating to wastewater lifting plants: EN 12050-1[220] EN 12050-2[221] EN 12050-3[222].

Non-sewered disposal

Waterborne sanitation is costly and requires extensive and reliable water resources. In some circumstances it may be important to develop and implement alternative non-sewered sanitation systems using little or no water, perhaps appropriate for isolated housing and small institutional buildings such as hostels. Systems may be designed that, if properly constructed and maintained, can contribute greatly to improving public health and personal comfort at minimum cost. The following topics need to be considered:

- health and nuisance problems associated with inadequate sanitation
- pollution risks to soil and groundwater

- sanitation technology options for no or low water-use systems
- sanitation options for unsewered buildings
- system design considerations
- on-site disposal of effluents
- assessing soil suitability[223]

In more recent years, disposal systems which do not rely on water (eg biological or composting toilets) have been monitored by BRE investigators. The option to use waterless installations is now included in the revised Part M of the Building Standards (Scotland) Regulations[190].

Materials
Plastics
Amongst the main plastics materials used in the 1980s and 1990s for underground pipework have been polyethylene – low, medium and high-density (LDPE, MDPE and HDPE), unplasticised polyvinyl chloride (PVC-U), polypropylene (PP), polybutylene (PB), acrylonitrile-butadiene-styrene (ABS), cellulose acetate-butyrate (CAB), and glass reinforced plastics, incorporating epoxy or polyester resins[77].

PVC-U pipes, which are identified by their orange-brown colour, are probably the most widely used type since they are suitable for most domestic installations and for surface water drainage. They are available with nominal outside diameters of 110 and 160 mm and in lengths of 3, 6 and 9 m. They should not be used for effluent at high temperatures, and become brittle at low temperatures. Push-in joints are available, either as separate couplings or integral sockets, and provide some flexibility. Although relatively expensive, PVC-U pipes have the advantage of being available in long lengths, are lightweight, and readily handled and assembled (Figure 4.12).

PVC-U pipes and their fittings are covered by BS 4660[224], BS 5481[225] and BS EN 1239[226]; ABS by BS EN 1455[227]; PP by BS EN 1451[228] and BS EN 1852[229]; GRP by BS EN 1636[230]; and modified PVC-U (MUPVC) by BS 5255[231].

Figure 4.12 PVC drain pipes being laid as replacements for salt glazed pipes cracked as a result of movements of the shrinkable clay ground

As well as plastics drain pipes, plastics inspection chambers and rodding eyes have been developed and have been in widespread use since the mid-1980s. Hepworth's Supersleve drainage system, which won the Building Innovation Award in 1985[232], is an example of a range of products that has gained wide acceptance.

Other European Standards are now being prepared and published (eg prEN 12666-1[233] for plastics piping systems).

Pitch fibre and pitch impregnated asbestos fibre
There will be many existing installations in these materials still in existence since they were very popular around the mid-1960s. The pipes themselves were flexible, mainly used in domestic installations in small diameters, and were made to the former BS 2760[234] which has now been withdrawn.

The original joints had machined, tapered spigots which made a drive-fit to machined pitch fibre couplings but joints using polypropylene couplings which seal to the pipes with rubber gaskets (snap joints) then became available. Apart from permitting longitudinal movement, these obviated both the need for machining the ends of cut pipes and the risk of splitting the couplings by over-driving (Figure 4.13).

Figure 4.13 Laying pitch fibre pipes in a common trench in 1960. The fittings are of cast iron

The withdrawal of production facilities was a commercial decision related to the availability of pitch and the fact that pipes in plastics were coming into production. There was very little wrong with the original product providing they were properly laid. Many installations are still giving excellent service.

Asbestos cement

Asbestos cement pipes, manufactured in diameters from 100–2500 mm and in lengths from 3–5 m were provided with flexible joints. They are occasionally to be found in existing drainage systems, but have the same disadvantages as concrete pipes. Asbestos cement pipes and their fittings are covered by BS 3656[235].

Concrete and fibre cement

Concrete pipes are suitable for use with normal effluent, but may be attacked by acids or sulfates in the effluent or in the surrounding groundwater. Concrete pipes are used mainly for sizes of 150 mm diameter and upwards; with these sizes, external wrappings of glass polyester laminates are available which reinforce the pipes and protect them from external attack. Concrete pipes are supplied either reinforced or unreinforced. Joints are usually either the spigot and socket type fitted with push-fit flexible rubber 'O' ring seals or are ogee joints for use with cement mortar or proprietary bitumen based compounds. Cement mortar joints are very unlikely to remain watertight and should be used only where some leakage into or out of the drain is acceptable (eg for surface water). Similarly, although using bitumen based compounds will provide some flexibility, it is difficult to guarantee the watertightness of this type of joint.

There are very many different diameters and lengths of precast concrete pipes ranging from 150 mm up to 1.8 m and more, and with many different wall thicknesses. Shallow bends are featured in the Standards, and various quality aspects such as hydraulic pressure, water absorption and crushing tests are described. Flexible joints may be of sliding or rolling ring patterns, with suitable constrictions in the sockets and spigots to prevent displacement of the ring.

Relevant British Standards include BS 5911-1[236] and BS 5911-2[237].

European Standards are now available, and include EN 588[238].

Vitrified and fired clay

Most drains constructed up until the 1950s, as already noted, will be fired clay, salt glazed until environmental considerations reduced the popularity of introducing dry salt glazing directly into the firing kilns. These practices produced vast amounts of pollution as smoke and fumes. Most of the pipes produced were of 4 in and 6 in diameters, and lengths were limited to 2 or 3 ft by the risk of distortion during firing. Jointing was by caulking with hemp yarn and forcing Portland cement based mortar into the socket. If the hemp was not fully inserted, there was a risk of mortar protruding into the pipe, which explains the old procedure of dragging a ball of sacking through the pipe immediately after laying and before testing. The joints between the pipes were obviously rigid and therefore extremely susceptible to damage from ground movement. Therefore it is rare to find a drainage system dating back to, say, Victorian times which does not contain numerous displaced or offset joints. In many cases, these defects will have occurred soon after construction as a result of the settlement of the building and the material used to backfill the drainage trenches.

During the 1960s and 1970s practice changed considerably, with the introduction of fired clay pipes which were no longer glazed. These pipes have proved to be no less suitable for a given usage than the older glazed variety. There is no significant difference in hydraulic resistance in service; the modern pipe with its greater length and more accurate jointing may often prove superior. Later manufacturing processes have also considerably reduced levels of pollution.

The performance of modern clay pipes has been vastly improved by the introduction of flexible 'O' ring seals and plastics couplings. Clay pipes remain popular because of their resistance to attack by a wide range of substances, both acid and alkaline. They are available with nominal bores of 100–800 mm, in lengths of 0.3–2.5 m and, depending on size, in three strength classes: standard, extra and super.

BS 65[239] now quotes a series of 12 preferred diameters (from 90–1000 mm) and 6 non-preferred. Lengths are not defined in the Standard, although deviations specified on straightness give practical limits on lengths.

European Standards are now available, and include the various parts of BS EN 295-1[240], BS EN 295-2[241], EN 295-3[242], EN 295-6[243] and EN 295-7[244].

Cast iron

Iron pipes were traditionally made from crude cast iron. However, ductile iron pipes, made from a modified form of cast iron, are now also available. Although iron pipes are not extensively used for drainage underground, their high strength makes them suitable for applications where special protection would is needed. Ductile iron pipes have the added advantage of being better able to withstand impact and distortion, making them particularly useful where pipes have to be laid at a very shallow depth under a road. Iron pipes can be supplied with flexible joints or with spigot and socket connections for caulking with lead or a proprietary jointing compound. The coating on these pipes gives good protection against corrosion and a reasonable life with average ground conditions and normal effluent, although care is needed during handling. Cast iron pipes are made in varying lengths, but the most commonly used length is around 4 m. Cast iron pipes and their fittings are covered by BS 4622[245] though this is now obsolescent; ductile iron pipes and their fittings are covered by BS 4772[246].

In addition to using cast iron for the drains themselves, the material, or cast steel, is used for manhole covers and gully tops. The British Standard covers design requirements, type testing, marking and quality control. Apertures for drainage purposes are defined, together with loads.

BS 497-1[247] and EN 598[248] are also relevant.

Main performance requirements and defects

Most of the drainage installations for existing buildings will have been constructed in accordance with the Model Building Byelaws dating from the 1930s and with the various editions of the BS Code of practice 301 and its successor, BS 8301[26]. These have now been superseded by BS EN 752-4 covering hydraulic design and environmental considerations.

Strength and stability

Cast iron pipes to BS 1211[249] and fittings to BS 78-2[250] used to be supplied with flexible joints which were effective in difficult, waterlogged ground conditions or where large movements were expected; they may still be found. Their strength was sufficiently high to permit their being laid at fairly shallow depths if reasonable care was taken in bedding.

Ductile iron pipes have a very much higher resistance to impact than cast iron, and can be struck and dented without fracturing. This makes them particularly useful in situations where there is risk of severe impact loadings; for example, at exceptionally shallow cover depth below a road.

Manufacturers used to suggest that pitch fibre pipes could be laid under new concrete roads at a depth allowing at least 150 mm between the top of the pipe and the underside of the concrete, the space between to be filled with small granular material. Under flexible pavements such as macadam, it was recommended that the pipes were at least 150 mm below the road formation level or 600 mm below the road surface, whichever was the greater. At shallower depths the pipes should have been encased in concrete and reinforced if necessary. Where the pipes were not used under roads and had less than about 500 mm cover, they should have been protected from mechanical damage by placing concrete slabs over them, with at least 75 mm of granular material thick between pipes and slabs.

Loads on pipes

The load on a pipeline depends on the diameter of the pipe, the depth at which it is laid, the trench width, the traffic, foundation or other superimposed loading, and the prevailing site conditions. The loading on the pipe itself is also influenced greatly by the type of backfill placed under and around the pipe. The choice ranges from placing the pipe directly onto the soil to total encasement in concrete. Flexible pipes are particularly dependent on the backfill material to prevent excessive distortion and, for rigid pipes, there is often a choice between using a high strength pipe on a weak bedding or a weaker pipe on a stronger bedding. The choice is usually determined by overall economy, although site conditions and material availability may also have to be considered. Especially with small jobs, it may be preferable to choose a pipe requiring minimum bedding requirements rather than have to provide the supervision needed to ensure compliance with a more strenuous bedding specification. In all cases it is important to ensure that the pipe is not laid on blocks or other supports that will create hard spots.

Design of bedding for rigid pipes

For rigid pipes there are five options for bedding which, in order of increasing levels of protection, are:
- Class D bedding – laying pipe directly onto bottom of trench
- Class F or Class N bedding – flat granular layer
- Class B bedding – granular support
- Class S bedding – granular surround
- Class A bedding – concrete cradle

Classes of bedding F, N, B and S are listed in Table 4.2.

A Class D bedding is suitable only in soil conditions where it is possible to trim the bottom of the trench accurately using a shovel (eg firm clays). It should not be used in soft silts and clays, soils containing large stones or flints, or in dense sands and gravels where a pick is required for excavation. The trench should be excavated to a depth slightly less than that required, so that hand trimming brings the bottom of the trench to the correct level. Over-dug places should be levelled using well compacted fill. Holes are formed in the trench bottom to give a minimum clearance of 50 mm between the sockets and the soil. Particular care is needed where the pipe length exceeds eight pipe

Table 4.2 Recommended aggregates for bedding rigid pipes

Pipe bore (mm)	Aggregate type	Nominal maximum aggregate size (mm): BS 882 grading[251] Class of bedding		
		S or B	F	N
100	Single sized	10	10	All-in aggregate with a max particle size of 10 mm, or coarse, medium or fine aggregate to BS 882 or 1201
	Graded	Not acceptable	Not acceptable	
50	Single sized	10 or 14	10 or 14	All-in aggregate with a max particle size of 14 mm, or coarse, medium or fine aggregate to BS 882 or 1201
	Graded	14–5	Not acceptable	
225	Single sized	10, 14 or 20	10, 14 or 20	All-in aggregate with a max particle size of 20 mm, or coarse, medium or fine aggregate to BS 882 or 1201
	Graded	14–5 or 20–5	Not acceptable	
300	Single sized	10, 14 or 20	10, 14 or 20	All-in aggregate with a max particle size of 10 mm, or coarse, medium or fine aggregate to BS 882 or 1201
	Graded	14–5 or 20–5	Not acceptable	

diameters to avoid uneven support along the length of the pipe.

For a Class F or Class N bedding, where accurate hand trimming is not possible, the trench should be excavated below the pipe invert level to allow a minimum thickness of 100 mm of well compacted granular bedding material to be laid on the bottom of the trench. For maximum protection (for a Class F bedding), a single size aggregate complying with the requirements of BS 882[251] should be used. For a Class N bedding, other fill materials can be substituted.

With a Class B bedding, for protection by greater depths of cover, the granular fill should be extended to the full depth of the pipe as shown in (Figure 4.14). The trench bottom should always be cut square, even if a vee-shaped trench has been excavated. The fill should be a single sized or graded aggregate; alternatively, a fill material with a compaction factor of less than 0.3 can be specified. Materials with compaction factors between 0.15 and 0.30 may also be used but require progressively more care during placement and compaction. This is to ensure reasonably uniform support along the length of pipes, particularly for pipes with length to diameter ratios of more than eight.

As a Class S bedding requires greater amounts of granular material, it is generally reserved for situations were the excavated material is unsuitable for backfilling around the pipe without careful selection. A fill material with a compaction factor of less than 0.15 may be specified as an alternative to the BS 882 materials.

Class A bedding is used where maximum strength is needed. It has a number of other advantages:
- minimising post-placement movement of the pipe, thereby allowing close control of the gradient
- allowing the trench sheeting to remain in place until after the bedding is complete minimising disturbance to the pipeline due to future excavations
- minimising the risk of subsidence to adjacent foundations

Pipes should be supported clear of the trench bottom by placing small blocks or cradles under each pipe, after which concrete is placed under the barrel of each pipe to give continuous support. Approved Document H1 of the Building Regulations[189] recommends that any concrete bed and surround shall not be less than 150 mm thick, and the bed or haunch shall extend at least 150 mm on each side of the pipe. A layer of fill 150 mm thick may be placed over the pipes for protection, but otherwise the pipeline should not be subjected to any additional load or vibration for 24 hours to allow the concrete to achieve adequate strength. Heavy rammers and traffic loadings should not be allowed within a metre or so of the backfilled trench for at least 72 hours, or longer in cold weather. Where the pipes are flexibly jointed, flexible movement joints must be provided through the full thickness of the concrete at each pipe joint, using a compressible material such as expanded polystyrene or impregnated fibre building board.

Design of bedding for flexible pipes

To provide adequate protection for flexible pipes, the drain must be fully surrounded by either selected granular material or, for maximum protection, concrete. The requirements of the granular material are given in Table 4.3.

The preferred granular bedding details for pipes in stable soils are shown in Figure 4.14. The pipe should be laid on a compacted layer of the selected material to provide a minimum clearance of 100 mm between the underside of the sockets and the bottom of the trench. The trench should be kept as narrow as practicable, but must provide a clearance of at least 150 mm on both sides of the pipe to allow the fill to be compacted. The selected fill should be used at least until the pipe is completely covered. Where the main backfill contains stones large than 40 mm or the trench is more than 2 m deep in poor ground, the

Table 4.3 Recommended aggregates for bedding flexible pipes

Pipe bore (mm)	Aggregate type		Alternative
	Single sized (BS 882)	Graded (BS 882)	
100	10	Not acceptable	Fill material with a max particle size of 10 mm and a compaction fraction not exceeding 0.30*
50	10 or 14	14–5	Fill material with a max particle size of 14 mm and a compaction fraction not exceeding 0.30*
225	10, 14 or 20	20–5 or 14–5	Fill material with a max particle size of 20 mm and a compaction fraction not exceeding 0.30*
300	10, 14 or 20	20–5 or 14–5	Fill material with a max particle size of 20 mm and a compaction fraction not exceeding 0.30*

* Compaction fraction for fill material used under roads where the depth of cover is less than 1.2 m should not exceed 0.15.

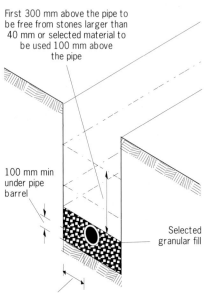

Figure 4.14 Class B bedding for a pipe should extend to the full depth of the pipe. To protect the top of the pipe, there must be no stones larger than 40 mm in the backfilling. If this is not practicable, the selected bedding should cover the pipe to a depth of at least 100 mm

4.2 Wastewater drainage

Table 4.4 Limits of cover for standard strength rigid pipes in any width of trench

Pipe bore (mm)	Bedding class	Field and gardens Min (m)	Field and gardens Max (m)	Light traffic loads Min (m)	Light traffic loads Max (m)	Heavy traffic loads Min (m)	Heavy traffic loads Max (m)
100	D or N	0.4	4.2	0.7	4.1	0.7	3.7
	F	0.3	5.8	0.5	5.8	0.5	5.5
	B	0.3	7.4	0.4	7.4	0.4	7.2
150	D or N	0.6	2.7	1.1	2.5	N/a	N/a
	F	0.6	3.9	0.7	3.8	0.7	3.3
	B	0.6	5.0	0.6	5.0	0.6	4.6

Table 4.5 Recommended cover depths and beddings for flexible pipes

Situation	Cover depth (below final surface in the case of roads) Min (m)	Max (m)	Bedding material
Fields and gardens	0.6	6.0	As specified in Table 4.3
		0.6	Where necessary, pipes with less than 0.6 m of cover should be protected by concrete paving slabs laid on at least 75 mm of granular or other flexible filling
Under roads	1.2	6.0	As specified in Table 4.3
	0.9	1.2	Single sized aggregate specified in Table 4.3 or fill with compaction fraction not exceeding 0.15

selected material should be extended upwards to provide at least 100 mm of cover for the pipe. The fill material should be placed and compacted equally on both sides of the pipe using hand rammers or light plate vibrators. Compaction may be unnecessary when using single sized aggregate, or when the compaction value of the granular material does not exceed 0.15.

In less stable soils (eg soft clays, silts or fine sands) and especially where there is a high water table, it may be necessary to increase the thickness of the bedding and surround by a factor of 2. In exceptionally soft ground with a high water table or running sands, expert advice should be sought.

Where providing protection from traffic loadings is needed, the pipe can be surrounded in a minimum thickness of 150 mm of lightly reinforced concrete. The concrete should have a minimum 28-day strength of 20 MN/m^2; or the pipe can be protected by placing a reinforced concrete slab over granular surround.

The trench will normally be backfilled using the excavated material. However, where the excavated material is difficult to compact and it is desirable to minimise settlement (eg where a road is to be constructed over the line of the trench), it may be preferable to import a more suitable material. When using flexible pipes, the first 300 mm of backfill should be selected to be free of stones larger than 40 mm, unless the selected fill material has been extended to provide 100 mm of cover.

In all cases the backfill material should be spread in layers no more than 300 mm thick and thoroughly compacted before placing the next layer. Heavy compactors should not be used until there is a minimum cover of 600 mm over the top of the pipe. Trench sheeting should be removed as the work proceeds, to avoid leaving unfilled voids at the sides of the trench.

Minimum and maximum depths of cover

Limits on the depth of cover for standard strength, rigid pipes laid under fields and roads are given in Table 4.4. Limits on the depth of cover for flexible pipes are given in Table 4.5.

Special precautions for laying pipes near foundations

To minimise the settlement of existing building foundations as a result of poor consolidation of backfill material, Approved Document H1 requires a pipe trench within 1 m of loadbearing wall foundations to be filled with concrete to the level of the underside of the foundations. The criteria were shown earlier in Chapter 2.1.

A drain is permitted to pass under a building provided it is surrounded by at least 100 mm of granular or other flexible filling, although additional pipes with flexible joints at both ends may be needed on sites prone to excessive settlement (so-called rocker pipes). Concrete encasement should be provided where the crown of the pipe is within 300 mm of the underside of the floor slab and this encasement should be integral with the slab.

Where a drain passes through a wall or foundations, it is necessary to provide either:
- an opening giving at least 50 mm clearance all round
- a short length of pipe built into the wall with its joints no more than 150 mm from the face of the wall or foundations and connected on each side to short lengths of pipe (ie less than 600 mm long) fitted with flexible joints

In both cases the point of entry will need to be packed to make it gastight.

Special protection against traffic loads

The less robust and more flexible drainage materials must be protected when laid near to a surface which is subject to traffic loads. This was usually assumed to be drains with inverts within 1.2 m of ground level; whereas below gardens and pavements, 0.9 m usually sufficed[26]. Although ductile iron was classified as flexible, it was usually satisfactory without surrounding in concrete.

Approved Document H1 prescribes special protective measures against ground loads.

- Where rigid pipes have less than the recommended cover in Table 4.4, they should be surrounded with concrete to a thickness of at least 100 mm; movement joints in the concrete surround formed using compressible material should be provided at each flexible pipe connection
- Where flexible pipes are not under a road (but nevertheless require some protection) and have less than 0.6 m of cover, they should be protected by concrete paving slabs laid as bridging above the pipes and by at least 75 mm of granular or other flexible filling
- Where flexible pipes are under a road and have less than 0.9 m of cover, reinforced concrete bridging should be used instead of paving slabs, or the pipes be surrounded in concrete

Hydraulic design

Hydraulic design for drains and sewers is covered in BS EN 752-4, and is not dealt with here except to record that there are a number of parameters which need to be considered, including:

- flow rates (range and peak)
- velocity of flow (including self-cleansing velocity)
- viscosity of fluid
- pipe diameter
- pipe roughness
- headloss coefficients caused by manholes and bends
- pipe length

Sometimes, flow simulation methods of design are used. See BS EN 752-4.

Gradients

Although it has been customary in the past to lay drains to minimum and maximum gradients to achieve self-cleansing velocity of flows carrying suspended solids, in practice a wide variety of gradients not meeting established requirements will perform well. On steeply sloping sites in certain areas of the country it has been customary for a number of years to lay drains at very steep gradients (eg 1 in 7); however there has been little evidence of erosion of systems or of undue numbers of blockages.

However, blockages have occurred. In the 1970s BRE Advisory Service survey described in Chapter 0, most blockages occurred in 100 and 150 mm diameter drains laid at gradients between 1 in 100 and 1 in 30, the range which was commonly recommended at the time, with peak flows of depths of between one quarter and one half of the pipe diameters. In a substantial sample of clayware drains, of those which had become blocked 4% occurred in gradients of between 1 in 500 and 1 in 71, 11% between 1 in 70 and 1 in 33, and 3% between 1 in 32 and 1 in 20. One 100 mm drain had been laid at a gradient of less than 1 in 1200; when observed, the drain was working well and there was no history of blockages, though the occasional WC flush seemed to assist in keeping it clear[252].

Building Regulations now recommend laying foul drains at minimum gradients of 1 in 40 for 75 and 100 mm diameters where peak flows are less than 1 l/s, and 1 in 80 where greater than 1 l/s. Where a minimum of five WCs are connected, a minimum gradient of 1 in 150 is recommended.

Unwanted side effects

Pipes have traditionally been bedded in natural or crushed rock aggregates of various grades. These are now increasingly in short supply, and the possibility of using recycled crushed demolition waste has been suggested. Recycled bedding material should be compatible with the pipes; for example, concrete pipes could be attacked by sulfates in plasters and mortars, whereas this risk is not present with fired clay pipes. Crushed waste consisting entirely of old brickwork contains too high a proportion of mortar and should not be used for this purpose[253]. See also BRE Digest 433[254].

Commissioning and performance testing

Blockages will occur in the best designed and constructed underground drainage systems, if only because of misuse. All parts of the drainage system should, therefore, be cleanable and access to the system should be provided to allow this cleaning to be carried out efficiently and economically[255]. It is important to provide alternative means of disposal for bulky items such as disposable nappies and sanitary towels.

Under the Model Building Byelaws, drains and private sewers were required to be:

- of adequate strength
- properly supported and protected
- capable of withstanding a reasonable hydraulic, smoke or air test under suitable pressure
- of adequate size
- of cast iron when passing through a building
- when laid under a building be of cast iron or protected with a surround of concrete not less than 6 inches thick
- provided with adequate means of access

In more recent years, building control officers have normally required drains to be tested to demonstrate conformance with the Building Regulations, though this is usually limited to a hydraulic test.

Although some local authority building byelaws required all manholes to be watertight, this would rarely have been achieved in practice. It is quite unrealistic to expect unrendered brick manholes full of water to be watertight, although clearly there would be some variability in both materials and workmanship. In most cases there will be a decreasing leakage rate with time as the joints or pores in the bricks silt up. In a test programme carried out by British Ceramic Research Association in 1986, one brick thick manholes constructed on sealed bases in engineering brickwork in a variety of bonds demonstrated that most leaks occurred at header joints and around mortar seals between brickwork and entry and exit pipes[256].

Retrospective testing of foul drains and sewers where there are reasonable grounds for suspecting that they may be defective, is possible under powers conferred on local authorities by Section 48 of the Public Health Act 1936[257]. Sewerage undertakers may also examine or test any drain or private sewer connecting with a public sewer under Section 114 of the Water Industry Act 1991[258]. Remedial work by the building owner may be required to be effected under Section 59 of the Building Act 1984[259].

Health and safety

There are four main aspects to consider from the point of view of health and safety:
- surcharging, specifically when it results in surface flooding
- pollution
- septicity
- vermin

Surcharging

Surcharging occurs when the drainage requirements exceed the capacity of the system, either through inadequate design or by accidental blockage, or, in combined systems, where periods of intense rainfall occur. Water levels in these cases fill the pipes and rise up the manholes. Where the surcharging overtops the manholes, this causes surface flooding which clearly is undesirable. Surcharging can sometimes have unfortunate consequences if any overflows occur in the immediate proximity to buildings. It may sometimes be necessary to incorporate non-return valves or anti-flooding valves in the system.

Surcharging is most common where basements contain water appliances. In circumstances where the risk is high, pumping the wastewater to levels well above the sewer connection may be necessary.

Pollution

The main criterion is that any discharge to an outfall should be that the water in the receiving watercourse or river or land infiltration system will not exceed its self-purifying capacity.

Septicity

All work on foul drains and sewers carries with it risks to health. One of the main problems is that of production of hydrogen sulfide under certain conditions. Hydrogen sulfide is toxic and potentially lethal, and can seriously interfere with sewage treatment. Additionally, it can oxidise to sulfuric acid, which will attack some materials used in the construction of drains. A watch needs to be kept, *inter alia*, on temperatures, biochemical oxygen demand, sulfate availability, acidity, ventilation, flow velocity and turbulence. Corrective measures may have to be taken (EN 752-4).

Vermin

Early in the 1990s, a postal survey of the 392 local authorities was conducted by BRE. Almost half of the authorities that replied reported a problem with rats in drains. The distribution of problems across the country followed no discernible pattern. In fact, the reasons for perceiving a rat problem varied greatly. In theory, sealed drainage systems should be effective in deterring rats from infesting drains. However the survey was unable to confirm this due to the unexpectedly low reported use of these systems in England and Wales. The survey showed that the traditional rat deterrent, the interceptor trap, is used extensively throughout the country (Figure 4.15). Due, though, to frequent problems with blockages, many local authorities are systematically removing traps that block persistently. A frequent cause of trap blockage is the rodding eye stopper falling into the trap. From the outcome of the survey it emerged that radical changes must be made in the design of the traditional interceptor trap to reduce the problem of rats.

The main requirements to limit rodent infestation are:
- systematic sewer baiting
- regular maintenance
- improved rodding eye stopper design
- improved trap design
- ventilation

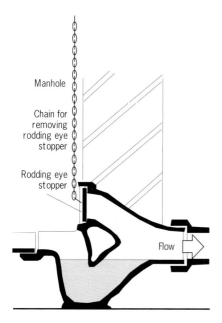

Figure 4.15 A clayware interceptor trap

It may be the case that deterring rats from entering houses through the drainage systems by means of a completely sealed drainage system, without manholes, will prove to be as effective as an interceptor trap but without the maintenance requirements[260].

Redundant drains and sewers

Vermin may be dispersed as a result of demolition or renovation. It is essential, then, that a property is disinfested before it is vacated. Demolition should follow immediately to avoid reinfestation. The remaining drains and sewers should be sealed to prevent access for rats.

Disused drains, sewers and cesspools should, as far as possible, be excavated and removed, the holes being filled with well consolidated hard rubble or the pipes filled with concrete which has been well rammed home. The open ends of land drains should be connected to an alternative means of drainage which takes them away from the site and leaves no possibility of access to a building.

During construction and connection of new or replacement drains, sewers should be kept closed as much as possible. Manhole and inspection covers should be removed only when work is in progress. If connection cannot be completed in a single day, the pipe ends should be temporarily stopped, and manhole and inspection covers replaced.

> **Interceptor traps**
>
> Interceptor traps, also known as intercepting traps, interceptors and disconnecting traps, are fitted between sewers and drains to deter sewer rats from entering domestic drainage systems. Six different designs of interceptor trap were compared dimensionally and hydraulically to determine which had a better performance in allowing 'stoppages' to pass through. The investigation determined some of the important parameters in interceptor trap sump design and revealed that, in terms of sump clearance properties, cast iron traps perform better than clayware traps. However, one cast iron trap costs the equivalent of about four clayware traps.

Durability

Drains may leak as a result of joint failure, ground movements or crushing from heavy vehicles. Inspection chambers built in a half brick skin or of non-durable mortar or bricks are unsatisfactory or may be porous, due to failure of internal rendering or degrading of unsuitable bricks; this may result in penetration of groundwater and, possibly, leakage of effluent into the ground. Where drains crack, tree roots can seek out water and can grow into the pipe and eventually block it.

Chambers or gullies can be damaged by impact and metal components may be corroded. Benching or rendering can break away and cause blockages or create difficulty of access.

Soakaways may silt up and overflow or, if built in heavy clay, never have functioned correctly. Soakaways built too close to buildings may increase the risk of rising damp. Since soakaways deal for the most part with rainwater or treated wastewater, they are included in Chapter 4.3.

Problems with drainage systems may be identifiable from maintenance records (if available) or from discussions with occupants. An accurate plan of the routes of all drains and access points, together with pipe sizes, will be found useful. All access points and inspection chambers should be checked for good flow characteristics, and visible parts of systems examined for cracks or other signs of deterioration. The possible effects of alterations to a system should be considered. If a more detailed investigation is required this may entail a levels survey of inverts, an air, smoke or water test, or a remote camera survey.

Clay pipes

Clay pipes are resistant to attack by a wide range of substances, both acid and alkaline, but certain substances could attack the joints; in these cases special jointing cements are recommended.

Provided there is no settlement and interference from tree roots in the vicinity, drains laid in glazed earthenware, and the manholes which connect them, ought to last 100 years.

Concrete pipes

Concrete pipes are suitable for use with normal effluents, but concrete may be attacked by acids or sulfates in the effluent or in the surrounding soil or groundwater. Where exposure to such conditions is likely, advice should be sought. Particular care is needed on sites where dumping has taken place as potentially injurious materials may be scattered within the soil. Pipes made with sulfate resisting cement give increased resistance to sulfate attack but not necessarily to acid attack. Various proprietary coatings are also available for internal or external application; for example, pipes of 225 mm diameter and upwards are available with an external wrapping of glass fibre laminate which reinforces the pipes and protects them from external attack.

Pipes made to the former BS 4101[261], now withdrawn, had ogee ends which were usually jointed with cement mortar. These joints were unlikely to remain watertight and their use should have been confined to drains carrying surface water only. From the 1980s, BS 4101 was replaced by various parts of BS 5911[206].

4.2 Wastewater drainage

Asbestos cement pipes
Substances that attack concrete may also attack asbestos cement and, where exposure to these substances has occurred, some deterioration can be expected. Investigations into the sulfate resistance of asbestos cement pipes have found that once they are exposed to an aggressive sulfate environment their fate is no different from that of any other cementitious product.

Some data from experiments in the USA indicates that, within six years of installation, normally cured pipes had deteriorated sufficiently to allow leaks and ruptures. After 10 years, normally cured samples of pipe had been completely destroyed while autoclaved samples showed no signs of deterioration[77].

Bitumen dipped pipes have been available, the coating giving increased resistance to attack, both internally and externally. The pipes were normally supplied with flexible joints and were available in one to three strength classes, according to size.

Pitch fibre and pitch-impregnated asbestos fibre pipes
Pitch fibre pipes first came into use in the USA in the early twentieth century, but were not generally available in the UK until they were manufactured here some decades later.

Pitch fibre pipes and fittings were deemed to be suitable for use with normal domestic effluents and with most trade wastes containing acids, alkalis or sulfates. Pitch fibre, pitch-impregnated asbestos fibre and polypropylene fittings to BS 2760-2 were considered unlikely to be affected by occasional releases of small quantities of oils, fats or organic solvents.

The cement in pitch-impregnated asbestos fibre fittings can be attacked by acids or sulfates but the pitch or bitumen impregnation was thought to provide good protection against this form of attack. The pipes disappeared from the market for commercial and not technical reasons when UK manufacture ended.

Cast iron pipes
Where normal effluents are carried, the coating on cast iron pipes gives good protection against corrosion and reasonable life in average ground conditions. Avoiding damage to the coating during handling is essential and any damage should be made good as soon as possible. Acid conditions, either in the effluent or in the surrounding soil (eg peaty soil), increase the rate of attack through any weaknesses in the coating. In some soils, sulfate reducing bacteria may accelerate corrosion. Certain vegetable oils, if carried in quantity in the effluent, could damage the inside coating.

The corrosion resistance of ductile and spun iron is similar to that of cast iron.

Plastics pipes
PVC-U resists attack by normal domestic effluents and a wide range of acids, alkalis and sulfates. It can be affected by certain organic solvents but occasional exposure to small quantities of these substances is unlikely to have significant effects. Where larger quantities are to be released (eg from industrial processes or laboratories) advice should be sought from the pipe manufacturer.

Because the wall stiffness of thermoplastics pipes decreases with an increase in temperature, the effluent temperature to which they will be exposed must be considered. Although discharges approaching 100 °C at the point of release can occur from domestic systems, the total quantity released at any one time does not usually exceed about 40 litres with a discharge rate of about 0.4 l/s. As plastics are poor conductors of heat, short duration discharges of this kind do not raise the temperature at the outer surface of the pipe to the same extent as would more prolonged discharges.

Push-fit joints are available for PVC-U pipes either as loose couplings or integral sockets, providing some telescopic flexibility. Solvent welded joints do not permit axial movement, but small changes of length caused by thermal movement can usually be accommodated by axial tensile or compressive strain in the pipe material.

In the mid-1990s, BRE investigators carried out a study of the factors affecting the durability of plastics pipes, especially in buried drainage applications. Previous investigations into the durability of buried plastics pipes showed that most have concentrated primarily on pressure pipes rather than drainage pipes. The later investigation found that there was no evidence of serious deterioration of pipe polymers in uncontaminated soil although studies continue.

Polyvinylchloride (PVC-U) is degraded by the action of oxidising acids. Nitric acid is particularly aggressive towards PVC-U, causing the surface to become brittle. On the other hand, PVC-U does not deteriorate under the action of neutral and alkaline solutions. The presence of oxidising salts may cause deterioration, but in most cases this is not a problem as the rate of deterioration is slow.

The resistance of PVC-U towards hydrocarbons is generally as good as, or better than, that of polyethylene. Organic compounds such as chlorinated hydrocarbons, anilines, ketones and nitrobenzenes cause PVC-U to swell and eventually to dissolve. Chemicals that do not cause PVC-U to soften do not permeate PVC-U; therefore no significant permeation occurs for alcohols, organic acids or aliphatic hydrocarbons[77].

PVC-U is not biodegradable, and soil burial tests have found no evidence of microbial attack after eight years. In contrast, plasticised PVC (PVC-P) is highly susceptible to microbial degradation.

PVC is essentially impermeable to alcohols, aliphatic hydrocarbons and organic acids.

Manhole covers

Traditionally, manhole covers were of cast iron, and these have proved to be remarkably durable provided they were not damaged by impacts. On the other hand, steel covers, while satisfactory for a time (Figure 4.16) are prone to corrosion (Figure 4.17).

Joints

Since continued watertightness of pipe joints depends on the correct choice of jointing materials, it is necessary to consider the type of effluent that is to be carried. In many cases in old installations the seal was made with a rubber gasket, either natural or synthetic.

Although these jointing products were suitable for use with normal domestic effluents, each had different properties that affected its suitability for use with a particular kind of trade waste such as oils, fats, organic solvents and copper salts. This was especially important for pipes carrying effluent that needed to be treated before the drainage authority would allow it to be discharged into the public sewer. Some joints for clay pipes are of plastics materials: much the same considerations apply to these and the manufacturers should be consulted about their use for trade wastes. Modern polymer jointing materials are on the whole much more durable than their predecessors.

Maintenance

Experience from 200-plus TV drain surveys and other drain studies has made it clear that it is unusual for domestic drains to be in perfect condition, especially if constructed by traditional methods and materials. Jointing is often far from perfect (eg with out-of-line pipework and infiltration of surface water). Defects in installation do not necessarily adversely affect drain performance; stepping at joints can be in the direction of flow, thereby just presenting a drop rather than a barrier to the effluent, and surface water may help to keep the drain clear. Back-falls or negative gradients due to pipe movement are common too but, apart from sometimes causing some buildup of fats and slime at the stagnant water area, do not appear to be a prime cause of drain blockage. However, if waste material which has accumulated in a 'dead zone' made by a backfall is allowed to dry out by lack of use of the drain, a blockage can form.

In all cases, external drainage systems should be accessible without the need to enter buildings to undertake rodding or periodic cleaning.

There may, though, be inaccessible lengths of drain with no inspection chambers at changes in direction. The positions of access points may be inconvenient or obstructed, or they may be lost or built over. Other design defects also include excessive or inadequate falls to drains, waste or rainwater pipes discharging above gully grids and 'flood level' fittings located inside a building.

Blockages in drains can result from bad installation practices; for example:
- branches joining the main run at narrow angles or against the flow direction
- small-radius bends at the bases of stacks, or elsewhere on a branch
- poor alignment of pipes and fittings resulting in badly swept direction changes

Figure 4.16 A 10 year old steel manhole cover

Figure 4.17 A 20 year old steel manhole cover which has suffered serious corrosion

The need for maintenance-free underground drains has long been recognised. Glazed drain pipes were one of the first building components to have a British Standard, and local authorities have always been responsible for testing drains during construction. Maintenance tends to be responsive rather than routine as drainage systems generally function for long periods without attention. Routine maintenance that is carried out is usually limited to checking open drains and inspection chambers. There are various opinions on how frequently this should be, but one suggestion is that it might usefully be undertaken every year or two. Grease traps and light liquids interceptors will need to be inspected and cleaned at least annually. Desludging wastewater treatment plants such as septic tanks will depend on usage.

For the purpose of future maintenance it is important to assess whether an existing drainage system can be rodded to clear blockages. This will require chambers or rodding eyes at reasonable intervals, located where there is sufficient space to work and where the resulting debris can be easily removed.

Simplification of future maintenance should be borne in mind when undertaking drainage improvement work. Improvements might include:
- replacing open gullies with sealed ones
- adding separate stormwater systems (where currently combined systems exist)
- re-routeing drains where they run (or may run) under buildings
- re-routeing drains to provide more direct routes to public sewers (for example to avoid traversing land of other properties)
- in combined systems, diverting surface water flows to soakaways or adjacent watercourses to reduce the load on the drains

Where settlement, seasonal ground movement or tree roots are known to be a problem, replacement of rigid drains with a modern flexible system may be worthwhile.

As already described, a study of the incidence and causes of blockages in some hundreds of 100 mm and 150 mm underground drainage systems was carried out by BRE investigators in the mid-1970s. Drains with a history of recurring blockage problems were examined with CCTV survey equipment, and other relevant design information collected. All but one of these drains showed specific physical defects in pipes, pipe joints or manholes. The performance of systems with few obvious physical defects was also studied to obtain data on the frequency of blockage in drains laid at different gradients. This work did not indicate that blockages were significantly more likely in flat gradient pipes than in the more widely used gradients of 1 in 40 to 1 in 70[252,255].

Manholes may need to be inspected for defects or buildup of fats and detritus. Covers as well as interiors will need to be cleaned and, if covers are to be airtight, the rebates re-greased. It is particularly important to avoid builder's rubble entering the manhole when properties are refurbished and the drains are to be reused (Figure 4.18).

When a private drain joins a public sewer, the owner of the drain is normally responsible for its maintenance and repair up to the point of connection to the sewer and not, as often supposed, only up to the building line or site boundary. Where the connection is by means of a junction, the owner's responsibility ends with the second pipe length from the junction. If a saddle connection is made, the owner's responsibility includes pipes up to the saddle and the saddle itself.

Cesspools and septic tanks
Regular emptying of cesspools has already been referred to. Maintenance frequency of septic tanks depends on many different factors, including design capacity, the volume of wastewater, the constituents of the wastewater and any substances that could destroy the bacteria which break down the effluent. Most require annual removal of solids. However, some systems have operated without such attention for 10 or more years.

Figure 4.18 Builders' rubble entering an uncovered manhole during refurbishment work

Work on site

Access, safety etc
At one time, rodding eyes could not be used as the sole means of access to a drain, but were used in conjunction with inspection chambers or manholes to provide access for the removal of silt or other debris[255]. However, since the tightening of the legislation on working in confined spaces, this rule no longer applies.

Storage and handling of materials
Storage and handling of plastics pipes has a crucial bearing on their performance. Not only can they become distorted if inadequately supported when stacked, but long exposure of some materials to the elements can affect long term performance[262].

Clayware pipes, while not subject to bending and distortion in the same way as plastics pipes, may easily fracture[263].

Workmanship
Workmanship should be in accordance with BS 8000-14[264].

Trenches
Local soft spots in a trench bottom should be stabilised by tamping in granular material. Large hard objects (eg rocks and large stones) and substantial roots should be removed and replaced by tamped granular material. In soft clays, silts or fine sands, it is advisable to spread a 75 mm layer of granular material or 50 mm layer of concrete, according to the type of bedding to be used, immediately after excavation to prevent the surface becoming disturbed and softened by trampling under wet conditions.

Concreting should not be permitted in frosty weather unless special precautions are taken. Partially set concrete must never be used and placing should be completed within 45 minutes of mixing. A 150 mm layer of fill may be placed over the pipe for protection but the pipeline must not be subjected to any further load or vibration until the concrete has acquired sufficient strength. As a general guide, backfilling should not commence until at least 24 hours after completion of concreting, and heavy rammers should not be used nor traffic loads permitted for at least 72 hours, or longer in cold weather.

The soft, relatively unstable conditions that can occur with a trench bottom in softened clays, silts, very fine sands or peat are unfavourable to flexible pipes because the buttressing effect of the sidefill, which normally helps the pipe keep its shape, may be considerably reduced; the effect is aggravated if there is a high water table. Under these conditions considerable care is needed when laying pipes at depths greater than 2–3 m, depending on the severity of the conditions. Where conditions approach those of running sand, expert advice should be sought.

Joints
Flexible joints must be assembled strictly in accordance with the manufacturer's instructions. Cleanliness is essential and if a lubricant is specified, only that recommended by the maker should be used.

Although in principle the flexibility of a joint could be impaired if stones get into the outer part of the annular gap, in practice this is not though to be a hazard, except that where the amount of movement at a joint is likely to be unusually large it would be wise to take special precautions to exclude the soil or bedding material, by a band of wrapping. The crucial point is where a drain crosses underneath the foundations of a building.

Where the pipes are bedded on or surrounded with in situ concrete, an annular gap of about 12 mm (in the length of the pipes) should be formed in the concrete at each pipe joint or the advantage of flexibility will be lost. However, if only a small amount of settlement is anticipated, the gaps can be smaller. The material used to form the gap should be shaped around the spigot and touch the end face of the socket. Care should be taken to prevent the concrete entering the annular gap between spigot and socket. Where increased localised movement is to be expected, for instance where the pipeline leaves a building, gaps should be formed in the concrete at both ends of the first pipe at least. The first joint should be kept close to the building so that the projecting pipe overhangs the face of the building as little as possible. The entry point should be suitably packed to avoid a route for the entry of gas[265,266].

If the site is made ground and the building is on piles, the soil surrounding the drains is likely to settle more than the building; in some other soils the building foundations may settle. In either case careful attention to detail is necessary at the points where drains pass from structure to ground.

Backfilling and protection
Normally the material excavated from the trench will be used for refilling, but if it is of a type difficult to compact and settlement of the surface must be minimised, as for example where a road is to be carried, it may be necessary to import other material. In any case a proper compaction technique should be used; it is essential that the material should be placed and spread in regular layers of not more than about 300 mm thickness and each layer adequately compacted before placing the next. Heavy compactors should not be used until there is at least 600 mm of soil over the pipes. Trench sheeting should be removed as the work proceeds so that the spaces are properly filled.

Although one of the important properties of plastics pipes is their flexibility, there may be some lengths of pipeline that need the protection of concrete even though this nullifies the advantage of flexibility. Some instances are sections of pipeline that pass at shallow depths under roads, situations demanding protection against mechanical damage, and to minimise the effects of settlement where pipes are laid close to and below building foundations or other pipelines.

Where laying pipes at depths less than 1–2 m is unavoidable, the following methods of protection may be considered:
- where the pipes are not under a road, precast concrete slabs can be laid over the pipes with a minimum thickness of 150 mm granular material between the slabs and the top of the pipes
- under roads, the pipes may be completely surrounded with 150 mm minimum thickness of in situ concrete; alternatively, ductile iron pipes could be used.

Rodding and cleaning

Although sewer and drain systems ought to perform well for many years without major disruption, from time to time problems do occur. They may be of two kinds:
- the slow buildup over the years of deposits within the system or the growth of tree roots
- sudden blockages caused by the introduction of unsuitable solids into the system

While the latter category will usually need to be treated by emergency measures, there is a lot to be said for planned regular maintenance to ensure that drains are kept in pristine condition, especially where past experience indicates potential risks. Planned maintenance may also in some circumstances be a condition imposed by the relevant drainage authority, particularly where there are risks to the environment.

Since flexible pitch fibre and plastics are somewhat less robust than traditional pipe materials, clearing blockages should be undertaken carefully. Careless use of mechanically operated machines or of sharp-ended hand-operated implements could cause damage. Suitable rodding equipment is available comprising small-diameter polyethylene tube, with a suitably designed cleaning head. The smoothness of the tube and absence of jointing ferrules reduce the risk of damage to the pipes. Winching a ball through the pipe may be found to be effective, and the use of jetting or of a hydraulic kinetic ram is also possible.

Repairs and rehabilitation

There will come a time when the condition of old sewers and drains no longer allows them to perform acceptably. Local collapses may occur, or there may simply be erosion of old masonry or deterioration of other materials. Repair or rehabilitation may be a possible alternative to renewal, and some general guidance is given in BS 752-5 and BS 8005-5[267].

With excavation

Perhaps the most satisfactory method of repair of both rigid and flexible drains is to determine where the defect is situated and then excavate to invert level and replace like with like. This may be the only practical solution where there has been a fracture and significant displacement of sections of the drain following subsidence or damage by heavy vehicles, or where the drain has been invaded by tree roots. In the past, identification of the sites of defects was usually effected by rodding with a suitable probe head until the obstruction could be felt, but in recent years has been carried out using CCTV cameras (Figure 4.19).

Without excavation (so-called trenchless technology)

Clayware and concrete drains which have cracked, or in which the joints have opened and lost their caulking, have frequently been repaired in the past by pulling a pair of rubber pistons sandwiching a wet cement grout from one manhole to the next; the grout is under pressure from the rear piston to which the pulling cable is attached, and is forced into any cracks or open joints. The system could not be used on branches without manholes.

A better alternative repair technique is to use a water based liquid resin to flood the drain. Since the resin liquid is of lower viscosity than cement grout it penetrates further into the defects and even into the surrounding bedding. The resin gel remaining in the bores is then removed before it fully sets.

Drains may also be found which have been repaired using lengths of flexible polyester resin-saturated lining sleeves which are inserted into the drain and pulled by cable or pushed by hydraulic pressure. The sleeves are held tightly to the inside of the bore by a temporary inflated air bag until the resin has cured. The lining reduces the bore of the pipe, but this will rarely be significant.

Figure 4.19 CCTV inspection of a drain

Where a reduction of bore cannot be contemplated, and the ground and original pipes are suitable, lengths of drain may be 'exploded' by hydraulic pressure pushing out the surrounding soil and allowing the water in the pipe to drain away. A replacement drain can then be pulled through.

Supervision of critical features

During construction or refurbishment, the building site should be kept as clean as possible and, if rat infestation is to be avoided, food scraps should be properly disposed of in covered metal bins. Otherwise rodent populations supported on scraps near the site huts may be the first occupants of the buildings and their drains.

So far as the efficiency of the drain is concerned, crucial points for checking include gradients, access points, jointing, builder's rubble and damage from site machinery[268,269].

Inspection

Observing the flow through an existing drainage system at inspection points may not reveal whether intermittent overflow problems can occur from blockages or surcharging. Theoretical calculations or occupants' comments may be more revealing about possible inadequacies in a system.

In first assessments of existing foul drainage installations, the tasks to undertake are:
◊ establishing whether the systems are combined
◊ establishing the drain runs
◊ exposing buried manhole covers
◊ identifying long runs serving single WCs

Whether the sewers or drains are sufficiently large to permit close visual inspections, or whether inspections must be carried out by CCTV, the problems to look for on a closer inspection are:
◊ defective manholes
◊ unnecessary bends or changes of gradient
◊ builder's rubble
◊ cracks or deformation in the walls of pipes
◊ erosion or chemical attack on walls of pipes
◊ incorrect or defective joints
◊ invasion by tree roots
◊ silt or grease deposition and mineral encrustation
◊ subsidence
◊ no breathing apparatus on site where required

Where a wash-out (void) is suspected adjacent to a defect in a pipe, joint or wall, this may require the use of geophysical apparatus to confirm its existence, or alternatively simple excavation to expose the site of any problem.

For sewage pumps and their housings, the problems to look for are:
◊ no gas testing facilities (explosion or health risks)
◊ damaged macerators
◊ sedimentation or clogged screens
◊ no spare fuel supplies
◊ no spare parts for pumps
◊ excessive noise
◊ vandalism

Chapter 4.3

Surface water drainage, soakaways and flood storage

It was noted in the previous chapter that drainage of surface water may be found to be combined with foul water drainage, or comprise separate systems. For design purposes, combined systems generally should have followed the rules for foul drainage. Where there are separate systems, design rules for surface water drains may be similar to but less onerous than for foul drains, although it will still be important that self-cleaning velocities of flows are achieved to reduce the deposition of silts within the systems.

Whenever alterations are made to the buildings and surface coverings on a site, new investigations must be undertaken thoroughly and competently so that all aspects of soil properties, geotechnology and hydrogeology are reviewed alongside the hydraulic designs of drains and soakaways.

Design procedures for urban drainage systems (sustainable urban drainage systems SUDS) are currently undergoing revision which, in turn, may lead to changes in building regulations[28,29].

The design of flood storage of whatever kind is a specialised engineering activity. Brief mention is made here of characteristics of some of the most common forms, but inspection and assessment of flood storage provisions should be made by competent persons.

Characteristic details

Basic description of types of systems

Surface water drains can fulfil a variety of functions. In their most familiar role they remove precipitation from the vicinity of buildings and surrounding road surfaces, hard standings, pavings and other impervious surfaces to appropriate disposal points, whether on or off site. However, surface water drains can also have other roles.

Shallow drains can increase run-off and thereby reduce seepage forces and lower groundwater levels. Drains of this type are relatively cheap and easy to maintain, but are also the ones most likely to fall rapidly into disrepair if neglected.

Deep drains can modify the seepage pattern within the soil or rock mass, thereby removing water and decreasing porewater pressures from the vicinity of potential slip surfaces. Drains of this type are generally more expensive than shallow drains, but are likely to be more effective and to require less maintenance.

Drains can also be installed specifically to dissipate excess porewater pressures generated, for example, by an increase in ground levels. These drains may be shallow or deep, and are required to operate for only a limited period of time.

Some drains may fulfil a combination of these three functions.

The major differences between foul and surface water drainage systems are mainly in the provision of access, intercepting traps and trapped gullies. The latter two items are not required unless the pipework eventually connects to a foul sewer. However, many estates provided

Figure 4.20 Excavating for a surface water drain

with separate rainwater systems which are eventually connected to foul sewers have intercepting traps at this final connection, thereby isolating the systems from sewer air and removing the need for trappage at the buildings. As a general rule, access to all surface water drains proves to be minimal, well below the requirements for foul systems; in practice it is often very difficult to find manholes for the systems since they are frequently grassed over or buried below paths and drives. Drain lengths are likely to be similar to those for foul systems given in Chapter 4.2.

Surface water ditches, trenches and swales

The simplest type of drain is the open one, either a shallow conduit close to the buildings (Figure 4.21), or, at a distance from the building, a ditch with steep banks. These are cheap to construct and are capable of carrying high levels of discharge for short periods. They have been in use for centuries and have commonly been used on natural slopes to carry away discharge from springs, to lower the maximum level of ponds, to re-route streams or to provide drained access to fields. Swales are long shallow channels with smooth banks which can serve similar purposes, but can also be designed to enhance local infiltration.

Ditches tend to be difficult to keep operational since high rates of discharge cause scour and low rates allow weed growth. Although at one time scour was commonly reduced by lining the ditches with concrete, this obviously increased cost and could be disrupted by even small movements. More recent thinking, however, is that lining with concrete can simply transfer the problem to elsewhere. The outfall end of the ditch may need careful detailing to ensure that the discharge, together with any entrained material, does not cause any problems.

Shallow gravel filled trenches can be used as an alternative to open ditches, and are often used in a herringbone or chevron pattern to intercept run-off on the face of a slope. To fulfil this function, they must be open at the top, not covered with earth, but need to be protected by a filter to prevent fines penetrating and blocking the pathways. Where high rates of discharge are anticipated, it may be necessary to concrete the base of the trench to protect it from erosion and to provide a perforated pipe to increase capacity.

Deeper rubble or gravel filled trenches (French drains) can be used to stabilise shallow ground slides by lowering groundwater levels. They tend to be particularly effective where there is a layer of highly disturbed soil or debris overlying undisturbed soil of much lower permeability. Where the trench penetrates into the undisturbed soil it may also provide some mechanical buttressing; in this form it is described as a counterfort drain. The effectiveness of the drains depends principally on their spacing, their depth and the ratio of the horizontal to vertical permeability of the ground.

Underdrains may be needed where soils do not drain quickly. They consist in essence of a perforated drain pipe laid in a trench, and covered with sand and gravel[212].

Sand-wicks

In the vicinity of a deep slip surface, vertical bored sand or gravel filled drains, known sometimes as sand-wicks, can be used to drain a perched water table to an underlying permeable stratum. A proprietary system consisting of sand filled geo-fabric tubes which can be lowered into small-diameter pre-drilled holes is also marketed for this purpose, and has the advantage that the strength of the fabric helps prevent the drains being severed by small down-slope movements. Alternatively, horizontal or near-horizontal bored drains can be installed from the toe of the face of the slope. Drains of this type are usually lined with a perforated or permeable material such as no-fines concrete rather than filling with a granular medium. For maximum effectiveness, deep drains should extend well beyond the point at which the reduction in porewater pressure is required – probably as much as two or three times the drain spacing.

Drains for special purposes

Drains that are used specifically to speed up consolidation following construction or a rise in ground level fall into two categories: drainage blankets and vertical wells. Drainage blankets are installed between lifts of fill or between fill and the underlying soil, and allow lower permeability fill materials to be used which would otherwise have to be rejected because of the risk of unacceptable buildup of pore pressure during placement. Vertical wells are essentially similar to the vertical drains (sand-wicks) described above except that it is unnecessary for a well to penetrate an underlying bed of higher permeability: excess pore pressures can escape at ground level if the tops of the wells are capped with a drainage blanket with suitable outlet details. Wells can be pumped out if necessary, although this is rarely done in practice other than as a means of carrying out an in situ permeability test.

Figure 4.21 An open drain taking the discharge from the gargoyles in the cloisters at Norwich Cathedral

4.3 Surface water drainage, soakaways etc

Land drains

Land drains have been used for many years for agricultural purposes. Traditionally the pipes, or the so-called tiles which preceded pipes, were of coarse fired clay laid with open joints (BS 1196[270]); water could then flow both in and out with equal facility. The problem is that they silt up, and are also vulnerable to blockage by roots. Drains may also be found which use plastics or concrete pipes BS 4962[271], BS 5911-110[272] and BS 5911-114[273].

Soakaways

Soakaways are pits or trenches used to collect stormwater from buildings and paved areas and allow it to percolate gradually into the ground without causing flooding. Traditionally their use has been restricted to areas remote from a public sewer or watercourse. In recent years, however, they have become increasingly common in fully-sewered urban areas to limit the impact of new developments on existing sewer systems.

Soakaways are basically of two kinds. Although similar in design, they differ in purpose, which is the accommodation of:
- surface and rainwater run-off
- treated wastewater

Whether a soakaway can be used for treated effluent will depend on consultation with the relevant authority. Effluents from septic tank installations can be drained away by either allowing it to disperse on the surface of grassland, or by using subsurface drains which allow it to disperse under the surface of the ground where the subsoil is permeable and sufficiently unsaturated. Of these two methods, the latter is much to be preferred, since it reduces the possibility of contaminants on the land surface.

Soakaways can be constructed in many different forms and from a range of materials. To function effectively, they must have an adequate capacity to store immediate stormwater run-off and must discharge their temporarily stored water sufficiently quickly to provide the necessary capacity to receive run-off from a subsequent storm. The time taken for discharge depends upon the soakaway shape and size, and the infiltration characteristics of the surrounding soil[90].

Traditionally, soakaways to cater for run-off for areas of less than 100 m² have been built as square or circular pits, which are either filled with rubble or lined with dry-jointed brickwork or precast perforated concrete ring units surrounded by suitable granular backfill. BS 8301[26], suggested that soakaways might also take the form of trenches that follow convenient contours. Trenches have larger internal surface areas for infiltration of stormwater for a given stored volume compared with square or circular shapes. The designer must therefore weigh the merits of the more compact square or circular forms of soakaway against the better rate of discharge from the trench for the soil type, available space, layout and topography of a site. Current recommendations for the design of soakaways are included in BS EN 752-4[91].

For drained areas in excess of 100 m², soakaways can be found of precast ring or trench type. They should not be deeper than 3–4 m if ground conditions allow. Limiting the depth means that the length must be increased, but then trench soakaways are cheaper to dig with readily available excavating equipment.

A simple catch pit or silt trap can be formed within a small shallow inspection chamber with the inlet and outlet at approximately the same level; this allows access for cleaning and rodding but with the manhole floor some 450 mm below the invert to trap and facilitate removal of the silt deposits (Figure 4.22). The chamber is designed to operate differently from a conventional inspection chamber in that it contains water all the time.

Perforated, precast concrete ring unit soakaways should be installed within a square pit, with sides about twice the selected ring unit diameter, and the space between the rings and the soil backfilled with coarse granular material. The need to oversize the soakaway pit for purposes of constructing the ring unit chamber may be used to advantage: the calculation of the size of the soakaway allows the volume of excavated ground below the discharge drain invert to be include in the total design storage volume.

The granular backfill must be separated from the surrounding soil by a suitable geotextile membrane to prevent migration of fines into the soakaway. If migration from the surrounding soil occurs, it can cause ground settlement around the soakaway sufficient to affect the stability of adjacent buildings. The top surface of the granular fill should also be covered with a geotextile membrane to prevent the ingress of material during and after surface reinstatement. The membrane should not be wrapped around the outside of the ring units as it cannot be cleaned satisfactorily or removed when it has become blocked.

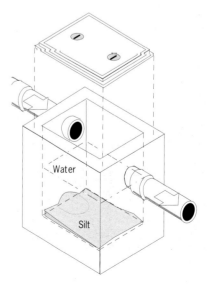

Figure 4.22 A simple silt trap

Drainage mounds

Where the drainage characteristics of the soil are poor, or where the water table is high, it may be possible to use a mound system for disposing of wastewater after treatment in a septic tank system, or for disposing of untreated rainwater.

A mound system consists essentially of a relatively long and wide mound of approximately 1 m height consisting of a thin bed of permeable soil at the base, surmounted by a deep bed of permeable sand into the top of which the water is distributed via perforated pipes. The whole system is covered with a geotextile membrane and a layer of topsoil (Figure 4.23). It is expected that this method will soon be permitted in building regulations and covered in a European Standard.

Detention tanks

Detention tanks are receptacles to store grey water or more usually stormwater on a temporary basis to reduce peak flows in combined systems. The main problem with them is the accumulation of sediment which needs to be removed periodically.

Flood storage

Buildings are inevitably surrounded with areas of hard paving which leads to an increase in rainwater run-off from a site when compared with the amounts which the original undeveloped land surface could accommodate. Alterations in the topography of the land surface associated with building developments will also create new drainage patterns for the site. If the development is of comparatively small scale, excessive run-off caused, for example, by the once in 150 year rainstorm and the subsequent flooding can usually be tolerated with perhaps a degree of surcharging of the drains. However, in large scale developments involving extensive car parking on local roads, when the amounts of run-off from intensive thunderstorms exceed the capacity of the existing surface water drains, as inevitably they will on occasions, the excess becomes more of a problem. The risk of local flooding in these circumstances can often be ameliorated by flood storage reservoirs. These reservoirs may or may not be suitable for amenity use, depending on, for example, size and local conditions including pollution.

Reservoirs can take one of two main forms, or perhaps a combination of the two:
- ordinary dry weather water flow passes through a storage basin – known as onstream
- ordinary dry weather water flow bypasses a storage basin – known as offstream

When water is normally present in the storage basin during non-flood conditions, it is called a 'wet basin'; when water it is not normally present, it is called a 'dry basin'. Of course, both are obviously impounded during flood conditions. Further information is available in the SUDS manuals[28,29].

Flood storage reservoirs may be provided with emergency spillways in addition to the controlled outlets.

Main performance requirements and defects

Outputs required

Rainfall amounts are dealt with in Chapter 1.3. These, together with the characteristics of the surfaces to be drained, will need to be considered in the design of both surface water drains and soakaways.

Drainage of paved areas

BS EN 752-4 contains a simplified method of calculating the surface water run-off for small development schemes. National Annex NE of BS 752-4 gives guidance on the drainage of surface water from paved areas, including access roads, but not trunk roads. The provision (position and number) of gullies will depend on the actual design of the paved areas; for example, the permeability of surfaces, or whether roads are cambered or have cross-falls.

The gradients of cross-falls or small cambers for footpaths are recommended to be a minimum of 1 in 40 and a maximum of 1 in 30. For other paved areas the gradient should be 1 in 60 minimum and for access roads 1 in 40 minimum. Where channels are formed adjacent to kerbs, the gradients recommended are 1 in 150 or 1 in 200 depending on the use of channel blocks and the class of surface.

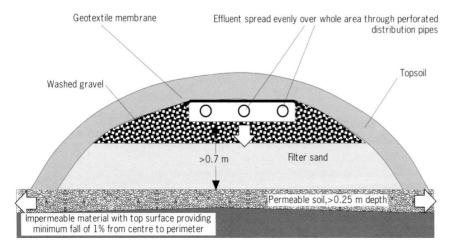

Figure 4.23 A mound disposal system

4.3 Surface water drainage, soakaways etc

Gullies

Water sealed traps are needed in gullies where discharge is to a combined sewer. Gullies, whether trapped or not, should be provided with grit interceptors, the actual capacity depending on such factors as use of the area and frequency of cleaning (BS 5911-230[274]).

Gullies at the ends of flows (terminal gullies) should normally be larger than intermediate gullies since the risk of flooding is greater at the ends of collecting channels. Kerbs provided with inlets are sometimes used as intermediate drainage points, though they are not very effective with steep channel gradients.

Gratings for gullies should be provided as close as possible to the kerb or some of the flow may be carried past. Spacing of gullies will depend on factors such as:
- design rainfall intensity
- areas to be drained
- gradients of channels
- width and depth of flows anticipated

Drains designed to accommodate summer storms are generally assumed to be able to accommodate run-off from melting snow.

Soakaways

The infiltration characteristics of the soil where a soakaway is to be sited are determined by excavating a trial pit, filling it with water and measuring the time taken for it to empty. In order for the infiltration characteristics to be truly representative of the proposed soakaway, it is important to observe the following points.
- The trial pit should be of sufficient size to represent a section of the design soakaway
- The pit should be filled several times in quick succession during monitoring of the rate of seepage to ensure that the moisture content of the surrounding soil is typical of the site when the soakaway becomes operative
- Existing site data should be examined to assess whether there are any geotechnical or geological factors likely to affect the long term percolation and stability of the area surrounding the soakaway. These will include variations in soil conditions, areas of filled land, preferential underground seepage routes and variations in the groundwater levels

Natural groundwater should not rise to the level of the base of the soakaway during annual variations in the water table. Local building control or planning authorities may be able to advise where fluctuations in groundwater level may cause a problem in the long term for any proposed depth of excavation.

The soakage or percolation trial pit should be excavated to the depth anticipated for the full-size soakaway; for areas of no more than 100 m^2, this will be 1–1.5 m below the invert level of the drain discharging to the soakaway, or a typical overall excavation depth of 1.5–2.5 m below ground level (BS 1377-2[58]).

If the test pit is deeper than about 3 m, it may be difficult to supply sufficient water for a full-depth soakage test. In these circumstances, tests may be conducted at less than full water depth, but the soil infiltration rates determined in this way are likely to be lower than those obtained from full-depth tests. This is because relations between depths of water in the soakage pit, the effective area of outflow and the infiltration rate can vary with depth, even when the soil conditions themselves are constant. The variation in infiltration rate with the depth at which the determination is made may be as much as a factor of 2.

If it is not possible to carry out a full-depth soakage test, the soil infiltration rate calculation should be based on the time for the water level to fall from 75–25% of the actual maximum water depth achieved in the test. The effective area of loss is then calculated as the internal surface area of the pit to 50% of the maximum depth achieved in the test, plus the base area of the pit.

In general, soakage trials should be undertaken where the drain will discharge to the soakaway. The use of full-depth, and of repeat determinations, at locations along the line of trench soakaways is very important where soil conditions vary; if the soil is fissured, infiltration rates can vary enormously. In these circumstances, a preliminary design length for the proposed soakaway should be calculated from the first soakage trail pit result and, if the design length exceeds 10 m, a second trial pit should be excavated at the design length distance along the line of the soakaway. In all ground conditions, a second trial pit should be dug if the trench soakaway (designed on the basis of one trial pit) is longer then 25 m; further trial pits are needed at intervals of 25 m along the line of a long soakaway. If more than one trial pit is used, the final design should be based on the mean value of the infiltration rates obtained from the tests.

Soakaways should not normally be constructed closer than 5 m to building foundations. In chalk, or other soil and fill material subject to modification or instability, the advice of a specialist geotechnical engineer should be sought as to the advisability and siting of a soakaway.

The recommended design procedure for sizing surfacewater soakaways depends, of course, on the anticipated rainfall to be dealt with as well as to the characteristics of the ground in which the soakaway is sited. This procedure is explained in detail in BRE Digest 365[90]. It is not repeated here, though the steps may be summarised as follows:
- excavation of a trial pit as described in Chapter 1.1 to determine soil infiltration prospects
- choosing the form of the soakaway (eg pit or trench)
- calculating design rainfall and the frequency of its occurrence from a rainfall map of the UK in conjunction with the use of tabulated information. A map giving the ratio of 60-minute to two-day rainfalls for a five-year return period is included in BRE Digest 365
- calculating inflow and outflow from the soakaway, and, therefore, the storage volume required

Flood storage
A flood storage reservoir should have sufficient capacity to accept the intended excess rainwater within the catchment area, and allow it to discharge over the period of time with which the existing watercourses and drains can cope*. Raised reservoirs (that is to say, reservoirs with part of their capacity above the surrounding land surface) with a capacity exceeding 25 000 m^3 come within the provisions of the Reservoirs Act 1975[276].

Unwanted side effects
Drains
Surcharging of surface water drains in periods of heavy rain are to be expected where the capacity is inadequate.

Soakaways
Care must be taken that the introduction of large volumes of surface run-off into the soil does not disrupt the existing subsurface drainage patterns; in this respect, it may be better to use extended trench or piped trench soakaway systems (BS EN 752-4). The effect of ground slope must be considered when siting soakaways to avoid waterlogging of downhill areas.

* The design of flood storage reservoirs is a specialised engineering task and is beyond the scope of this book. For those readers wishing to investigate further, a suitable reference is to be found in *The design of flood storage reservoirs*[275].

Health and safety and environmental aspects
Rainwater run-off from urban areas can be very variable in quality, sometimes heavily polluted, sometimes relatively clean, but never wholesome. Pollutants may include suspended solids, both organic and inorganic in origin, oils, heavy metals and coliforms[275].

There are three main considerations with respect to environmental issues:
- Standards for protecting river aquatic life
- Standards for protecting bathing waters
- Standards for protecting amenity use

For further information on these matters, see BS EN 752-4.

Flood storage
There are obvious safety implications with flood storage reservoirs, particularly those on land to which the public has access. Although gently shelving banks rather than steeply shelving ones may provide a margin of safety, they can be a double-edged sword, inviting paddling by children. Planting of suitable vegetation on gently shelving banks may provide some discouragement to adventurous children, and specialist advice will be necessary for this.

Durability
Drains
Since materials commonly used in the construction of surface water drains are mostly identical with those used for foul drains, similar durability considerations apply to attack from aggressive ground conditions (see Chapter 4.2). On the other hand, surface water should not normally contain very aggressive agents – unless from contaminated land or from, for example, roads and vehicle hardstandings – and durability problems should not arise.

Figure 4.24 An elementary error. Pea shingle from an adjacent area has more-or-less completely blocked these drainage grilles

4.3 Surface water drainage, soakaways etc

Maintenance
Drains
The following items may need to be inspected regularly, including taking action on:
- manholes – defects or built-up of detritus
- gullies and grilles – cleaning, grit bucket emptying and flushing (Figure 4.24)
- drainage ditches – debris removal and vegetation removal
- land drains – rodding to clear debris

In the short term, drains can be a very effective method of stabilising a slope, but as a long term solution they can suffer from failure to carry out periodic maintenance to ensure their continued functioning.

Soakaways
Soakaways can provide a long term, effective method of disposal of stormwater from impermeable areas as large as several hundreds of square metres; but long term maintenance and inspection must be considered during their design and construction. With catch pits, vehicle mounted suction emptying and jetting equipment can be used, so suitable access covers must be provided.

Maintenance of soakaways has always presented problems. With rubble filled soakaways, in particular, the problem can often be locating the soakaway in the first place. All soakaways should be provided with some form of inspection access, so that the point of discharge of the drain to the soakaway can be seen. This access will identify the location and will allow unwanted detritus to be cleared from the soakaway.

Little monitoring of soakaway performance is done, but this could be the best way of obtaining information about changes in soil infiltration rate and in warning of soakaway blockage in the long term. The inspection access should provide a clear view to the base of the soakaway, even when the soakaway is of the filled type, as shown in Figure 4.25. For small, filled soakaways, a 225 mm perforated pipe provides a suitable inspection well. Lined soakaways are easier to access for inspection and cleaning and this advantage should always be used. Trench-type soakaways should have at least two inspection access points, one at each end of a straight trench, with a horizontal perforated or porous distributor pipe linking the ends along the top of the granular fill. It may be convenient with a trench soakaway to have several drain discharge points along the length of the trench, each connected to the soakaway via an inspection access chamber.

The risk of pollution reducing the quality of groundwater must also be considered. The limited evidence presently available suggests that roof surface run-off does not cause damage to groundwater quality and may be discharged directly to soakaways. Those pollutants entering a soakaway from roofs tend to remain in the soakaway or in its immediate environs, attached to soil particles. However, paved surface run-off should be passed through a suitable form of oil interception device prior to discharge to soakaways. Maintenance of silt traps, gully pots and interceptors will improve the long term performance, and installing wet-well chambers within soakaway systems will assist in pollutant trapping and extending operating life.

Work on site
Workmanship
Drains
Requirements for workmanship are covered in the code of practice BS 8000-14[277]. Since materials and practices are very similar to those for foul drainage, the relevant sections of Chapter 4.2 are appropriate.

Soakaways
The trial pit for soakaways should be 0.3–1 m wide and 1–3 m long. It should have vertical sides trimmed square and, if necessary for stability, be filled with granular material. However, when granular fill is used, a full-height perforated vertical observation tube should be positioned in the pit to allow the water levels to be monitored using a dipping tape. A suitably dimensioned pit can be excavated by a back-hoe digger or mini-excavator. Narrow, short pits use less water for the soakage tests, but may be more difficult to trim and clean prior to testing. The dimensions of the pit should be carefully measured before trials begin and, for safety reasons, no one should enter the pit.

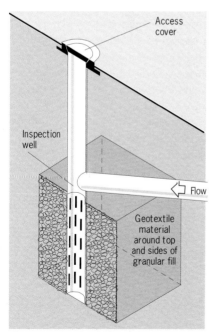

Figure 4.25 A small filled soakaway with a perforated inspection well extending to the base of the soakaway. This provides access to the discharge drain outlet as well as enabling inspection of the condition of the fill at depth

Determining the soil infiltration rate will require a great deal of water, so a water bowser may be needed. The inflow should be rapid so that the pit can be filled to its maximum effective depth in a short time (ie to the proposed invert level of the drain to the soakaway). The inflow should not cause the walls of the pit to collapse. BRE Digest 365 suggests that the pit is filled and allowed to drain three times to near empty; each time, the water levels and times from filling should be recorded so that the drops in water levels can be calculated as functions of time. The three fillings should be on the same or consecutive days. Extraneous water (eg rain) should not be allowed to drain into the pits during the test.

The Building Standards (Scotland) Regulations[93] require a percolation test somewhat similar to the above to determine the soakaway provisions for a small septic tank installation, though there are differences in the procedure. A 300 mm square hole, 300 mm deep, is excavated at the bottom of each of at least two trial pits, not less than 5 m apart. The tops of these holes should be below the inverts of the proposed soakaways (represented by the trial pits). The holes are filled with at least 300 mm depth of water and allowed to seep away overnight. On refilling the next day, the times taken for the levels to drop from 75% full to 25% full are recorded. The test is repeated three times. The percolation value is obtained by dividing the time in seconds by 150 to give the time taken for the water level to drop by 1 mm.

The area of the soakaway, whether it is a pit or a trench, is then calculated from:

$$A = p \times V_p \times 0.25$$

where A is the area of the subsurface drainage trench in m²
p is the number of persons served by the tank
V_p is the percolation value obtained in seconds per millimetre

If the effluent is to receive secondary treatment, or is greywater, the area may be reduced by 20%.

Inspection

Drains
The first task in any investigation is establishing whether the foul and surface systems are combined.
The problems to look for are:
◊ poor siting of rainwater gullies
◊ manhole covers hidden underground
◊ systems not designed for appropriate rainfall patterns leading to surcharging
◊ invasion by tree roots (Figure 4.26)
◊ silt or grease deposition and mineral encrustation
◊ subsidence

Figure 4.26 Tree roots entering this clayware drain through joints have completely filled it (Photograph by permission of John Bricknell)

Soakaways
Whether the surface water runs to soakaways or to mains must first be established.
The problems to look for are:
◊ reduced effectiveness
◊ soakaways hidden under ground or pavings
◊ soakaways are too near to buildings

Flood storage
Reservoirs coming within the provisions of the Reservoirs Act 1975 are required to be inspected every 10 years. Since the inspection is solely the province of qualified civil engineers, it is not appropriate to include a check list here.

Chapter 4.4　Other utilities

This chapter provides a few points relating to electricity, cable TV and telecommunications, and gas mains which lie between the site boundary and the building. Although much of the work required is carried out by specialist teams, one of the main potential problems which seems to affect all utilities is that of ensuring adequate reinstatement of trenches and the pavings lying above them.

Routes followed by overhead cables may, of course, be perfectly obvious (Figure 4.27) but those buried underground may be much more difficult to locate. It may even be necessary to trace and expose the service entry points within the building (Figure 4.28).

Characteristic details

Electricity
Small installations in rural areas, and in a few urban areas too, are still served by overhead wires, either single or three phase, carried for the most part on preservative treated wooden poles. The last few metres may be carried down the pole and buried underground.

Cables for external use must be waterproof, especially if they are to be routed underground. Plastics, including both polyethylene and polyvinyl chloride, have largely superseded bitumen wrapping for sheathing underground electricity mains.

Cables should be laid in a sand or other suitable bedding before backfilling. The cables themselves are normally coloured black for low voltage installations and red for high voltage. Interlocking fired clay or concrete tiles were often used in the past on top of the sand bed to provide an early warning to those excavating at a later date. In more recent years highly coloured marker tapes have been used.

As recorded in *Building services*[20], single phase 230/240 volt, 60, 80 or 100 amp, 50 Hz earthed neutral installations, in accordance with BS 7671[278], are the usual types of supply to individual dwellings. These installations all come within the definition of low voltage installations (extra low voltage is up to 50 volts AC or 120 volts DC, low voltage up to 1000 volts AC or 1500 volts DC between conductors). Dwellings built during the 1980s and 1990s, especially those with electric space heating provision, have 80 or 100 amp installations although 60 amps installations can be found in older properties. A few dwellings, served in the main by overhead lines, are restricted to 30 amps.

Cable TV etc
The 1990s have seen an extensive programme of installation of cable distribution systems, mainly in urban and residential areas from a variety of competing commercial firms. Cables are normally laid within the public highway limits, terminating in a buried access trap at the site boundary. CCTV for security purposes has also, in some circumstances, required cabling to be run below ground to avoid the risk of tampering.

Figure 4.27 Electricity cables to rural locations are normally of the overhead kind. Responsibility for adequate and safe performance normally lies with the utility company

Figure 4.28 Rehabilitation work under way on a Victorian house. Excavation has been used to find the routes of the old services installations

Telephone lines in rural areas, and in some urban areas too, will often be found strung overhead from wooden distribution poles to terminals at eaves level on dwellings. It is only since the 1960s that underground supplies in some areas of the UK have come into general use. Telecommunications cables are normally coloured grey, and TV cables green.

Many telecommunications copper cables are 'jelly filled' (ie grease filled and sheathed, for example, in polyethylene).

Gas

It was noted in *Building services* that some 85% of dwellings in England now have mains gas supply, with the percentage much smaller in rural areas. Over the whole of England, some one in eight dwellings have no gas supply of any kind[4]. The perceived risk of explosion since Ronan Point in 1969 has been a deterrent to providing gas supplies in multi-storey blocks.

Until the 1960s, the gas used first of all for lighting and later for cooking, and distributed mainly in urban areas was manufactured from coal, with a by-product of coke. Since the 1960s the coal gas producing industry has disappeared in favour of natural gas piped from the North Sea.

The main legislation relating to the safety of gas installations is contained in the Gas Safety Regulations 1972, 1984, and 1990. Many of the Approved Documents to the Building Regulations (for example B, E, F, G and J) are concerned with the wider issue of gas safety.

For those areas outside the range of mains distribution there is the option of using propane or butane, stored in small pressure cylinders that can be manhandled or distributed by tanker vehicles for transfer to large fixed storage tanks.

Gas may be supplied at different pressures according to location and type of buildings served, and it may be necessary to provide space for pressure reducing plants near the buildings to be serviced.

For buildings built before the 1970s, incoming service pipes were normally in steel or even wrought iron, protected by bitumen wrapping. In recent years, though, yellow coloured plastics piping has become almost universal.

Pipe diameters for small installations could be as small as 25 mm, although larger domestic installations, or those at some distance from the mains, might need to be larger. Special arrangements are made for the design of larger installations.

Main performance requirements and defects

Avoiding danger from buried electricity cables

There are about 250 accidents a year involving injury to workers inadvertently touching live buried electricity cables or striking them with contractors' equipment. Many of them are fatal, but many also result in burns caused by arcing. Obviously there is a paramount need to obtain as much information as possible before excavation work is undertaken. Although cables are normally buried at least 0.5 m deep, they can be found at much shallower levels[279].

Durability

Electricity and telephone transmission poles of redwood, spruce or Corsican pine, treated with creosote, normally have a service life of around 60 years; many, though, last much longer than this, sometimes up to 100 years (Figure 4.29). Poles are inspected at least once every 10 years, but frequency may be increased if the consequences of premature failure are likely to be severe. Poles in remote or severely exposed areas also may have different inspection schedules. At the time of inspection, poles are assessed for likely further durability, and are either replaced or left in service. Restrictions may be placed on linesmen climbing suspect poles until replacement becomes feasible.

Work on site

Workmanship

Electricians undertaking work should be suitably qualified, and be employed by firms who are members of the Electrical Contractors Association (ECA) or registered with the National Inspection Council for Electrical Installation Contracting (NICEIC).

Inspection

Many problems will be self-evident when a system fails. There is little that the non-specialist can do to foresee technical problems apart from those relating to the durability of supports and enclosures, and, depending on levels of competence, reviewing compliance with regulations.

The main problems to look for are:
◊ inadequate consolidation of fill in utility trenches leading to settlement of paved surfaces
◊ loose or sagging brackets supporting overhead electric wires
◊ non-registration of CCTV cameras in accordance with requirements of the Data Protection Act 2000
◊ buried cables and pipes that are unidentified or unmarked

Figure 4.29 Redundant transmission poles removed for investigation of durability

Chapter 5 Walls, fencing and security devices

This fifth chapter deals with freestanding boundary walls, fencing, site lighting, and other security devices such as controlled entry gates and CCTV. Also included are notes relating to walls used to retain earth embankments. In many cases embankments and various kinds of barriers are used to mitigate the noise of traffic from roads, particularly in residential areas. For the purposes of this book therefore, the performances of various kinds of barrier are included in Chapter 5.1.

Throughout history, measures have been necessary to provide a degree of protection to properties, whether by fortification in medieval times, or, more recently, by simple enclosure against casual housebreaking and burglary. On the other hand, those measures may simply be there to establish ownership and privacy. Even the Victorian byelaw housing in the most densely developed urban areas needed their enclosing yards and 2 m high brick walls (Figure 5.1).

Figure 5.1 Byelaw housing in an inner city area. The ginnel provided a convenient route for access for disposal of domestic refuse, as well as a route for the drains and piped water supplies. It is interesting to see that the yards on the north side of the terrace (left side of the picture) have had their walls whitewashed to increase available light in what would otherwise have been a dreary location

Chapter 5.1 Embankments and retaining walls

Where building on sloping ground cannot be avoided, particularly where the ground is potentially unstable, there are a number of techniques that can be used to improve the stability of the slope. These include:
- regrading
- anchors
- retaining walls
- grout injection
- vegetation
- drainage

These techniques are described briefly in the following sections, although it is emphasised that slope stabilisation is a specialised subject. In general, the application of these techniques should be left to geotechnical engineers with relevant knowledge and experience. As well as taking steps to improve the stability of marginal slopes, care must be taken in the design of the building, including its foundations, services, drives, pathways and any landscaping, to ensure that the effect on slope stability is minimised.

One situation where a retaining wall was frequently needed in the past was the ha-ha, or sunken ditch and wall, which kept livestock out of a mansion garden without a wall or fence interrupting the view. Some of the more aggressive conditions affecting materials are to be found in these walls. Indeed, for many years the former BRS used the local ha-ha for exposure trials of brickwork (Figure 5.2).

Where the natural slope of the ground was stable, however, it would not always be necessary to build a retaining wall (Figure 5.3).

Embankments are required in many civil engineering works including highway and dam building. These are specialised forms of construction with specific performance requirements and there is a large body of technical literature dealing with them; these applications are outside the scope of this book. In this chapter, a brief description is given of some relevant matters concerning the use of embankments in landscaping works which are usually concerned with reducing the visual and acoustic impact of major roads on residential areas.

Characteristic details

Basic description of slope maintenance
Regrading
The most obvious and often most effective way of improving the stability of a slope is to regrade the slope using cut and fill techniques. However, the practicality of this approach is likely to be limited in areas that contain existing buildings. There are three possibilities:
- grade to a uniform, flatter angle
- apply filling at the toe of the slope, creating a step or berm in cross-section
- reduce the overall slope height while keeping the profile unchanged

Reducing the overall height of the slope tends to be the most effective solution where the potential failure plane is near the surface. Conversely, adding material to the toe, or the removal of material from the crest, tends to be more effective for correcting deeper-seated failures.

Figure 5.2 The ha-ha on the BRE site at Garston used for assessing the performance of brickwork in aggressive conditions

5.1 Embankments and retaining walls

Figure 5.3 Excavation for this ha-ha has revealed a limestone seam to create a more-or-less vertical barrier to livestock

Retaining walls fall into categories according to the means used to provide the necessary resistance to the applied soil forces:
- gravity walls
- reinforced walls on spread foundations
- embedded walls
- anchored or propped walls
- piled or cantilevered retaining walls
- sheet pile walls

These basic types are described below. It is also common to have hybrid designs which combine the features of two or more of the basic types.

The choice between a cut or a fill solution is normally constrained by the need to maintain either the crest or the toe at a specified position; where this is not the case, a combination of cut and fill is normally adopted to obviate the need to import or remove material. Cut slopes can normally be left a little steeper than filled slopes to reflect the greater strength of the undisturbed material, but, where necessary, fill slopes can be steepened by importing a suitable material.

For natural slopes with irregular and complex slip surfaces, cut and fill solutions are complicated by the risk that placing fill in one area may reduce the stability of another.

Small retaining walls

Retaining walls are used to support soil at slopes steeper than those at which they would be naturally stable. Typical uses for small walls are to provide near flat 'terraces' in a sloping site for gardens or access paths. Small is defined here as typically up to 2 m, or occasionally up to 3 m. Retaining walls are generally constructed from masonry or concrete and may be reinforced or unreinforced (Figure 5.4). It is recommended that the services of a suitably experienced, chartered civil or structural engineer is engaged for the design and construction of any walls exceeding a height of 3 m and for any walls over 2 m that incorporate reinforcement.

Soil can also be retained using other methods:
- sheet piling
- piles
- stone filled baskets (gabions)
- stacked box-like structures containing granular soil (cribwork)
- burying reinforcement in the sloping surface of the soil itself (reinforced earth)

Further details can be found in BS 8002[280] which also covers the design and construction of retaining walls of up to 8 m in height.

Retaining walls are normally specifically designed for each separate application, and a combination of techniques may be found in certain cases. For example, BRE investigators were involved in testing a piled retaining wall that, in addition to lateral loads, was also used to carry vertical loads from a large building situated above. Instruments used during the tests allowed the load carried by the pile base and shaft to be separately assessed, and deflection and bending moment profiles to be obtained for the piles subjected to lateral load, thus verifying the design criteria[281].

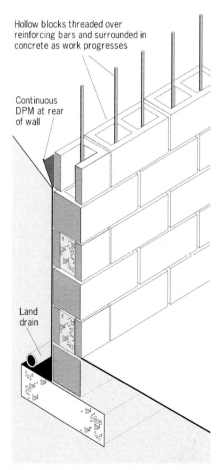

Figure 5.4 A typical small reinforced hollow concrete block retaining wall

Gravity walls
Constructed from stone, brick or concrete, gravity walls are the oldest form of retaining wall. They rely on their dead weight to provide adequate resistance to the applied soil forces. Masonry walls may be designed on the basis that small tensile forces are allowed to develop in the mortar, or on the gravity principle whereby no tension is permitted within the wall. Indeed, for some applications employing large massive blocks of stone or concrete, such as harbour walls, walls may be built without using mortar. Higher walls, especially those constructed using brickwork, are likely to be reinforced.

Reinforced walls on spread foundations
By using spread foundations, the weight of some of the retained soil contributes to the stability of walls. Walls of this type are normally constructed using reinforced concrete so that the vertical stem can be relatively thin. In a cantilevered wall the stem is supported solely by its connection to the base. Alternative forms of construction are the counterfort wall where the stem is strengthened by vertical slabs built onto the retained face of the wall, and the buttressed wall where the support is added to the outside face.

Embedded walls
Embedded walls depend mainly for their stability on mobilising passive earth pressure on the buried section of the wall. They are normally constructed using contiguous (ie overlapping) piles or some form of interlocking pile as described in Chapter 3.1.

Anchored or propped walls
Ground anchors or props can be used to improve the stability of any retaining wall, although they are most commonly used with embedded walls. Props are generally only used as temporary supports in applications where there are two retaining walls opposite each other (eg during construction of a basement or cut-and-cover tunnel). Anchors, on the other hand, are usually installed on a permanent basis. Walls may have several levels of anchors or props, although they are most effective when installed near the top of the wall.

Piled or cantilevered retaining walls
These can be used to stabilise slopes by transferring loads from the unstable soil or rock to the underlying undisturbed ground. The forces and moments which act on the wall can be very large and can only be calculated if a full stability analysis is performed incorporating the properties and stress conditions in the ground. Unless the wall is supported by prestressed anchors, a certain amount of deformation will be needed to mobilise its resistance.

Sheet pile walls
Sheet pile walls can be used where movements are likely to be small, or as a very temporary expedient, but in the majority of cases their resistance is too small to provide permanent support. They do, however, have the advantage that, unlike concrete walls, their resistance is mobilised immediately upon insertion. Additional capacity can be provided by using piles with a high-modulus section.

Grouting

Injection of cement or chemical grouts into the ground has been successful in arresting movement in highway or railway slips where there are significant voids or zones of very weak ground into which the grout can penetrate. The benefit of grouting clay soils is less obvious. Ion exchange in the porewater can cause a change in mineralogy, which can produce an improvement in shear strength; grout may also be able to penetrate slip surfaces producing a cementing action. However, the benefit to the soil mass as a whole is likely to be limited by the relatively low permeabilities of the soil.

There is also a possibility in permeable soils that grout may block natural drainage pathways and, in urban areas, may intrude into buried pipes. These risks should be carefully considered before grout is used.

Figure 5.5 An anchored piled wall

Gabions and cribs

A gabion is a cage of durable wire filled with stones, or formerly of wicker filled with earth, to form a type of fascine to retain earth embankments. In its modern form, the technique has been widely used in the construction of embankments and ditches for motorways, but can increasingly be found used as earth retaining structures in the hard landscaping of building sites. Indeed, in the 1990s, gabions began to be used on a small scale in the structures of buildings, and even used for loadbearing purposes.

Most kinds of durable stone, and even crushed concrete, have been used to fill the cages. Mesh size is usually related directly to infill size. Cobbles or flints may be tipped into the cage, but stone rubble needs to be packed by hand on the outer face to achieve an acceptable appearance.

Other earth retaining structures sometimes seen include shaped precast concrete units (cribs) stacked one above the other and interlocked to provide pockets of soil exposed on the outer face within which plants are encouraged to grow. It is understood that timber structures which achieve the same result have become available.

Rock and soil anchors

The stability of slopes can be improved by anchoring the unstable surface soil or rock to the ground below the slip surface (Figure 5.5). Anchors fall into two broad categories:

- unstressed anchors which rely simply on a dowelling action to increase the resistance to sliding
- stressed anchors which increase the effective stress acting on the slip surface and can also offer a component to counterbalance the disturbing force

Unstressed anchors are normally formed by excavating or drilling a hole or holes of appropriate diameter through the potential slip plane and filling it with a suitable cementitious material.

Stressed anchors are normally formed by grouting a steel bar or tendon into a pre-drilled hole using cement mortar, concrete, epoxy or other chemical grout. In soft rocks and soils, the pull-out resistance may be increased by enlarging the end of the drill hole using a reaming tool. In 'more competent rocks', a mechanical device is often used to expand the anchor in order to grip the sides of the drill hole. Having formed the anchor, the bar or tendon is tensioned at the surface using jacks or by simply turning a nut on a threaded bar. Where no re-tensioning is anticipated the remainder of the drill hole can then be filled with grout to protect the length of bar or tendon between the surface and the anchor. However, in some cases, it may be necessary to re-tension the anchors periodically as the ground consolidates under the load applied by the anchor. In these circumstances, it is necessary to provide corrosion protection to the exposed length of steel in the form of grease or a PVC sheath.

The design of an anchoring system is a complex process requiring the estimation of the total force that is needed to stabilise the slope. As a general rule, however, stressed anchors are likely to be most effective if they are inclined at an angle to the slip surface equal to the mobilised angle of shearing resistance. They are also more likely to be effective if they are placed near the toe of the slope, since anchors placed near the crest would not prevent a tension crack developing, thereby allowing the lower part of the slope to fail in isolation.

Other possibilities for stabilising sloping ground are described in Chapters 1.5 (tree planting) and 4.3 (drainage).

Main performance requirements and defects

Strength and stability

Embankment design

A combination of geotechnical and economic parameters will normally control the design of embankments.

- The geotechnical performance requirements of embankment barriers are quite simple: the embankments must be stable over long periods. Small consolidation settlements of the fill of the foundation of an embankment will not usually be of concern
- For economic reasons, it is desirable to use local materials and, if possible, to obtain the fill from excavations at the site where the embankment is being constructed. Although construction costs will be linked to the volume of fill to be placed – and therefore the steeper the slopes, the cheaper the embankment – there may be environmental and maintenance considerations which make flatter slopes more acceptable

The two major, interrelated elements of the design are the selection of a fill material and the choice of the gradient of the embankment slopes. Granular fills can generally be safely built to steeper slopes than can clay fills. However, where embankments are built on soft ground, the low strength of the ground may be the factor controlling embankment stability.

Embankments should be designed to provide an adequate factor of safety against slope instability. A landscaping embankment is unlikely to be more than 5 m high, and where a small embankment is built on firm, level foundations using well graded and well compacted granular fill, slopes of one vertical to two horizontals should be satisfactory. If a small embankment is built on firm, level foundations using a well compacted sandy clay fill, slopes of one vertical to three horizontals might be adopted. Where the embankment is higher than 5 m, or is being built on soft or sloping ground, or is being built of heavily over-consolidated clay, or there is some unusual feature, it is recommended that specialist geotechnical advice is sought.

The minimum acceptable width of the top of the embankment is likely to be governed by the practicalities of construction with earthmoving machinery, but it is unlikely to be smaller than 3 m.

Embankment construction

The method of deposition and spreading of the fill will depend on the earthmoving plant which is available. The fill should be placed in horizontal layers which are sufficiently thin to permit adequate compaction through the full depth of the layer with the compaction plant that is used. Many different types of compactor can be used in embankment construction; for example, pneumatic tyred rollers, vibrating smooth-wheeled rollers and vibrating sheep's-foot rollers. Each type of equipment is produced in a wide range of sizes: there are small, pedestrian operated vibrating rollers and large, towed vibrating rollers. The selection of compaction plant will depend on the type of fill to be compacted, the scale of the operation and, of course, the availability of the equipment.

Suitable types of compaction plant for the main types of fill are described in BS 6031[282]. It also gives guidance on many practical aspects of earth moving such as site clearance, wet weather, freezing conditions and haul roads.

A detailed specification for layer thickness for different types of fill and different types of compaction plant is available in the Department of Transport specification for highway works. However, this specification has a more exacting situation in mind than that of a landscaping barrier and therefore some relaxation of the specification should be tolerable.

Retaining wall design

This section deals with the general principles of design for masonry or concrete retaining walls. As already noted, it is recommended that the services of a structural engineer or other suitably qualified professional are engaged for any walls incorporating reinforcement or exceeding a height of 3 m.

The choice of concrete mix for the foundations of a retaining wall is important if they are to be strong and durable. If the soil conditions generally are dry, a wide range of concrete mixes, including those based on Portland cement, can be used. If the soil is consistently wet or damp, the foundations will need to be resistant to any sulfate salts in the water and the choice of mix should be governed by the concentration of salts present. If the soil is contaminated by industrial waste, an appropriate mix should be used. If precast units are to be used, the effects of contaminants on the units will need to be checked with the manufacturer. If there is any doubt over the stability of the soil, the water table, or the salt or acid content of the soil, advice must be sought from a geotechnical, civil or structural engineer, or similarly qualified person.

There are three primary modes of intact failure for a small retaining wall illustrated in Figure 5.6:
- overturning about the toe
- sliding
- bearing failure

Additionally, the wall may crack, usually by a horizontal shear failure in the brickwork.

An adequate factor of safety must be provided against each of these failure modes and, in addition, the structural strength of the wall must be adequate to ensure that the wall is not over-stressed. Furthermore, any slope of the ground in which the wall is constructed must be adequately stable against both short and long term failure. Where appropriate, the stability of the slope should be assessed by a competent professional using standard methods of analysis.

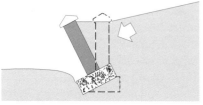
Overturning failure

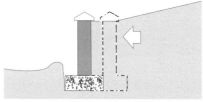
Sliding failure

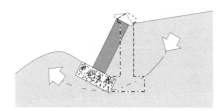
Bearing failure

Figure 5.6 Typical modes of failure in small retaining walls

The factor of safety against overturning is calculated by estimating the limiting values of the forces acting on the wall and taking moments about the most critical point, which would normally be the toe. The factor of safety, which is the ratio of the resisting moment to the overturning moment, must be at least 2.

The vertical thrust on the base of the retaining wall should not exceed the allowable bearing pressure for the soil. The calculation of the allowable bearing pressure should be based on settlement or ultimate bearing capacity calculations, with due allowance for the inclination and eccentricity of the loading and with the incorporation of suitable factors of safety: 2 for granular soils and 3 for cohesive soils.

Designing small retaining walls to remain initially stationary against lateral movement is unlikely to be practical. Indeed, to generate the sliding frictional resistance at the base of a wall's foundations requires some lateral movement. This movement can be limited by placing loose granular backfill between the wall and the retained soil or by using a proprietary compressible material such as lightweight expanded polystyrene which compresses slowly under pressure.

The construction of a wall may cause a change in the moisture equilibrium of a clay soil. Where there are large trees, the soil may be desiccated and the construction of a wall in their vicinity may cause some reversal of this desiccation by cutting through roots and by introducing a pathway for moisture to get into the soil. Construction is therefore likely to be followed by a period of swelling which, in a severely desiccated clay, may continue for many years. During periods of dry weather the clay may shrink away from the wall allowing a gap to open up and, during subsequent wet periods, the soil will swell back again to more or less its original volume. The changes that take place in the soil are largely reversible. However, there is a risk that backfill material or general debris will fall down the cracks and prevent the soil reoccupying its original space. When this happens, each cycle of shrinking and swelling may increase the earth pressure on the wall and slowly cause it to move forward. The provision of compressible material between the back face of the wall and the clay surface may help to minimise this effect.

For large retaining walls, methods of calculation have been developed to ensure adequate factors of safety against the common modes of failure. For small walls, calculations are not normally necessary and typical sections of masonry wall are accepted as suitable for the relatively small disturbing forces that will act on them.

If any of the following apply to a retaining wall, the advice of a chartered civil or structural engineer, or similarly qualified person should be obtained.

- The wall is higher than about 3 m above the top of the foundations
- There is supporting backfill on which banked up soil, stored materials or buildings are to be placed close to the wall
- The wall is higher than about 2 m above the top of the foundations supporting backfill on which vehicles or other heavy items will stand or pass close by
- retaining soil with a slope adjacent to the wall is steeper than 1 in 10
- The wall is supporting a fence of any type other than a simple guard rail
- The wall retains very wet earth, peat or water (eg a garden pond)
- The wall forms part of or adjoins a building
- The wall is not constructed of bricks or blocks, or is of dry masonry construction
- The wall is in an area of mining subsidence or other unstable ground
- The water table lies within 0.5 m of the underside of the wall foundations

Marginally under-designed retaining walls can be made safe by raising the ground on one or both sides to reduce the effective height.

Retaining walls in difficult ground conditions

Where the site has difficult ground conditions (eg abutting a river bank) special retaining structures will usually be necessary. These might involve piling, frequently of interlocking steel sheets forming a complete wall in itself. Provided the site conditions have been adequately assessed, structural failure of such a wall will be rare (Figure 5.7), though corrosion will ultimately take its toll.

Figure 5.7 Failure of a sheet pile wall

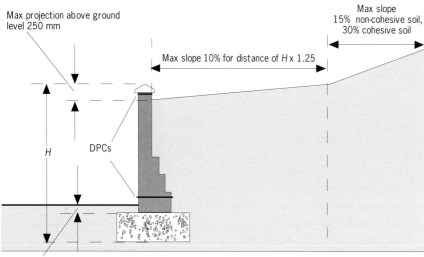

Figure 5.8 A common form of unreinforced retaining wall

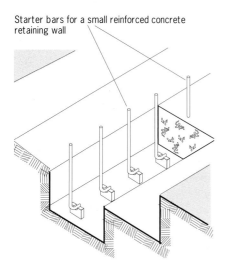

Figure 5.9 Starter bars for a small reinforced concrete retaining wall

Retaining wall construction
BRE Good Building Guide 27[283] provides suitable dimensions for some small masonry walls.

To avoid ground collapse, excavation of the ground prior to construction of a retaining wall needs to be undertaken with care (BS 6031 and BS 8004[68]). Faces higher than 1.2 m should be sloped back to a safe angle (a 45° or 1 in 1 slope is recommended). Exposed faces in cohesive soils should not be left unsupported for longer than necessary as the risk of instability increases with time. The ground may be excavated progressively and the wall constructed in panels.

Again, particular care should be taken when setting out the wall footings. With the exception of piered walls, the bases of all retaining walls must be constructed centrally on their foundations (Figure 5.8). BRE Good Building Guide 27 provides advice on locating wall footings. It also gives detailed advice on providing structural reinforcement to small walls. For the connection between the wall and its concrete foundations, at which considerable movement can occur, reinforcing bars should have a bend to optimise anchorage as illustrated in Figure 5.9. The orientation of the bend in the bar is not important.

Backfilling and drainage
The retaining side of the wall should preferably be backfilled with lightly compacted non-cohesive material. The use of intermediate to high-plasticity clay, or highly organic soils should be avoided (see BRE Digest 383[64]). A drainage layer of free-draining material such as coarse aggregate, clean gravel or crushed stone should be incorporated next to the wall. If the general backfill consists of fine grained material, it should be separated from the drainage layer by a geotextile filter fabric. This will prevent the drainage layer from becoming clogged. Advice on the selection of a suitable geotextile fabric can be obtained from the manufacturers.

The drainage layer must be able to discharge through weep holes in the wall (Figure 5.10). They should be at least 50 mm diameter (even 75 mm diameter holes may not be excessive) and spaced horizontally at not more than 1 m intervals. Weep holes should be incorporated near the base of the wall below, or within, the low-level DPC. To prevent water from reaching the foundations, concrete should be placed behind the wall below the weep holes (Figure 5.11). The infill concrete should have the same mix proportions as that used for the foundations.

After construction, it is important to advise the owner to keep the weep holes clear and free from obstruction.

Case study
Behaviour of an anchored diaphragm wall in stiff clay
An anchored diaphragm wall in stiff London Clay, supported by four rows of anchors, was instrumented by BRE investigators so that displacements, both surface and internal, could be monitored during and after excavation of an 8 m deep cutting. Pore pressures and anchor loads were also measured. The effects of the anchors were twofold:
- appreciable horizontal displacements occurred beyond the limits of the anchors
- the anchored zone moved as a block, with horizontal and vertical movements becoming quite large (up to 50 mm and 30 mm respectively). Horizontal strains in this zone reached up to 0.34 % (extension). Nevertheless, the component of displacement along the anchors was of the order of only 2–3 mm, and the loads remained nearly constant, indicating satisfactory performance of the anchors. Therefore, the successful installation of ground anchors does not necessarily preclude the possibility of high horizontal and vertical movements[284].

5.1 Embankments and retaining walls

Dampness

High-level and low-level DPCs are recommended in all walls that are not frost resistant (Figure 5.11). See BRE Good Building Guides 14[285] and 17[286], and BRE Digest 380[287] for general advice on DPC selection.

High level DPC detailing is more complex in staggered walls and flexible DPCs may be difficult to incorporate unobtrusively. This is because of the large proportion of overlaps necessary and the need to bed on both surfaces. If it is intended to use a flexible DPC at high level, it is advisable to check detailing with the brick and DPC manufacturers. Any flexible DPCs incorporated must have good bonding properties.

The preferred minimum solution for a low-level DPC is to build up from the foundations with DPC Type 1 or 2 clay bricks (or equivalent) to one course above the top of the weep holes. A more practicable solution is to form the total construction from the top of the foundations to approximately 200 mm above finished ground level with DPC Type 1 or 2 clay bricks (or equivalent). Flexible materials or slate should not be used for DPCs at low level.

One of the main problems in relation to the durability of materials in retaining walls in the past has been the lack of adequate provision for prevention of groundwater percolating through the retained soil and into the facing material. This leads to saturation of the brick or stone facings, and in turn to a greater risk of frost action. This is why so many retaining walls are provided with weep pipes. Weep pipes should ideally protrude sufficiently to ensure that water drips clear of the wall surface, though this may be a problem with battered facings.

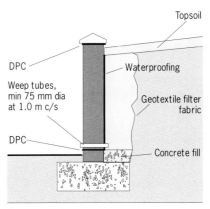

Figure 5.11 A retaining wall needs to be properly detailed to resist water penetration

Case study

Damp penetration through a retaining wall

Road salt was being stored on ground against a wall. Over the years chlorides dissolved from the stockpile, moved through the ground into the wall, and evaporated from the wall resulting in a salt deposit of around 10 mm thickness on the exposed face. The BRE Advisory Officer's proposed remedy was to excavate the earth away from the wall and to provide vertical asphalt tanking. The subsequent drying out of the wall and surface treatment would need to be monitored.

The case illustrated the enormous amount of water that can be carried through a wall which has no DPM.

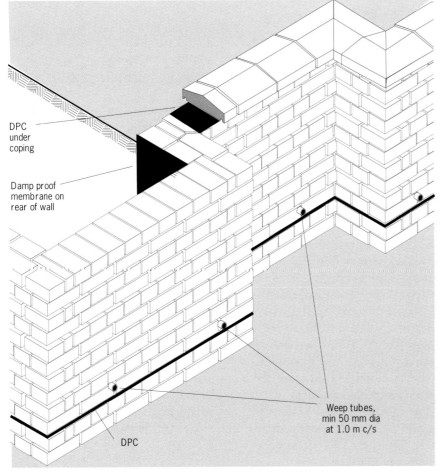

Figure 5.10 Weep holes or pipes are an important feature of the design of retaining walls where there is a high risk of water penetration from the retained soil. For low risk situations, a land drain buried behind the wall, above the footings, may suffice

5 Walls, fencing and security devices

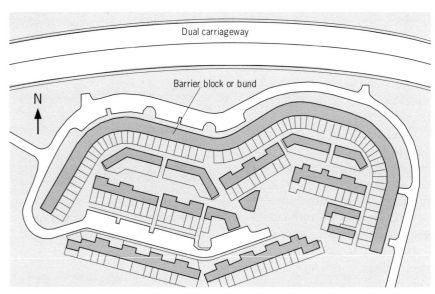

Figure 5.12 An embankment, or barrier block of dwellings, with minimum apertures on the traffic side, can be used to mitigate noise from road or rail traffic

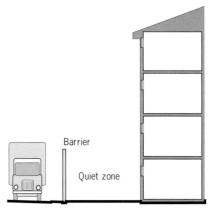

Figure 5.13 A quiet zone may often be created behind a barrier

Unwanted side effects
Gabion walls and crib walls are self-evidently permeable to water. Earth fines, under certain conditions, can be washed out from the retained banks to be deposited within the cages or cribs, providing support for plant life. Surface growths of moss and algae can be expected.

Traffic noise, barriers and bunds
In urban areas, particularly following the implementation of the Noise Insulation Regulations 1975[288], it has been common practice to provide earth embankments or other kinds of noise barriers, such as walls, fences and tree planting, to mitigate the effects of traffic noise (Figures 5.12 and 5.13). Since 1975, most schemes to which the Act applies will have been designed to meet predictions of traffic noise calculated in accordance with the two editions of *Calculation of road traffic noise*[289].

Common noise descriptors were described in *Walls, windows and doors*[101], which also deals with the sound insulation properties of the building fabric. The same criteria cannot, though, be applied universally to assess the value of noise barriers at site perimeters.

The calculation of entitlement to protection, and the prediction and measurement techniques are not dealt with here, but the following points may be noted.
- The main criterion for protection selected was 68 dB(A) $L_{A10,\,18\,hour}$. It is expected that this criterion will be amended in the future to reflect new thinking on units of measurement
- At least 1 dB(A) increase was required in the relevant noise level over the prevailing level
- Noise levels are measured at a point 1 m in front of the most exposed part of an eligible room
- Traffic flow is based on maximum flow on a normal working day
- The texture of road surfaces, whether concrete or bitumen based (and in the case of bitumen whether impervious or pervious) affects the noise produced

Noise from sources on the ground can be reduced by increasing the separation between the source and the building, but this is often impractical as the separation must be doubled to reduce the noise by 4–6 dB(A)[10]. Noise barriers can reduce the noise by 5–10 dB(A), but they must be:
- without holes (through which noise can pass)
- close to either the source or the building
- as tall as possible
- either longer than the building or returned at the ends

There is no doubt that the most effective barrier to noise is a raised embankment. The calculation procedures assume that performance of barriers can be enhanced with absorbent types of ground cover including vegetation. The actual performance achieved, however, will depend on many factors, and further advice from specialists may need to be sought.

5.1 Embankments and retaining walls

Durability
Traces of wicker cages used in former times may occasionally still exist where used below water tables. Steel wired gabion cages will depend for their longevity on the protection given to the wire. The most durable will be stainless steel, but ordinary steel wire will need to be protected by galvanising or other suitable covering.

The following steel wire specification is recommended:
- in sheltered locations, galvanised steel (galvanised after bending)
- in exposed locations, austenitic stainless steel
- in very acidic or corrosive environments, austenitic stainless steel should always be used
- the material used for tying wires and connectors must at least equal the quality of material used in the bars

The durability of precast concrete units depends entirely on the quality of the concrete, but, if hydraulically pressed from suitable mixes, they should achieve lives in excess of 30 years.

Frost resistant cappings or copings sized according to BS 4729[290] or other copings to BS 5642-2[291] should be used. The cappings or copings should incorporate an overhang and drip if the wall is not in a very sheltered location or is not going to be built of frost resistant bricks. Before a particular wall type is selected, the availability should be ascertained of a suitable combination of coping and DPC for the site conditions (see BRE Good Building Guide 17). If there is a risk of children playing on walls, or of vandalism, a wall type should be selected which allows the addition of interlocking capping (see BRE Good Building Guides 17 and 19[292]).

To accommodate cappings and copings it may be easier to finish piers or sections over staggers higher than the rest of the wall. Special bricks or concrete units are often available for cappings and copings at piers.

Work on site

Workmanship
For masonry accepted good workmanship practice should be followed (see BS 5628-3[161] and BS 8000-3[293]). A simplified construction sequence is given in BRE Good Building Guide 14. The following precautions are recommended:
- if the bricks are frogged, they should be laid frog up
- joints should preferably be finished with a bucket handle profile
- if the wall is to be rendered, the mortar joints should be raked back 10–12 mm (not necessary on blockwork walls)
- the render should be appropriate to the environment conditions. (See Chapter 10.2 of *Walls, windows and doors*[101])
- excessive loads on fresh mortar should be avoided, lifts not exceeding 1.5 m per day
- new masonry should be protected from frost, rain and wind
- 28 days should be allowed for the mortar to set before backfilling

---

Inspection

The problems to look for are:
- ◊ fissures or 'crumpling' in soil adjacent to retaining walls
- ◊ saturated masonry indicating no vertical DPMs
- ◊ no weep holes
- ◊ no rear land drains
- ◊ no DPCs at bases of walls
- ◊ no DPCs at heads of walls
- ◊ copings not anchored
- ◊ copings not durable
- ◊ non-durable bricks or blocks
- ◊ no handrails or raised parts of walls where pedestrian access is possible

Chapter 5.2 Freestanding walls

Freestanding walls, not forming part of a building, are widely used for boundary demarcation, landscaping, screening, security, and noise barriers (Figure 5.14). By their nature they are exposed to the weather and to wind loadings on both faces. If only a small part of a wall becomes unstable, this can lead to progressive collapse of the whole wall with a risk of serious injury to the occupiers or passers-by. These walls are not subject to building control except in the Inner London Area where approval is required for walls exceeding a height above ground level of 1.83 m.

Figure 5.14 A low boundary wall to a front garden in Yorkshire, adjoining a public pavement and built in a similar stone to that of the house

Characteristic details

Basic description

Most modern freestanding walls are constructed using brick or blockwork. In some parts of the UK where typical wind speeds are low, walls of up to 3.25 m in height can be built in sheltered areas using unreinforced masonry. However, reinforcement is likely to be needed for most walls exceeding a height of 2.5 m and, in certain parts of the UK, reinforcement may be needed for walls of no more than 0.4 m (Figure 2.8 in *Walls, windows and doors*[101]). Diaphragm wall construction can be used as an alternative to using reinforcement, but is unlikely to be economical unless the height of the wall exceeds 5 m. As a general rule, walls taller than 2.5 m should not be constructed without the advice of a structural engineer or other suitably qualified professional.

Piers will be needed where gates are to be hung, the design depending on the circumstances.

Main performance requirements and defects

Strength and stability
Walls
The structural design of a freestanding wall is based primarily on consideration of the wind loading that the wall will have to withstand. Wind loading is assumed to be proportional to the wind speed and design values – incorporating a suitable factor of safety – which are based on published wind speed data for the UK. The regional values have to be adjusted to allow for the local topography and the ground roughness (ie the size and frequency of any features that will act as windbreaks). The wall then has to be designed to ensure that the wind loading does not cause over-stressing within the wall itself or that it does not overturn about its base.

A simplified, rule-of-thumb approach to the design of unreinforced walls is described in BRE Good Building Guide 14[285] which is based on dividing the UK up into four wind exposure zones as shown in Figure 5.15. Within each exposure zone, proposed wall locations are categorised as being either sheltered or exposed. Sheltered locations are typical of urban areas, but may include other environments where there is considerable local interruption of wind flow. Exposed locations are typical rural areas or other areas where there is a clear view over open country. Where there is any doubt over whether a site can be classified as sheltered, specialist advice should be sought or, alternatively, the site should be classified as exposed.

For a given exposure, the rules-of-thumb can be used to determine maximum permissible wall height for common wall thicknesses. These are summarised in Table 5.1. The wall height should be measured from the lowest ground level to the top of the capping or copping. In the absence of other information, wall thicknesses that are intermediate between those listed should be built to the height given for the next smallest thickness.

5.2 Freestanding walls

Table 5.1 Rules of thumb for wall thickness and maximum height above ground

Zone	Wall thickness	Wall height limit (mm)	
		Sheltered	Exposed
1	Half brick	725	525
	One brick	1925	1450
	One-and-a-half-brick	2500	2400
2	Half brick	650	450
	One brick	1750	1300
	One-and-a-half-brick	2500	2175
3	Half brick	575	400
	One brick	1600	1175
	One-and-a-half-brick	2500	2000
4	Half brick	525	375
	One brick	1450	1075
	One-and-a-half-brick	2450	1825

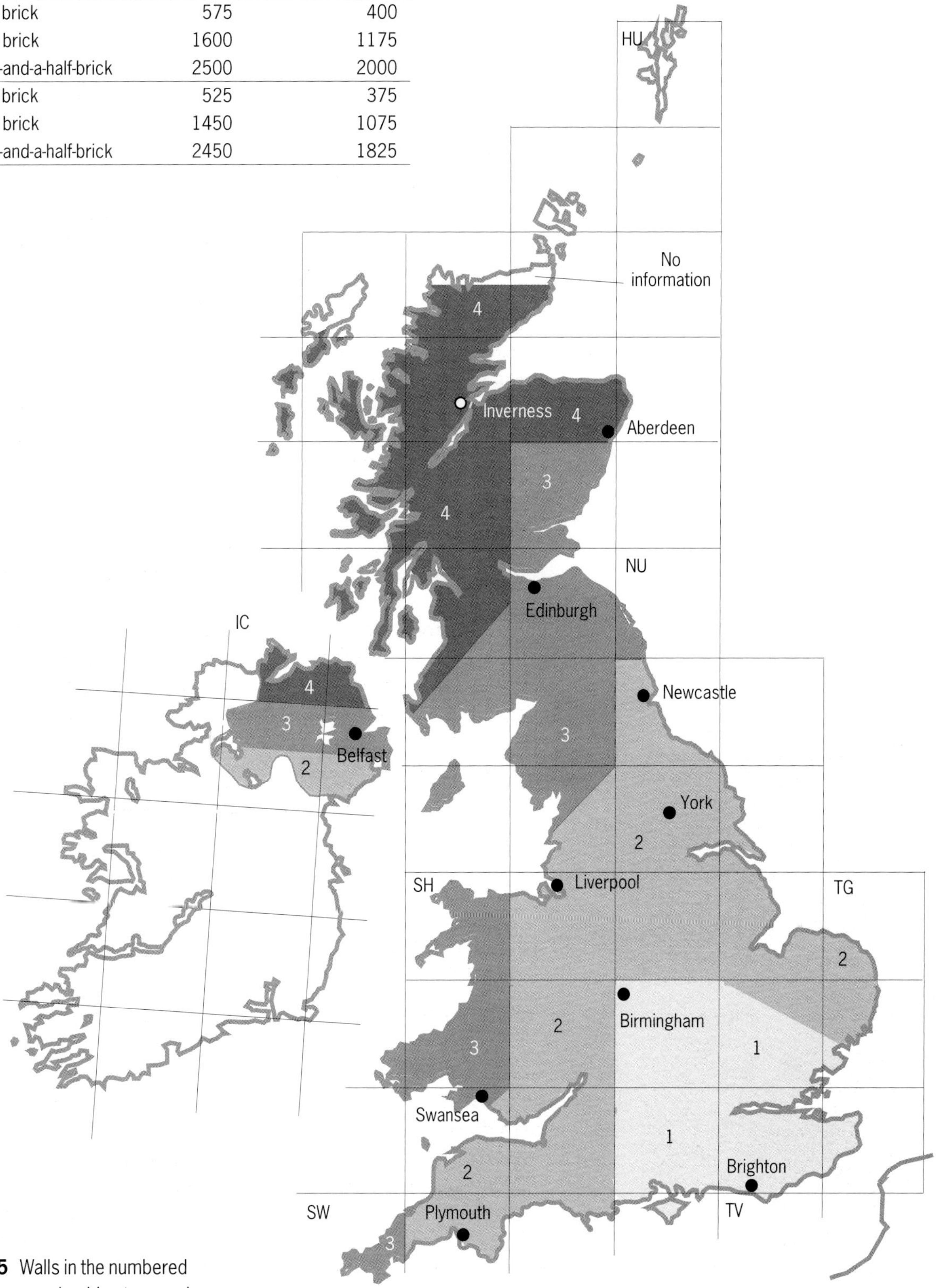

Figure 5.15 Walls in the numbered wind speed zones should not exceed the heights shown in Table 5.1

The figures given in Table 5.1 are for an average slope of up to 1 in 20. For slopes of between 1 in 10 and 1 in 20 the wall heights should be reduced by 15 %. Formal design procedures should be adopted where the slope is greater than 1 in 10. Special consideration should also be given to the need for formal design where the proposed wall may be subjected to influence by one or more of the following:
- vehicle impact (eg adjacent to a vehicle access area)
- pressure of a large number of people (eg adjacent to a public right of way to a stadium)
- excessive vibration (eg from heavy traffic)
- higher than normal wind loading (eg close to a medium or high rise building, on the crest of a hill, or near an extensive hill or mountain range)
- an atypical loading (eg having to support a large gate or door, or where the difference in ground level between each side of the wall exceeds twice the wall thickness)
- excessive ground movement (eg as a result of settlement of made ground, or shrinkage and swelling of clay soil)

For construction of one brick walls, a compromise between strength and appearance is achieved by laying the bricks to a particular pattern or bond; two bonds which are often used for freestanding walls are English garden wall bond and Flemish garden wall bond (Figure 5.16). As an alternative, walls of equal strength can be built using half brick (ie 103 mm) construction by staggering the designs on plan or by using piers.

Where the height of the wall exceeds the values given in Table 5.1, it will be necessary to adopt a formal design procedure incorporating reinforcement into the construction as required. The foundation width should be selected to ensure that the resultant thrust on the underlying soil from the wind loading and the dead weight passes through the middle third of the foundations.

Existing freestanding walls should be checked (even if plumb and undamaged) for theoretical stability, especially if adjacent to footpaths, play areas or on exposed sites[294]. The possibility of vehicle impact on walls around parking areas should be considered, together with the provision of suitable kerbing and edging. Existing walls can be thickened or buttressed if improved stability is required.

Foundations for freestanding walls
The foundation requirements for freestanding walls are generally less onerous than those for buildings. For walls not forming part of a building and not exceeding 2.5 m in height, a foundation depth of 0.5 m is normally considered adequate for most ground conditions. For higher walls in cohesive soils, the foundation depth should be increased to at least 0.75 m. Consideration should also be given to deeper foundations where walls are to be founded on highly shrinkable soils and in close proximity to large trees or where large trees have been recently removed. In these circumstances, the required foundation depths should be based on the recommendations for house foundation depths (eg using those given in the NHBC Standards). For most applications it should be possible to found a freestanding wall at, say, half the depth recommended for house construction.

Foundation width is more likely to be governed by the need to resist the overturning moment generated by wind pressure rather than allowable bearing pressure. Nevertheless, some consideration of allowable bearing pressure may be necessary for construction on soft clays, loose sands and made ground. Widths for concrete strip foundations are tabulated on a rule-of-thumb basis in BRE Good Building Guide 14. These tables are not repeated here, but, as examples:

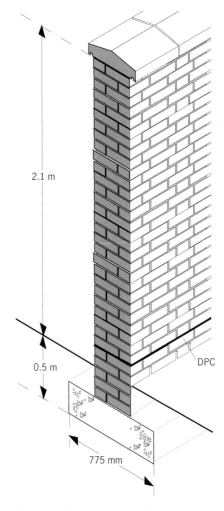

Figure 5.17 In an exposed location within zone 2 (Figure 5.15), a wall of 2.1 m height would need a foundation width of 775 mm

Figure 5.16 A Suffolk boundary wall built with a fairly soft brick in Flemish bond with a plinth just above pavement level. Ivy has a detrimental effect on durability and should be cleared

5.2 Freestanding walls

- a half brick wall of 525 mm height above ground in exposure zone 4 (Figure 5.15), but in a sheltered location within that zone, would need a foundation width of 375 mm. If the same wall was in an exposed location in zone 4, the foundations would need to be a little wider, 400 mm
- a one and a half brick thick wall 2.5 m high in exposure zone 2, but in a sheltered location, would require a foundation width of 525 mm. In an exposed location within the same zone a wall of 2.1 m height would need a foundation width of 775 mm (Figure 5.17)

Wherever possible, construction on made ground should be avoided unless the fill has been properly compacted. Where construction on uncompacted made ground is unavoidable, estimation of the allowable bearing pressure should not exceed 25 kPa for newly placed fill or 50 kPa for older fill. Moreover, it may be necessary to remove any local soft spots or pockets of organic material and backfill using lean mix concrete. It may be necessary to accept a certain amount of settlement and foundations should be reinforced longitudinally to limit the distortion to the wall.

Security

Low walls tend to attract human climbers. The worst configuration is where a wall provides a step between the ground and the roofs of flats, garages or sheds. If a wall is intended to keep out trespassers, it will need to be at least 3 m high with no foot holds in the masonry nor hand holds on the coping. To deter the criminal, broken glass set in cement mortar or freely revolving spikes have been traditional solutions. Razor wire is a more recent invention – a sad comment on current society. However, using glass, spikes or razor wire to deter entry can have legal implications for property owners.

Figure 5.18 A brick which is insufficiently durable for a freestanding wall is block bonded into an adjacent stretch of wall in another type of brick, but there is no movement joint. The wall has therefore created its own movement joint in the form of a crack

Figure 5.19 A boundary wall of a prestige building. Although the coping is substantial in size, it does not respect the movement joint beneath, with risk of subsequent distress

Figure 5.20 An urban brick boundary wall in stretcher bond. The design of the coping makes it inevitable that rainwater run-off will disfigure the front face

Durability

To ensure durability it is important that the specification for a freestanding wall should include the following:

- the right type of brick or block (Figure 5.18)
- the right type of mortar and, where appropriate, render
- providing of adequate dampproofing
- correct positioning of piers and movement joints (Figure 5.19)
- providing adequate cappings or copings (Figure 5.20)
- protection from vehicle impact, where appropriate

Guidance on how to comply with these requirements can be found in BRE Good Building Guide 14 and in the Brick Development Association publication, *Design of free standing walls*[295].

Freestanding walls are often cracked or leaning, too tall or too slender, or with mortar breakdown; and they can include frost damaged bricks, blocks, stone or mortar, and missing copings. Existing boundary walls should be robust and have a reasonable life expectancy; if not, they should be repaired or replaced.

The longevity of freestanding walls of brick or stone will depend on many factors, but a well built wall of suitable materials, and equipped with DPCs at the foot and under the coping, should last nearly as long as the external walls of buildings built in similar materials – say 100 years. Some copings are prone to vandalism, especially those placed over DPC materials to which bedding mortar does not adhere, and proprietary anti-vandal features may need to be specified in replacement work.

Case study

Problems with external rendered finishes on screen walls

There had been considerable worsening of the defects from inadequacies in the dampproof arrangements at wall-head and at ground level on an exposed site in Scotland. A BRE Advisory Officer found that the rendering on the walls had been carried down to ground level and, in places, inadequate drainage had been provided at the base of walls. The beginnings of sulfate attack were seen in the form of horizontal cracks in the rendering, but there was no tilting of the walls. However, some doubts were expressed by the Advisory Officer on the long term satisfactory performance of parts of the brickwork since the bricks used were not of the special quality recommended for wet conditions. Because of this the main objective of the remedial measures suggested, including replacement of the render and the provision of improved cappings and DPCs, was aimed at stopping further sulfate attack by minimising the penetration of rain into the walls.

Cracking in freestanding walls

Masonry walls can be subjected to a number of factors which result in cracking; any cracks that appear may or may not require attention[296]. The following rule-of-thumb criteria apply to walls of adequate thickness for their height and where no significant other defects are present:

Cracking which can normally be ignored or which have only cosmetic significance:
- single hairline cracks and cracks up to 5 mm wide, substantially vertical, in either mortar or through both masonry and mortar, which are not near piers or changes of section
- more than one crack, each up to 5 mm wide as above but separated by a distance equal to or greater than the wall height

Cracking which requires monitoring or further investigation:
- single hairline cracks up to 5 mm wide, substantially vertical, in either mortar or through both masonry and mortar within two brick lengths of a pier or change of section
- multiple hairline cracks in masonry, mortar or render
- more than one crack each up to 5 mm wide as above, but separated by a distance less than the wall height
- single predominantly vertical cracks over 5 mm wide, extending no more than 600 mm horizontally
- single horizontal cracks less than 600 mm long through the complete wall thickness but not in a pier

Cracking which normally indicates a need for repair (Figure 5.21):
- single raking or vertical cracks over 5 mm wide within two brick lengths of a pier, wall end or change of section
- single cracks adjacent to a pier or change of section
- single cracks within two brick lengths of a gate pier
- single cracks over 5 mm wide extending more than 600 mm horizontally
- single horizontal cracks more than 600 mm long penetrating the full wall thickness
- single horizontal cracks completely through a pier

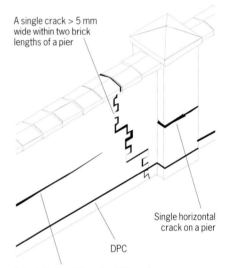

Figure 5.21 Examples of cracking in freestanding walls which will normally require repair

5.2 Freestanding walls

Work on site

Inspection

Walls should be pushed to see if they are 'live' (ie can be moved by hand pressure alone – but care is needed if further damage or personal injury is to be prevented), checked for plumb, and their thickness and pier arrangements noted. If no distress is visible in any of the existing adjacent landscape features, problems will probably arise only as a result of exceptional conditions (flood, storm, subsidence), changes due to rehabilitation (in particular, increased traffic loads) or to natural unchecked growth of trees and other plants. Existing distress should be diagnosed – cracks may require monitoring to establish whether movement has ceased and if any remedial work is required.

The problems to look for are:
- part or complete collapse of wall structures
- piers inadequate
- poor foundations
- poor bonding
- tree growth near walls
- sulfate or salts damage
- lack of movement joints
- walls out of plumb (beyond the middle third)
- weak or friable mortar eroding
- copings poorly fixed
- cracking
- no DPCs at bases of walls and under copings
- walls rendered on both faces

Chapter 5.3 Fencing

This chapter deals with fencing in the immediate vicinity of buildings. It does not deal with fencing for agricultural land or prisons.

There is an enormous variety of fencing in existence and this chapter can only cover some of the more important considerations, though it pays particular attention to durability, an area in which BRE has been extensively involved.

From Victorian times, where no visual screening was required, urban fencing has chiefly consisted of wrought or cast iron standards restrained by upper and lower rails, and caulked into holes formed in the copings of dwarf brick walls. The flame-cut remains of many thousands of these railings can still be seen – a reminder of the scrap iron drives of the 1939–45 war. Fortunately, railings of some prestige buildings were spared this fate (Figure 5.22) and they continue to grace the urban scene.

Figure 5.22 Decorative cast iron railings at the British Museum

Characteristic details

Types of fencing
The main types of fencing are covered in the various parts of BS 1722.

Wire
- chain link
- rectangular wire mesh and hexagonal wire netting
- strained wire
- anti-intruder fences in chain link and welded mesh

Wood
- woven wood and lap boarded
- cleft chestnut pales
- close boarded
- wooden palisades
- wooden posts and rails

Steel
- mild steel (low carbon steel) continuous bar
- steel palisades
- open mesh steel panels

Steel guard rails and fences are also covered in BS 7818[297]. The standard covers guard rails for carriageways, footways, bridleways and cycleways.

Heights of fencing
The height of a fence will normally be related directly to its purpose, as well as to the building type being fenced. Heights quoted in British Standards include 0.9 m, 1.2 m, 1.4 m, 1.8 m, and 2.15 m, which, for those fences composed of mesh or link, corresponds with the width of the material in the roll.

The lowest standard heights are used for domestic garden fencing at the fronts or sides of individual dwellings. It is possible for an intruder to vault a fence 0.9 m high. Where the property adjoins other uses such as public land, railways, commercial or industrial property, the height requirements will often be dictated by security considerations and may need to be increased. For the highest risks, where intruders need to be prevented from access, the heights of fences may be increased still further by 340 mm extension arms, which may be straight or cranked.

Guard rails for pedestrians and cyclists are normally around 1 m, but those for horse riders need to be 1.8 m.

Where the fence or rail is built on sloping ground, it may either follow the slope, or be stepped. In the latter case the posts will need to be longer than for the same kind of fence following the slope. To quote one example from the Standards, BS 1722-14[298] indicates that fences may be stepped for slopes up to 20° or follow the ground surface for slopes over 20°.

Steel wire fencing

The simplest kind of fencing where no visual privacy is needed but where some protection is desirable is that made of wire netting on posts of steel, wood or concrete. Increased durability is provided by chain link instead of wire netting. The essential difference between these, of course, is that wire netting is twisted together in manufacture to form a hexagonal pattern, whereas chain link is simply adjacent wires looped together to form an interlocking diamond pattern.

Sizes for wire fencing depend very much on its purpose. Straining posts and struts are normally used at all ends, corners and significant changes in levels, but also on straight runs. Intermediate posts will normally occur at around 3 m centres.

Occasional examples may be found where the lowest 300 mm or so of wire has been buried below ground level either as an anchor for added security or as a precaution against burrowing animals.

Chain link netting

For chain link, the mesh size is normally 40 or 50 mm, but the wire diameter can be 2.5 mm (medium), 3 mm (heavy) or 3.55 mm (extra heavy) (BS 1722-1[299]). Gates in chain link fencing are normally of steel framing, and infilled in exactly the same type of material as the remainder of the fence.

Wire netting

This type of fence, including chicken and sheep netting is usually to be found in agricultural applications, but some varieties have been used in domestic situations (BS 1722-2[300]). There are various sizes of mesh, and heights, depending on the type of animals to be constrained. Materials include posts of rolled or hollow section steel, or of timber or reinforced concrete; wire of spring or high tensile steel mesh is also available. Straining posts are normally positioned at all corners and ends, and also at not more that 150 m centres on long runs. Intermediate posts are normally at not more than 3.5 m centres.

Strained wire fencing

Strained wire fences can be supported on posts of rolled or hollow section steel, timber or reinforced concrete (BS 1722-3[301]), as with wire fencing described above. Although the wires are strained tightly, the distances between intermediate posts is frequently too great to entirely eliminate sag in the wires. Droppers of wood or steel are therefore introduced, which act as spacers for the wires. These droppers can be either the full height of the fence, or part of the height, in which case they need to overlap (Figure 5.23). The wires are maintained in tension for the life of the fence, and the Standard specifies a tension of not less than 1600 N four days after tensioning. Old fences constructed in the early part of the twentieth century will often be found to have straining posts with built-in ratchet tensioners, whereas many later fences will simply have separate ratchet tensioners or eye bolts.

Anti-intruder fences

Although no fence can be considered to be proof against determined intruders equipped with bolt cutters, fences of a more robust construction do offer a good standard of protection against the casual intruder.

The type of fence described in BS 1722-10[302] is mounted on substantial concrete posts, with main posts cross-section 150 × 150 mm and intermediate ones 125 × 100 mm. Alternatively steel rectangular hollow or circular hollow section posts or rolled steel sections of similar robustness can be used. Zinc or plastics coated chain link or welded mesh consists of 3.0 mm diameter wires, and the tops of the posts carry cranked or straight extension arms with anything up to three strands of barbed wire or a razor wire.

Wood fencing

The British Standards for the particular types of wood fencing give various criteria for wood quality in new construction; for example, knot area ratio is defined together with control over splits, particularly splits at the ends of posts. Defects such as past evidence of insect attack or rot are also controlled in newly produced supplies, for example not more than 20 pinholes in a 0.3 m length of post or rail.

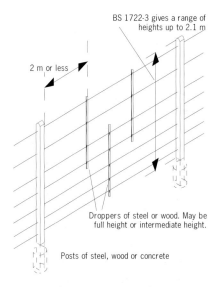

Figure 5.23 A strained wire fence to BS 1722-3

Figure 5.24 Lapped wood fencing panels. In this situation, lining a public footpath, the panels will probably not survive more than a few years

5 Walls, fencing and security devices

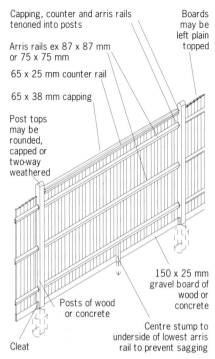

Figure 5.25 A typical wooden close boarded fence to BS 1722-5

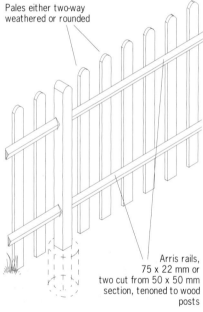

Figure 5.26 A typical wooden palisade fence

Woven wood and lap boarded fencing

Although giving a good standard of privacy, some of the least robust of fences are used almost exclusively in domestic situations (Figure 5.24 on page 213). Heights available range from 0.6–1.8 m. Posts may either be of square cross-section in timber or plain or slotted, or slotted reinforced concrete (BS 1722-11[303]).

The slats or boards for woven fencing are usually 75 × 5 mm thick, held in place by rectangular sawn battens. Lapped boards may be square, waney or feather edged, and may be slightly more robust than the woven kind.

Cleft chestnut pale

The cleft chestnut fence is a traditional type of fence, much used for general purposes, though mostly in domestic roles (BS 1722-4[304]). The wiring encircles the pale during manufacture, giving a robust construction from which individual pales are not easily removed.

Heights can be from 0.9–1.8 m, the Standard specifying six standard heights altogether. The wiring will be at heights varying between 450 and 750 mm, depending on the total height of the fence and the level of security required. Pales spacing is 50, 75 or 100 mm centres, the closer ones used for the higher security situations. Intermediate posts are normally spaced between 2.25 and 3 m for concrete, and slightly closer for wood, again depending on the level of security required.

Close boarded fencing

This type of fence has been in wide use for many years, following more or less the same design as described in BS 1722-5[305]. Wood is usually either treated softwood, or oak left untreated (Figure 5.25).

Wooden palisade

The wooden palisade is a largely decorative type of fence used mainly in domestic situations, but sometimes seen elsewhere (BS 1722-6[306]) (Figure 5.26). Fences with inclined palisades are sometimes met with, and occasionally they may be found with palisades laid alternately each side of the rail.

Wooden post-and-rail fencing

This type of fence is not entirely stock-proof. Heights are usually around 1 m or just over (BS 1722-7[307]). Posts can be sawn rectangular, cleft or left in the round. Main posts are usually 75 × 150 mm at 2.85 m centres where the rails are morticed and 1.8 m centres where they are butt jointed. Thinner posts may used at intermediate positions – these are called prick posts – and are usually around 38 × 87 mm in section.

The rails, either two, three or four in number, depending on circumstances, are usually rectangular and may be either sawn or cleft. Where relatively slender posts are used, the rails are usually scarf-jointed, although wider posts give the opportunity for butt-jointing.

Steel panel and pale fencing

Continuous bar fencing and hurdles
These types of fences have been in widespread use for many years normally for farming purposes, and largely for the containment of the larger farm animals. In the early twentieth century, these fences were typically made of wrought iron. Many examples dating from this time have survived for upwards of 75 years and are still capable of giving further service. The Standards refer to light, medium or heavy duty quality (BS 1722-8[308] and BS 1722-9[309]).

5.3 Fencing

Steel palisade
Even the ordinary qualities of steel palisade fences offer an enhanced standard of security over wire fences (BS 1722-12[310]). They are available in heights of up to 2.4 m, and, where greater security is needed, in heights of 2.4 m, 3.0 m and 3.6 m. Steel posts are normally set at 2.75 m centres. Pales for normal use are 2.5 mm thick steel, and for greater security, 3 mm thick. They have been available in a variety of profiles, but are usually of vee or other shaped and rolled profiles to give greater strength. A very great variety of tops is available, including plain, rounded, rounded and notched, single pointed, and triple pointed with several different profiles within each category.

High security fencing pales may also be embedded 350 mm into the ground or 150 mm into concrete to discourage burrowing underneath. Finishes are metal coatings to BS EN 22063[311] or hot dip galvanising to BS 729[312].

Open mesh steel panel
This is a basic type of fence for both ordinary use (at 2.4 m high) and for various standards of security, including high security and maximum security (at 3.0 m high) (BS 1722-14[298]). All are available with extension arms, both straight and cranked.

The mesh is constructed from very thick wires, up to 5 mm thick, spaced at 12.5–75 mm centres and welded into panels, giving a solidity of from 12% to a maximum of 58%.

Sheeted steel
Fences sheeted in corrugated or other forms of shaped galvanised steel will occasionally be seen. The galvanising of corrugated sheets is by dipping after manufacture so that cut-end corrosion ought to be eliminated. They will not necessarily behave in similar fashion to the same materials used on pitched roofs, since the conditions of service will be different (see the section on durability later in this chapter).

Gates
For prestige buildings where a reasonable degree of security was (and still is) required, the traditional solution has been to provide gates in decorative wrought iron (Figure 5.27) hung from brick piers.

In recent years, designs have become more utilitarian, with materials for smaller gates being of timber or wrought iron, and larger gates for vehicle access more or less confined to steel tube, bar and sheet (Figure 5.28).

Main performance requirements and defects

Specification considerations
When preparing a specification for a fence, once the basic pattern is established it will be necessary to decide on the following:
- the materials to be used (particularly for posts)
- the excavation for posts
- the degree of security to be achieved
- the type of infill required (remembering the potential for foot and finger holds)
- the life expectancy and therefore the degree of protection to be afforded to the fencing materials

Many permutations are covered in British Standards. The topping of the security fencing will also need to be considered; for example, barbed wire, barbed tape or rotating spiked vane (BS 4102[313]).

Strength and stability
Installation Standards normally refer only to what is called normal ground conditions, and where harder or softer ground is met with, alternative installation features may be called for. Where fences offer a virtually solid barrier to the wind, it will be self-evidently necessary to consider the effects of strong winds and to carry out the necessary calculations to ensure stability. Some British Standards give test procedures for deflection under both wind and dead loads.

Figure 5.27 A pair of wrought iron gates made for a country house by a skilled blacksmith around 1910. Although wrought iron is reasonably durable, unless gates of this kind are regularly maintained, they will gradually deteriorate. The main problem is corrosion of the inside surfaces of the bands which hold adjacent scrolls together

Figure 5.28 Controlled access vehicle and pedestrian gates in steel bar and tube with bracing in sheet steel

Wire fencing

Wire fencing relies for its strength on the frequency of posts, and the use of taut straining wires between the straining posts which are usually strutted. They are largely transparent to wind forces.

So far as the wires themselves are concerned, BS 443[314] covers wires of different tensile strengths from 0.33–10 mm diameter, and coatings of from 45–305 g/m².

Wood fencing

Wooden fences are not transparent to wind forces. Even if erection has not been faulty, they deteriorate and become, with the passage of time, increasingly liable to damage in strong winds.

Steel panel fencing and guard rails

Various performance tests are specified in British Standards. For example, steel palisades are required not to deflect more than 8 mm under a load of 2.5 kN for normal fencing, and not more than 10 mm under 3.5 kN loading for security fencing.

So far as guard rails are concerned, there are four main classes, shown in Table 5.2.

Durability

Existing fences should be robust and have a reasonable life expectancy; if not, they should be repaired or replaced.

Fences may have corroded, rotted or have insufficiently deep posts. Concrete posts carbonate and spall. Preservative treatment of wood used in fencing is covered in BS 5589[156].

Wire fencing

Chain link fencing is normally available with plain zinc, zinc alloy or plastics coatings (BS 1722-16[315]). The wire itself in high tensile steel or low carbon steel.

The longevity of wire mesh seems to depend very heavily on its siting. For example, in an urban location, a fence seen by BRE investigators virtually disappeared in 35 years, whereas, in a rural situation, a 35 year old fence is still functioning with just a little surface corrosion where the galvanising on stirrups has been rubbed away.

There seems to be very little written information on the durability of steel wire used in fencing, although from the 1960s to the 1980s a USA-led investigation did carry out research at a site in England[316]. Various types of wire including single strand 12.5 gauge (say 2.5 mm) barbed, 9 gauge (say 3.75 mm) chainlink, and 7 wire strand were investigated over approximately 20 years. The English site, monitored for 13 years, had high levels of sulfur dioxide contamination. Loss rates of the coatings were measured, and varied considerably over all the sites, depending in large degree on the presence of industrial emissions in the atmosphere. Metal loss rates were lowest with hot dipped aluminium coated wire, with hot dipped zinc next lowest, then electroplated zinc, and the highest loss rates were experienced with aluminium powder clad wire. The English site had the highest loss rates for aluminium coatings, though not for zinc coatings. All wires suffered a loss in breaking strength over the exposure period, some by as much as 60%. It was noted that copper-containing steel wire was more resistant to corrosion than copper-free steel wire.

Figure 5.29 Forty year old woven wood fence panels, probably larch. Although rot-free, they become very brittle with time and are easily damaged

Since loss rates are fairly consistent, doubling the thickness of protection will yield approximately double the life. A 940 g/m² zinc coating can be expected to give around a 30 year life, depending on levels of pollution.

In the experience of BRE investigators, unprotected mild steel tying wires and stirrups are commonly found, and these are usually the first parts of a fence to show rusting. When these eventually part, the mesh or chain link detaches from the horizontal straining wires, and collapse of the mesh or chain link follows.

The durability of posts and standards needs to be considered as well as that of the wire. It will rarely be practicable to renew corroded wire without at the same time renewing posts of wood. Even preservative treated softwood posts will rot in time. In one wire fence examined by a BRE investigator which had around fifty preservative treated intermediate posts approaching 35 years of age, approximately half had rotted off at ground level (but see the next section). The fence, though, was still in position because the straining wires were strung between secondhand railway sleepers used as straining posts. These sleepers had been steeped in creosote before their railway use.

Table 5.2 Criteria for the performance of guard rails		
Quality	Design loads, longitudinal members (N/m)	Maximum deflections (mm)
Light	500	50
Normal	700	50
Heavy	1400	40
Special	2800	40

5.3 Fencing

Figure 5.30 Non-preservative treated oak fence to BS 1722 after 20 years life

On the other hand, the wrought iron standards of one rural wire fence erected in the early twentieth century, and examined by a BRE investigator in 1999, were still in serviceable condition with only surface rust discoloration. However, the few remaining galvanised wires were heavily corroded and the galvanised steel sheet droppers had almost entirely disappeared.

Wood fencing
Wood fencing, if specified according to environmental and working conditions, ought to last at least 30 years, though the less robust materials may easily suffer from vandalism (Figure 5.29).

Oak is no guarantee of longevity, in spite of its reputation. It all depends on how much heart wood is included in each unit. In a close boarded, non-preservative treated oak fence built in 1970 to the then available edition of BS 1722, and examined 20 years after construction, one post out of eight had rotted at ground level. On the other hand, the remainder of the fence, apart from the capping which had slightly deteriorated, was in reasonably good condition, and assessed as well able to last a further 10 years (Figure 5.30).

BRE investigators have carried out a substantial amount of research into the longevity of timber, buried in the ground, which had not been treated with preservative. The results of the research have given a good indication of the likely durability of fencing posts. The experiments involved driving 610 × 50 × 50 mm stakes 380 mm into the ground. They were examined at intervals of 6–12 months, and struck with a wooden mallet to see if they broke[317].

Typical results are shown in Table 5.3

In interpreting the results of these tests, deterioration is directly related to minimum thickness so that a 100 × 100 mm stake should give double the above lives. On the other hand, a 50 × 100 mm stake would only be expected to give the same as a 50 × 50 mm stake.

Preservative treated timber may be expected to give significantly longer lives. Generally speaking, the stronger the preservative, the greater the durability, the most effective formulations being creosote or those containing only copper, chromium and arsenic, either as salts or as oxides[318].

In some field tests, BRE investigators compared the performance of copper-chromium-arsenic (CCA) (3% w/w to BS 4072[319]) and creosote (at 82 °C to BS 144[320]) in hardwood and softwood stakes exposed externally in a similar way to the untreated stakes described above. To a considerable extent, the durability achieved depends on the penetration of the preservative, which may be poor with hardwoods. Some of the results are shown in Table 5.4

The research from which the above data was produced is still ongoing and has shown evidence that creosote treated spruce still maintains its integrity after 49 years.

Steel panel and pale fencing
The durability of steel fencing depends on the degree of protection afforded to the steel and on the local pollution levels. A detailed examination of the different forms of pollution and their effects on metals is to be found in Chapter 1.9 of *Roofs and roofing*[97]: the major influences affecting the durability of steel fencing above ground include sulfur dioxide near industrial areas and chlorides in coastal situations. Below ground, the risks on brownfield sites are mainly from industrial pollution, and on greenfield sites from the nature of the soil and subsoil.

Table 5.3 Natural durability of timbers in the ground

Name	Mean life to breaking (years)
Western hemlock (*Tsuga heterphylla*)	6.0
Scots pine (*Pinus sylvestris*)	7.4
European larch (*Larix europea*)	12.0
Leyland cypress (*Cupressocyparis leylandii*)	14.9
European oak (*Quercus robur*)	26.8

Table 5.4 Durability of treated timbers in the ground

Name	Mean life to breaking (yrs) CCA treated	creosote treated
Ash (*Fraximus excelsior*)	13.5	>24.0
Elm (*Ulmus procera*)	10.0	12.7
Birch (*Betula* spp)	11.6	22.7
Scots pine (*Pinus sylvestris*)	>24.0	>24.0
European larch (*Larix europea*)	>24.0	>24.0
Norway spruce (*Picea abies*)	>24.0	>24.0

The Ministry of Agriculture Fisheries and Food publishes a map every few years illustrating the average atmospheric corrosivity rate for 10 km grid squares of the UK. This map is based on a zinc reference database but corresponding corrosivity rates for other metals (steel, aluminium, copper and brass) can be calculated using Dose Responsive Function relationships. This may help to assess long term durability of corrosion protection or of the proposed solution to a corrosion problem in a particular location[321].

Galvanised steel fencing components for use in normal conditions of exposure should be protected by a minimum of 350 g/m² zinc coating on both sides or an aluminium alloy coating of 135 g/m². In a marine environment, a minimum zinc coating of 275 g/m² each side with factory applied organic coatings to BS 5427[322] is suggested, or a zinc coating of 350 g/m², a coat of primer and two coats of paint. These should give lives of at least 10 years depending on the exposure conditions. For lives in excess of 10 years, a heavier coating of zinc would be needed (eg 610 g/m²) which can only be applied to a preformed sheet. As stated in *Principles of modern building*[9], the durability of corrugated sheet steel depends on its effective and continuing protection. BS EN ISO 1461[323] may be appropriate. This Standard also covers coatings containing not more than 2% of metals other than zinc. Powder coating on galvanising ought to give a life of around 30 years before repainting becomes necessary.

Coating thicknesses may sometimes be given in g/m² and sometimes in mm. For zinc, 1 mm is approximately 7 g/m².

Stainless steel used in fencing can retain its bright finish for many years given appropriate local conditions. Where a dull finish is required, matt finish stainless steels are available.

Maintenance
Wire fencing
It is possible to restore some protection to wire which has been damaged, perhaps in erection of the fence. The damage should be protected with two coats of zinc-rich paint to BS 4652[324].

Concrete posts
So far as posts are concerned, those of concrete can be subject to severe deterioration. Steel reinforcement is normally protected against corrosion by the alkaline environment of fresh concrete. However, carbon dioxide gradually penetrates from the concrete surface inwards, neutralising the alkalinity as it progresses (carbonation). There is never much cover available in slender concrete posts, and when carbonation reaches the reinforcement, the steel is much more vulnerable to corrosion. In bad cases, spalling of the concrete cover can result (Figure 5.31) and repairs are not feasible.

Wood fencing
Maintenance of the original treatment of wood fencing may be impractical if it has been treated with preservative. Recoating may be possible only if the original treatment can be identified; creosote is one of the most straightforward systems to identify.

Deteriorated paint systems on wood fences will need to be prepared and renewed in the same way that wood surfaces are treated on buildings. Reference should be made to Chapter 9.4 in *Walls, windows and doors*, although some of the main considerations may warrant repeating.

Solvent-borne paint systems
Traditional paints containing lead pigments were very durable, but their use is now restricted to certain historic buildings. Most paints for external use on wood are now based on solvent borne alkyd resin systems, and a three coat system, consisting of primer to BS 5082[325] or BS 5358[326], and undercoat and top coat to BS 7664[327], is still the most popular finish.

Microporous paint systems
Some paints, both solvent and waterborne, are available which are described as microporous or breathing paints. There is no evidence that the long term moisture contents of wood under these paints differs significantly from those under conventional systems, though there may be a short term benefit. Durability will be expected to be similar to that obtained from more conventional systems.

Moisture-permeable paint systems
Waterborne paints are more permeable to moisture than conventional paints and have high levels of film extensibility which is retained during weathering; therefore these tend to be more durable than solvent borne alkyd paints. They are slow drying in adverse weather conditions.

Figure 5.31 Spalling of the concrete cover to steel reinforcement. At worst, the original cover in this example cannot have been much more than around 15 mm

5.3 Fencing

Wood stains
Exterior wood stains consisting of lightly pigmented dispersions of resins and fungicides are available in zero-build, low-build (10–15 μm) and high-build (30–40 μm) versions. The zero-build and low-build versions are moisture permeable and tend to erode rapidly. High build stains are more resistant to moisture transmission, and the substrate is therefore less affected by moisture induced movements.

Varnishes
Varnishes of unpigmented resin solutions give coats of up to 80 μm thicknesses. These have in the past tended to be of relatively short durability, although improved durability formulations have been developed. They do not, however, prevent long term colour changes to the underlying substrate.

Creosote
Creosote has long been used successfully on agricultural and garden buildings, and has a good track record. However, it fades, and cannot be coated over with alternative treatments.

Steel panel fencing

In general the British Standards for steel panel fencing do not specify the maintenance requirements of the fences. Nevertheless it is important to make regular provision for maintenance after installation. Where rust begins to show, patchwork painting with a zinc rich paint may be sufficient, until visual unacceptability dictates overall painting with a suitable system compatible with the original coating.

Chapter 9.2 of *Walls, windows and doors* deals with painting systems for metals, including suitable undercoats and topcoats. Primers for metal are listed in BS 6150[328] and include:
- one and two-pack treatments
- red lead primers
- red oxide primers

One and two-pack treatments
Two-pack pretreatments, used as washes or etching primers, are typically polyvinyl butyral resin solution with phosphoric acid (as a separate component) and a zinc tetroxychromate pigment, and are usually yellow in colour. One-pack systems are typically polyvinyl butyral/phenolic resin solutions with tinting pigments and are usually blue in colour. The main function of these primers is to improve the adhesion of paint to non-ferrous metals. They can provide temporary protection to blast-cleaned steel and sprayed-metal coatings. The two-pack types generally give superior performance but may be less convenient to use. They do require, however, the application of a normal type of primer on top of the treatment. Most pretreatment primers may be used in conjunction with conventional and specialist coating systems.

Red lead primers
Red lead primers to BS 2523[329] are linseed drying-oil type binders with red lead (types A and B) as the sole pigments; they have an orange–red colour. These are primers of traditional type for iron and steel, though more especially for new construction when the steel is exposed to the elements for long periods before completion of the paint systems. They are more tolerant of indifferent surface preparation than most other metal primers, but are slow in hardening. However, even the low-lead content paints are now being phased out of production in the UK.

Red oxide primers
These are typically drying-oil or resin type binders with red oxide pigmentation and are usually red–brown in colour. These primers for iron and steel are quicker in drying and hardening than red lead primers to BS 2523, and are therefore more suitable for use in maintenance work, or when early handling or recoating is necessary. They should not be left exposed for too long without top coats.

Work on site

Workmanship

Fence erection is a skilled task, and people involved ought to have had some training and experience under supervision. British Standards in general refer to level 2 NVQ or Scottish NVQ for fencing work.

Inspection

Fences should be rocked gently to check their stability.
The problems to look for are:
◊ mechanical damage (eg from vandalism)
◊ instability of posts
◊ corrosion of metals
◊ wood rot, especially at ground level
◊ spalling of concrete

Chapter 5.4 Exterior lighting and security devices

There are four main aspects where exterior artificial lighting within the curtilages of building sites is important:
- pedestrian wayfinding and safety
- access by vehicles and parking
- prevention of crime and the operation of CCTV
- amenity lighting, for example the floodlighting of buildings and other structures (Figure 5.32).

Well designed lighting schemes can have a major beneficial effect on the night-time appearance of buildings, and the spaces and landscaping around them (Figure 5.33), accentuating both character and appearance. Accordingly, responsibility for the design of exterior lighting installations should be given to competent lighting engineers.

The lighting of sports facilities is not covered by this book.

Characteristic details

Lighting
Pedestrian areas
Safety and security issues are paramount for pedestrian walkways and, particularly, for steps. Bollard-style fittings are available for lighting paths (Figure 5.34) and steps can be lit by recessed luminaires set into the side walls of flights. For low level lighting of footpaths, it may be possible to use fluorescent sources.

Access roads
The preferred light sources for minor roads are low pressure sodium giving predominantly yellow light, or metal halide or high pressure sodium, giving a more complete spectrum.

Prevention of crime
Areas around buildings carry different levels of risk to pedestrians and vehicles, depending on local topography and crime rates. Generally speaking, the greater the levels of illuminance, and avoidance of dark areas and shrubbery, the lower the risk of muggings and thefts from parked vehicles. Personal safety and crime deterrence can be enhanced by CCTV cameras and intruder detection devices.

Closed circuit television
Exterior lighting provides a useful security function for vulnerable buildings; if an area needs to be protected by CCTV, the exterior lighting levels must be adequate for the cameras. Unobtrusive night surveillance can be achieved with infrared CCTV and dark-filtered tungsten halogen illumination sources.

Figure 5.32 Effective floodlighting at Tower Bridge

Figure 5.33 Artificial lighting providing for the needs of pedestrians and vehicles may also serve security, commercial and amenity needs

5.4 Exterior lighting and security devices

Figure 5.34 A footpath light fitting – especially useful where the integrity of the footpath surface is questionable

Intruder detectors
Properly sited and maintained external lighting, which is operated by a time switch or sensor, will act as a deterrent to a potential intruder. Luminaires or lanterns should be positioned where they can cast a pool of light around front doors, windows etc, preferably at high level and where they are out of reach of intruders. Cables must be concealed to protect them from tampering. Luminaires or lanterns should be positioned so that they do not allow shadows which could conceal an intruder or annoy neighbours, who may object to being awoken by badly positioned security lights.

The type of intruder detection devices often used externally to buildings are those which transmit a narrow beam of energy, either infrared or microwave, to a receiver. These systems are most frequently used just inside site perimeter walls and fences as additional protective features. When that beam is interrupted by an intruder, the signal triggers an alarm.

Since the beams will only travel in straight lines, a set of three or more overlapping paired receivers and transmitters needs to be used to protect a large area. Correspondingly more pairs need to be used if the area to be protected is irregular in shape. Narrow LED infrared beams may need to be stacked one above the other to ensure that a sufficient height is covered. Microwave beam transmitters usually project a wider beam, and so fewer may be needed.

More sophisticated systems are sometimes specified for buildings or contents having very high risks. Systems include, for example, underground tubes which are sensitive to differences in hydraulic pressures when the ground is traversed by an intruder. They can be tuned to ignore small animals.

Amenity lighting
Floodlighting of buildings depends entirely on the effect to be achieved, though to be effective the levels must be set in consideration both of the general levels of artificial lighting in the immediate vicinity and of the reflection factor of the materials of the building fabric. (Figures 5.35 and 5.36).

Good modelling of architectural features by artificial lighting requires selective siting of luminaires – a uniform coverage with no appropriate use of light and shade is unflattering to most buildings. Consideration also needs to be given to the selection of suitable fittings which do not mar the appearance of the property during the day.

Figure 5.35 A simple lighting effect for a church tower

Figure 5.36 One of the ground mounted floodlights for the church tower shown in Figure 5.35

Types of lamps

Exterior lighting schemes will normally use one or more of the following lamp types:
- tungsten halogen – gives a warm light with good colour rendering
- tubular or compact fluorescent – light varying from warm to cool with moderate colour rendering
- low pressure sodium – a warm light with very poor colour rendering
- high pressure sodium – a warm light with average colour rendering
- high pressure mercury – a cool light with average colour rendering
- metal halide – an intermediate or cool light with average to good colour rendering

Metal halide lamps are suitable for floodlighting where a cool, modern look is required and high pressure sodium lamps can give warm floodlighting to stone or brick buildings. A number of buried, and therefore unobtrusive, uplighters and portico downlighters are available that use compact fluorescent lamps to illuminate the entrances and ground floor façades of buildings. High efficiency discharge lamps can also be used in conjunction with fibre optics for detailing, and cold cathode tubes are available in a variety of colours for accentuation.

Table 5.5 Categories for illumination of roads, footpaths and verges

Use	Lux requirement
High	average 10
(eg with high crime rates)	minimum 5
Moderate	average 6
	minimum 2.5
Minor (eg residential)	average 3.5
	minimum 1

Main performance requirements and defects

Outputs required

Irrespective of the actual average values achieved for artificial lighting, it is important for most situations to achieve even coverage so that there are no dark areas. Also, using multiple lamps to provide for lamp failure may need to be considered. This policy may not, however, be acceptable for floodlighting of buildings for particular aesthetic effects – uniform coverage can look bland. So far as the average levels are concerned, these will often depend on the building type and its location (Figure 5.37). There are special considerations governing lighting in the vicinity of airports, railways and waterways.

The luminous efficiency of different types of lamps – the rate of conversion of power consumption in watts to light output in lumens – varies considerably, and manufacturers' literature should be consulted.

Pedestrian areas

Recommended lighting levels range from 5 lux for secondary pathways, to 10 or 20 lux for main pathways. Illuminance levels on steps should be slightly higher than the surrounding area, perhaps up to 50 lux, but glare should be avoided. There might also be a case for lighting treads and risers separately. Paths leading to commercial areas, shopping precincts and hotels may even be 50 lux or higher in certain cases, and railway and coach stations will be 150–200 lux.

Access roads

The required values of lighting for access and service roads are covered by BS 5489-1[330] and BS 5489-3[331].

Three categories of use for illumination requirements are given for roadways, footways and verges in Table 5.5. Roads which provide access from local distributor roads to residential, business or industrial areas for light traffic are normally lit from 30 minutes after sunset to 30 minutes before sunrise.

Figure 5.37 An old gas street lighting column and lantern has been adapted to illuminate this pedestrian access. The horizontal bar was to support the top of the lamplighter's ladder

Light columns can be spaced at appropriate intervals, though around 30 m is quite common for situations where crime rates are low. Columns should not be placed nearer than 0.8 m from the carriageway for roads designed for traffic speeds of up to 50 km/hr. Mounting heights are normally 4, 5 or 6 m, should not overhang the road, and normally be placed on the outside of bends. It is also advantageous to site them so that they throw light on footpaths running normal to the road.

Preventing crime

For car parks the major lighting requirement is for security. For this reason, car parks, especially large open-air parks, are often treated as roads and are lit using the road lighting Standard, BS 5489. Using relatively energy efficient lighting such as high or low-pressure sodium lamps normally leads to well lit and uniformly lit spaces but the overall illuminance levels should not exceed those recommended for street lighting in the neighbourhood. It is also important to use well designed

5.4 Exterior lighting and security devices

Figure 5.38 Swan-necked floodlight fittings on the Swan Hotel! The lighting arrangements here serve both commercial and amenity requirements

luminaires or lanterns that minimise upward light spillage.

This type of lighting can be bland and unattractive, and some efforts have been made in recent developments to add interest to parking areas by integrating landscape design with lighting arrangements, but avoiding too many strong pools of light and areas of dark shadow, particularly the abrupt cut-offs at the car park boundaries. Reflected light from tree canopies should provide sufficient light for recognition purposes.

For both external intruder devices and CCTV, one of the main considerations is their performance in adverse weather conditions, particularly fog. Modern designs of high power LED infrared transmitters can normally cope with fog, and with rain and condensation on apparatus, but the effectiveness of CCTV in these conditions should be checked with the manufacturers.

The Chartered Institute of Building Services Engineers recommends values of standard maintained illuminance for pedestrian and car parking areas external to buildings which are of high and low crime risk (Table 5.6).

Pedestrian subways are particularly crucial areas, and are normally lit to 150–200 lux depending on such factors as risk of mugging and length of subway[332].

Amenity lighting

For floodlighting buildings, values of between 30 and 100 lux on the building's surface are usually set. Different strategies may need to be used for buildings of different character, for example those with very strong horizontal or vertical modelling. Account needs to be taken of the reflectance values of building surfaces since they can vary widely: from the highly reflective gault brickwork at about 0.8, to dark brickwork or granite at around 0.2.

In many cases floodlighting will need to be positioned close to or actually on the building as spaces available to the owner for this purpose will be limited (Figure 5.38). Balconies or cornices lit by adjacent fittings can produced unsightly shadows and supplementary lighting may be needed on the balcony or over the cornice to overcome the problem, although particular architectural features at the top of the building can usually by lit effectively with accent lighting at close range. Light-coloured window reveals can be lit to provide effective modelling of the more plane façades. The beam angle (ie the spread of light) of the floodlight and the aiming angle should be chosen to give good illumination of the building with minimal overspill to the sides or top. Tall buildings are sometimes lit with diminishing brightness towards the top, in order to emphasise apparent height.

Other smaller structures in the urban environment (eg statues and fountains) are usually lit to around 50 lux, but those items which need to be read by artificial light, such as notice boards and clocks, will need to be lit to 200–500 lux depending on local conditions including the brightness of the surrounding area[332].

Colour filters are sometimes used to create special effects, and these will call for increased outputs from the lamps. Blue filters need the strongest outputs, with amber the weakest.

Energy efficiency

A number of high efficiency light sources are available for quality exterior lighting at relatively low cost to both the building owner or occupier and the environment.

Security floodlighting in car parks using high power tungsten halogen lamps is particularly poor from the point of view of high glare and deep shadows. These lamps are often used with passive infrared sensors so that the lamp is only switched on when it is needed. Although this may appear to be an energy efficient strategy, the high power requirement of the lamps (250–1500 watts are typical) means that if the lamp is only on for a few minutes each hour, it can use more energy than an energy efficient lamp running continuously. Unfortunately most energy efficient lamps are not suited to this type of security system as they are slow to warm up and cannot re-strike rapidly.

Dimming or switching off car park lighting at night, say between midnight and 6 am, and during the day, is recommended where security is not an issue.

Table 5.6 Recommended illumination for external crime risk areas

Area	Lux
Pedestrian:	
low crime risk	20
high crime risk	100
Car parking:	
low crime risk	20
high crime risk	50

Unwanted side effects

Levels of illumination should be carefully chosen to distinguish a building from adjacent buildings but without creating excessive glare which could distract the viewer. Inappropriately sized and situated exterior luminaires can cause glare to pedestrians and vehicle drivers in the vicinity of illuminated buildings. Careful positioning of lights is essential, too, to avoid unnecessary overspill into the night sky causing light pollution, or away from the building causing a nuisance to neighbours.

Commissioning and performance testing

It is a fairly simple matter to measure available light on both horizontal and vertical surfaces. Measurements must be taken at the appropriate heights above ground.

Lamps should be selected which have appropriate re-strike times where the system is switched by, for example, passive infrared (PIR) sensors. This is particularly important where security is a requirement. Some lamps will not re-strike for up to 20 minutes. Another factor is the so-called run-up time; that is to say, the time taken by a lamp after switching on to achieve its normal output – some lamps may take up to two minutes to achieve normal output. The manufacturer's literature should be consulted.

Although BS 4737[333] applies to intruder detection devices installed within buildings, it may still assist with specifying equipment mounted externally but performance and quality standards will need to be agreed with individual manufacturers (Figure 5.39). Specifiers should consider the likelihood of false alarms being triggered by birds and animals.

Durability

Apart from the adequate functioning of intruder detection apparatus, the main issue to consider is impact damage to luminaires, either through accident or by vandalism. Transmitters, receivers and luminaires must be of robust construction.

Maintenance

Lighting fitments should be chosen so that each lamp is easily accessible for cleaning and re-lamping, or if accessibility is poor, the lamp with the longest life should be chosen provided it fulfils all the other specified requirements. Expected lamp life can be obtained from the manufacturer's literature. The appropriate strategy for replacement – whether all lamps or just selected lamps – will depend on the chosen housing as well as on the lamp average life.

Lantern cleaning intervals will vary according to local conditions and pollution levels, but annual cleaning is common.

Work on site

Inspection

The problems to look for are:
◊ light levels not complying with those recommended
◊ inappropriately positioned luminaires giving light and dark areas
◊ luminaires not vandal resistant
◊ tungsten halogen security lighting switched by PIR (long restrike times)
◊ light pollution or overspill

Figure 5.39 Column mounted CCTV equipment

Chapter 6

Hard and soft landscaping

Figure 6.1 The York stone paved courtyard of the Victoria and Albert Museum

Steps may be hazardous due to loose stone treads, chipped or cracked nosing, unequal risers, treads not level, pitch too steep, width too narrow, or surfaces liable to become icy and slippery. Paths become uneven and dangerous as a result of ground settlement, heave and tree root disturbance (Figure 6.2). Vehicular areas can be similarly affected. Paths may be graded too steeply, too narrow for pram wheels or wheelchairs, and not routed to suit users' needs.

Public areas that are not maintained invite vandalism, rubbish and dumping which can rapidly degrade a development or neighbourhood.

This sixth chapter deals with the areas of the site immediately surrounding the building; that is to say, the upper surface of what is normally understood by the term building curtilage. The need is to provide a secure all-weather access to buildings for both pedestrians and for vehicular traffic, or simply to provide pleasant places for relaxation (Figure 6.1).

All paved surfaces

Pavings are exposed to rain, frost and sunlight, and unless bricks, stone, tiles and mortar are frost resistant they will have a considerably reduced life. Poor maintenance can be assessed visually, but effects which may not be apparent at the time of inspection may also need to be considered: leaf fall affecting yard drainage, for instance.

Figure 6.2 This concrete flag pavement was probably founded on inadequate bedding and now presents a considerable hazard for pedestrians

6 Hard and soft landscaping

Figure 6.3 Hard standing for fire engines. This particular solution offers approximately 30% grassed area

Access to buildings for firefighting vehicles

For certain kinds of buildings, it is essential to provide suitable access for firefighting vehicles and escape ladders (Figure 6.3). A common solution which avoids using large areas of tarmacadam adjacent to a building is to use interlocking cellular grids of suitable material with the interstices filled with grass. Since the grass is flush with the surface of the grid, mowing should present no problem. Various proportions of grid material to grass are available, with different vehicle loading capacities.

Car parking

Although this book does not include specifications for public highways covered, for example, by the Roads (Scotland) Act 1984, the Highways Act 1980 or the New Roads and Street Works Act 1991, or any Regulations made under these items of legislation, some aspects of roads, parking areas, and their construction are briefly mentioned.

Dwellings

Nearly two thirds of all dwellings in England have a garage, car port or designated parking space, with most of these having garages. The remainder rely on street parking, being most common in city centres. However, two thirds of all garages are not used for housing the car, but for storage (EHCS[4]).

Other buildings

Local authorities in many areas have specific requirements for car parking provision for non-domestic buildings in their areas. Planning requirements will generally dictate the level of parking provision appropriate to the site. In many cases these requirements have changed in recent years in line with revised thinking on road congestion in towns, and efforts to persuade people back to public transport. However, in spite of considerable efforts by designers, it is arguable that provision for the motor vehicle in many cases makes a negative contribution to the immediate surroundings of buildings (Figure 6.4) and car parking situated too close to a building can be something more than a nuisance when exhaust emissions are carried into the building through open windows and when soot deposits disfigure the walls (Figure 6.5).

Access roads can be adopted by the local authorities provided they are constructed to specific design standards. This will allow the maintenance of the roads to be undertaken by local highways departments or by their contractors.

Figure 6.4 Car parking provision at an office building in a rural area

Figure 6.5 Cars parked too close to a building leave exhaust marks on the fabric. Oil which has dripped from parked vehicles onto paved surfaces permanently damages those surfaces (at worst) or disfigures them (at best)

Chapter 6.1 Pavings

This chapter sets out the various forms of paving: small element paving, bound materials and unbound materials.

Traditional materials for pedestrian area pavements, until the introduction of pressed concrete flags, ranged from gravels, brick pavers or paviors, through York and Caithness flags, to rounded cobbles and stone (usually granite) setts.

After the Romans, having laid down paved roads, left the UK, and before the widespread introduction of broken stone macadam in the nineteenth century, most roads would have consisted of natural gravel which was inadequate even for the horsedrawn traffic of the day. There would have been just a few areas in towns and round prestigious buildings of stone setts (Figure 6.6) – a durable, picturesque, but rough surface that is still specified for a few areas today.

Tarmacadam, that is to say, broken stone or slag macadam with a tar or pitch binder, was introduced in the 1880s, although its use did not become widespread until the 1920s.

Characteristic details

Pavements can be classified by structural type into rigid or flexible. Most paving is to some extent flexible, having the ability to deform under loads and regain its shape to some extent when that load is removed. Rigid construction is only generally used in areas of heavy load or where the ground conditions are poor.

They can also be either permeable or impermeable to rainwater, draining, in the former case, through the paved area, and, in the latter case, from the perimeter of the surface, to disposal or treatment and reuse.

Hardcore and bases

Base materials for pavings and roads will consist of a variety of specifications, including hardcore, hoggin and cement or lime-bound aggregates, or even soils as found and won on site. By themselves they do not form a permanent weatherproof finish and must be further protected from the ingress of water, both from below and from above. Normally, hardcore bases are specified only with the lightest of loadings in car parks etc. The specification of cement-bound bases will normally differ according to whether the top surface is flexible (eg concrete block or bituminous) or non-flexible (as with concrete slab finishes) (BS 1924-2[334]).

For general landscape paving, hardcore recycled from site demolition is frequently used. This is acceptable provided the hardcore is clean and free from contamination. Crushed concrete is another recycled material often specified as appropriate for footpath bases.

Dry bound macadam bases are formed by laying coarse aggregate in layers 75–150 mm deep according to requirements, rolling between each layer, then covering with fine aggregate (say 25 mm) vibrated into the interstices of the coarse aggregate; finally rolling the finished surface with a heavy roller until fully compacted.

Coated macadam bases normally comprise 20–40 mm sized crushed rock or slag aggregates coated with bitumen binders.

In all cases, overall thicknesses of bases must relate to the anticipated traffic.

Figure 6.6 Coursed stone setts laid to camber

6 Hard and soft landscaping

Flags, kerbs, edgings, quadrants and channels

These are taken together because they are covered in the same British Standard (BS 7263-1[335]).

Flags are available in seven standard sizes, ranging from 600 × 450 mm to 300 × 300 mm. They are either 50 or 63 mm thick, depending on strength requirements. Most are made by applying hydraulic pressure to comparatively dry concrete mixes which develop high strengths early in the curing process and allow early handling. A number of surface finishes are available including, for example, tactile pedestrian crossing flags with raised buttons. The traditional bedding procedure is to lay concrete flags on lime mortar spots on a bed of gravel topped with sand, with dry mix brushed into the narrow joints.

Precast kerbs range in lengths from 450–915 mm, and are available to various radii from 1–12 m as well as in straight lengths. Heights are from 150–305 mm, and profiles available include bullnose, splayed, battered and dropped. Quadrants to similar profiles are also available. Precast concrete kerbs and edgings are the most frequently specified products. It is generally preferable to specify the narrowest kerb that will provide sufficient retention. This avoids a visually over-dominant edge to paved surfaces.

Square or dished channel inverts are available in a number of sizes, with different depths of depressions.

Kerbs in natural stone are available to BS 435[336], with finishes in the following grades:

- fine picked
- fair picked
- single axed
- rough punched

Pavers

Pavers (or paviors to use the old term) are commonly available in a variety of materials. Those of fired clay used to be available to the nominal brick size of 9 × 4½ in on plan, and were made in vast numbers from Victorian times until around the 1930s (Figure 6.7).

Concrete pavers or blocks, on the other hand, are of more recent manufacture (BS 6717-1[337]). Sizes are 100 × 200 mm, but the Standard also permits any shapes filling a 300 mm square. Thicknesses are not less than 60 mm and can be up to 100 mm, though some made in the 1980s may be found to be thinner. Access roads and car parks are usually formed of 80 mm thick blocks. Brick kerbs are commonly used in conjunction with concrete block paving.

Calcium silicate pavers may also be found.

BS 7533[338] suggests design lives of 20 years or 40 years for pavements of clay blocks or concrete blocks; an algorithm is given for the design of these pavements. The different situations are described including the lowest category of strength for areas serving up to four dwellings.

Setts and cobbles

Setts are normally of granite, dressed roughly to shape and laid on consolidated sand beds in similar fashion to concrete blocks, but in a variety of traditional patterns.

Cobbles are large stones rounded by wave action. They are normally laid touching, or nearly touching, on a bed of gravel or concrete, and tamped into position until the tops form an acceptably level surface (Figure 6.8). Grout is then brushed between the units. The cobbles will naturally vary in character, and may be sorted as to both size and colour before laying in order to achieve particular patterns.

Figure 6.8 Cobbles set between brick pavers provide access in this mews area for both pedestrians and light vehicles

Wood block paving

Wood block road paving, laid in bitumen in similar fashion to flooring but commonly cut from end grain blocks, used to be an alternative road surface in Victorian times where it was necessary to achieve quiet conditions (eg near hospitals) as an alternative to laying straw on the cobbles. The wood ameliorated the noise from the steel rims of the wheels of horsedrawn carts before the introduction of pneumatic tyres. The surface was highly susceptible to flooding and at least one West Midlands town centre, as late as the 1930s, lost most of the paving in its market place for this reason. Needless to say, it was not renewed. Wood blocks were also laid, at one time, adjacent to and between tramlines in city centres. Wood paving is hardly ever seen now.

In situ concrete pavings

Sawn or cast movement joints are crucial to the performance of in situ concrete pavings. To prevent differential settlement of adjacent slabs separated by these joints, it is normally necessary to include sleeved stainless steel bars (BS 2499[339]).

Figure 6.7 Fired clay pavers of two different thicknesses, 50 and 63 mm, made in the West Midlands in the 1920s

6.1 Pavings

Tarmacadam

Coated macadam wearing courses of 6–14 mm crushed rock or slag bound with bitumen are normally specified to BS 4987-1[340].

Definitions of tarmacadam normally include:
- dry and wet bound macadam bases
- coated macadam bases and wearing courses
- hot rolled asphalt bases and wearing courses
- surface dressings of tar, crushed rock and gravel

Top dressings are covered in a number of Standards (eg BS 3690[341] which lists 10 viscosities for different uses). Dense tar surfacing is covered in BS 5273[342].

Asphalt

Pavements consist of hot rolled asphalt bases and wearing courses to BS 594-1[343] and BS 594-2[344]. Mastic asphalt pavings may also be found (BS 1447[345]).

Gravel, clinker and hoggin

Hard binding gravel of 20 mm maximum size and with a low clay content is normally to be found in surfacing which consists of well consolidated layers, each of around 50 mm thickness to a total thickness of 150–200 mm. Well graded and dust-free clinker may also be found. Hoggin is a naturally occurring mixture of fine gravel, sand and clay. All three surfaces are raked to falls and consolidated by rolling, and may receive a top dressing of fine gravel or pea shingle (Figure 6.9).

Main performance requirements and defects

All paving, and particularly that comprising small elements, is very vulnerable to poor design detailing. This is particularly apparent at corners and junctions.

Strength and stability

Horizontal or inclined features, such as pavings and steps, rely for support on the ground beneath. In the event of failure in the sub-base, the consequences are usually limited to some vertical movement of paving units leading, in turn, to risk of tripping.

Adequate edge restraint is a key factor in ensuring stability. Concentrated loadings from turning vehicles can be significant and it is important that the design specification is appropriate for all likely usage.

Pavings that are too thin or laid on inadequate bases, or uncompacted or ungraded backfill, may collapse, particularly when subjected to unanticipated or unauthorised loads. These cases typically will occur when flags or cobbles are laid on gravel bases which have not been properly compacted. Vehicular traffic then easily fractures the flags or disturbs the cobbles, which, in turn, present opportunities for vandalism.

Concretes and flags

Many tests have been carried out by the BRE over the years for investigating certain properties of concrete products; for example, the transverse and flexural strength and water absorption characteristics of commercially produced precast concrete flags, generally in relation to the various revisions to relevant British Standards. These properties are relatively easy to specify and to test. Pavers used for domestic situations, hard landscapes, footpaths and private drives commonly have only half the strength of pavers used for roads, lorry parks and farmyards (BS 6717-1).

Tarmacadam

In addition to causing long term damage to houses and other structures on shallow foundations, slope creep tends to produce noticeable ripples in roads and other asphalt surfaces. However, as the soils that experience slope creep tend to be shrinkable clays, the effects are often difficult to distinguish from clay shrinkage.

In south east England the summers of 1989, 1990 and 1991 had below average rainfall. At the end of October 1991, considerable cracking of tarmacadam paths and road surfaces was noticed in north west London, particularly in the Mill Hill

Figure 6.9 A domestic driveway consisting of a hard binding gravel base dressed with fine gravel

Case study

Slipperiness characteristics of paving flags

The BRE Advisory Service was asked to test the slipperiness characteristics of some concrete paving flags after an accident had occurred outside a public building in an urban location. The small regular pimples which result from the manufacturing process had been completely worn away on the flag on which the accident was alleged to have occurred (the subject of Test A), but two similar neighbouring flags which appeared to be slightly less worn were also tested (Tests B and C). One adjacent flag which had a completely different surface finish, not wet pressed like the others, was also tested (Test D).

It is misleading to suggest that the pimples found on the surface of paving flags are deliberately produced to enhance slip resistance in wet conditions. They result from the vacuum de-watering during the production process, and are not a requirement of BS 7263-1[335].

The tests were carried out using the Transport Research Laboratory (TRL) portable skid tester – the pendulum tester – and according to procedures described in BS 8204-3[347] (Figure 6.10). The slip resistance values obtained were the arithmetic means of at least five readings, and were corrected for temperature. For those tests carried out in the wet state, the slider was shod with soft rubber. The corrected values are shown in Table 6.1.

Experimental work over many years has amply demonstrated that values below 20 indicate that the surface is dangerously slippery, 20–40 marginal, and above 40 satisfactory. The Advisory Officer concluded that the flags tested in this investigation met the normally accepted criteria for safety.

For good slip resistance in wet conditions, the surface walked on must have some surface texture so that the water film is broken. Hydrodynamic lubrication then does not occur and good contact between footwear sole and surface should be achieved. The surface roughness required to avoid slipping problems in the wet appears to be in the region of 8–10 μm.

Figure 6.10 Concrete flags under test using the TRL pendulum tester

Table 6.1 Slip resistance values for tests on concrete paving flags

Test	Condition	Slip resistance value
A	Dry	56
A	Wet	43
B	Wet	55
C	Wet	45
D	Wet	81

---

area. The cracking was caused by the shrinkage of the ground due to the growth of nearby mature and semi-mature trees. An attempt was made to relate the height, distance and species of a tree to nearby surface cracks in tarmac roads and the results were consistent with guidelines published in BRE Digest 298[105].

Unwanted side effects

Good design detailing is essential for paved surfaces to work well, and a key criterion for all paved surfaces is good surface water drainage. Poor design and construction is frequently encountered and often leads to inadequate shedding of surface water. This then leads to standing water, the damaging effects of which can be cumulative. Cross-falls or crowning are essential to shed water and prevent puddling.

A common fault seen by BRE investigators in a number of cases, including even prestige buildings, is to specify pea shingle of the same or similar size to the apertures in adjacent covered drainage slots. The shingle wedges in the slots, and grossly interferes with drainage (Figure 4.24 on page 190).

Health and safety

The design and construction of any paved area must ensure an absence of trip hazards. Changes in level must be clearly indicated and be in 'expected' locations.

Slipperiness of paved surfaces

A comprehensive review of factors governing slipperiness of floor surfaces of various kinds was given in Chapter 1.5 of *Floors and flooring*[66]. Although some of the factors affecting external surfaces are similar to those affecting floor surfaces, there may be other factors at play. For one thing, the surfaces are much more likely to be wet or, worse still, icy, and moss will grow, for example on concrete surfaces. Criteria are given in BRE Information Paper IP 10/00[346].

Flags, as indeed most products made of concrete, can become polished under continuous use. BRE Advisory Service has carried out a number of special investigations where flags have reportedly been involved in accidental slipping by pedestrians. This is also discussed in Chapter 1.5 of *Floors and flooring*.

Some surfaces, like concrete, can be improved by pressure hosing away vegetable and fungal growths, or by shotblasting the surface to provide a texture. As with flooring materials, other surfaces can be improved by coating with epoxy resin into which hard aggregate like carborundum or crushed flint is sprinkled. A second coat of resin is usually applied to partially close in the surface. Most treatments to existing pavings will alter their appearance.

Ramps to footpaths need to be considered from the point of view of pedestrian safety. Paved areas outside entrance doors need to be laid to a minimum fall to prevent water accumulating and then freezing. This is conventionally accepted to be 1%, but, because of the almost inevitable deviations in construction, and the need to reduce the likelihood of backfalls and subsequent ponding, the nominal fall should be increased to at least 2%. Ramps or slopes should not exceed 6%, and preferably be shallower than 8%.

One further risk area which needs to be considered is the size of gaps in drainage grilles. There is some difference of opinion on this matter. In public areas, particularly at entrances to buildings, consideration should be given to limiting the mesh clear-gap size to 8 mm, though up to 15 mm seems to be quite common. The smaller dimension should allow snow, rainwater and most stones picked up on moulded rubber shoe soles to fall through, while at the same time preventing narrow heels of shoes and umbrella ferrules from penetrating the gap.

Security
Extensive footpaths on estates, particularly where they are hidden from view by shrubbery, can provide opportunity for assault, and poorly lit areas (eg in car parks) can offer opportunities for vandalism and theft. For access roads, routes which incorporate features forming physical or psychological barriers, or which give the impression that the area beyond is private property (eg rumble strips or changes in the colour of road surfaces), can suggest that entrance would be unwanted or unauthorised[348].

Durability
Materials used in bases which are susceptible to frost action should not be used within 450 mm of the final surface which largely eliminates them for use under footpaths. The use of permeable materials with high void contents can prevent the worst effects of frost heave; the frost lenses simply expand into the large voids within the material rather than accumulating within the small pores which would lead to the upper surface lifting.

Sunken pavings and hardstandings are often the result of using inadequately prepared ground during construction. Settlement may indicate drain collapse, wash-out, poor backfilling, or the effect of tree roots. Alternatively, pavings may have been overloaded by regular or occasional traffic (such as household refuse collection vehicles or fire engines). Drains should be checked before reinstating damaged or defective surfaces.

The clay bricks used for pavings do not always prove to have adequate frost resistance. When laid as pavings, they will be in a more or less continually wet environment. Bricks, identical to those which perform entirely satisfactorily in the walls of buildings, can decay rapidly when used as pavings alongside those very same buildings, sometimes failing after one or two hard winters. Clay units sold specifically as pavers should be satisfactory, but even these have been known to give unsatisfactory service. One area of clay pavers laid at BRE had around a 5% failure rate due to spalling of the top surface from frost action during a 10 year service period (Figure 6.11).

Figure 6.11 Fired clay pavers interspersed with concrete flags. Many of the pavers subsequently failed as a result of frost action during a 10 year service life

Concrete flags and pavers laid in the 1980s and 1990s rarely seem to suffer from frost action, although those which are defective in manufacture will be vulnerable. Output from works in specific areas of the UK seem to have been particularly involved in the past, although cases of failure now occur less frequently. Flags and other pavings intended for pedestrian use only will fracture easily when traversed by vehicular traffic and will need bollard protection (Figure 6.12).

In 1983, a number of concrete slabs containing a wide variety of different aggregates were placed on one of the main site roads at BRE and the durability and skid resistance were monitored at ages of 2 and 10 years. The mechanical strength of the aggregates varied between 20 kN and 290 kN. Some of the slabs cracked in half, probably due to foundation problems, but this did not seriously impair serviceability; surface spalling occurred on one slab only. A comparison of the skid resistance for the different slabs found that the resistance improved with higher silica contents of the fine aggregate.

Concrete pavings should give a life of at least 30 years, though it does depend on the degree of misuse, and movements in the sub-base caused by tree and shrub growth. Natural stone paving may last a little longer, but it has been known since the early 1930s that laying stone paving on an impermeable base could lead to premature failure due to frost action on the saturated stone[349].

Interlocking concrete block pavings are suitable for light traffic where speeds are not in excess of 60 km/hr and for hardstandings.

Tarmacadam and asphalt footpaths should give working lives of upwards of 40 years, although wearing courses may need to be replaced earlier. The lifetime of roadway surfacing depends on traffic density and wheel loadings, and softening caused by oil contamination.

Maintenance

Inspections of external paved areas need to be carried out at least annually, and arguably more frequently. Assessments should be made for replacing paving units which have broken or otherwise deteriorated.

Work on site

Workmanship

Compaction of hardcore, granular or cement-bound bases should be carried out with plant that suits the traffic anticipated. Suitable plant ranges from smooth, tyred or vibrating rollers, or vibrating compactors, tampers or rammers, and depends also on the area to be compacted. The degree of compaction – the number of passes of the plant – is determined by the chosen equipment and the thickness and nature of the material. Compaction is sometimes found to be inadequate.

Deviations on finished surfaces should not normally exceed 15 mm under a 3 m straight edge. This figure may be compared with tolerances on concrete floor slabs given in Chapter 1.5 of *Floors and flooring*.

Deviations in the sub-base under concrete blocks should not normally exceed ± 20 mm, and in other circumstances ± 30 mm. Some form of edge restraint prevents sub-base displacement during compaction.

The correct curing of cement-bound bases and in-situ concrete slab pavings is important.

Figure 6.12 The scars and damage to the concrete bollard on the left show just how necessary protection can be

Inspection
The problems to look for are: ◊ using inappropriate unit sizes for the function of the paving ◊ poor detailing around posts and street furniture ◊ poor detailing at changes in level (eg ramps) ◊ poor compaction of sub-bases ◊ poor curing of cement-bound bases and in-situ concrete pavings ◊ base materials susceptible to frost action used within 450 mm of final surfaces ◊ missing DPMs or geo-membranes ◊ missing movement joints in in-situ concrete pavings ◊ proximity of trees and other vegetable growth leading to disruption ◊ absence of edge restraints to concrete block paving ◊ poor levelling of service covers ◊ excessive cutting of units for awkward areas ◊ uneven laying with large gaps between paving units ◊ ponding of surface water

Chapter 6.2

Trees, plants and grass

This chapter sets out the soft landscape elements frequently associated with buildings. Trees, plants and grass are important elements in providing the settings for buildings (Figure 6.13).

The scope of the chapter is restricted to the garden areas and amenity land immediately surrounding buildings and car parks. It does not include highway verges or agricultural or horticultural applications, nor are sports surfaces included.

The climatic benefits of tree planting and shelter belts are included in Chapter 1.4. Roots of trees or shrubs that are too large or inappropriately sited can cause damage to masonry and pavings as well as to buildings; this topic is covered in Chapter 1.5.

Characteristic details

Plants: general considerations
It is important that the selection of plant material is appropriate to the site conditions. Landscape architects are the professionals best qualified to provide advice on the selection and specification of plant material. The specification of plant species native to the British Isles is to be encouraged as this has ecological benefits in that a greater diversity of insects and birds will be able to utilise the plants. The aim of many planting schemes is to include a range of foliage size and shape, and provide colour interest that will change with the seasons. A balance of evergreen and deciduous species also allows for year round variation.

There are nearly 2,000 plants widely available for use in landscaping and these are now commonly specified according to *The national plant specification*[350].

See also *List of trees and shrubs for landscape planting*[351].

Trees and shrubs
Semi-mature trees may be available for transplanting, although certain species are more suitable than others. Deciduous trees are on the whole somewhat easier to establish than evergreens. Among the species which transplant and re-establish most readily are certain varieties of acer, though most native species also transplant and re-establish reasonably easily.

The largest semi-mature trees which are generally available are usually root-balled. Extra-heavy standards are the next largest size trees normally sold; they are 4–5 m tall and have a trunk girth of 12–14 cm. There is then a range of sizes of trees down to light standard size. Below this, trees and some shrubs are available as whips and transplants.

The dimensions of trees are given in BS 3936-1[352]. Standard trees range in height from half standards at 1.8–2.1 m to tall standards at 3.0–3.5 m, though other sizes and descriptions, such as whips and seedlings, are also specified in BS 3936-1.

Newly planted trees require support until the root system establishes and grows sufficiently to hold the tree up by itself – usually 3–4 years. Traditionally, a single stake was placed next to the tree; this can cause problems due to rubbing damage from the stake or tree tie. Staking practice changed somewhat in the final years of the twentieth century. In former years

Figure 6.13 The gardens of Culzean Castle, Ayrshire

6 Hard and soft landscaping

Figure 6.14 A newly transplanted tree supported by twin stakes and a flexible rubber tie

stakes were driven in to stand vertically close to the root ball. It is now thought that allowing the tree a degree of movement encourages the root growth to develop faster. Stakes positioned in the ground further away from the root ball are now favoured, and in recent years many trees have been planted between two stakes. The tree is tied onto a extended rubber tie that is then fixed to both posts (Figure 6.14).

Protection of the trunk should be sufficiently durable to last the required time of staking which frequently depends on the age of tree at transplanting.

Trees planted as transplants or whips are frequently protected by tree shelters to provide a favourable micro-climate for early growth. Tree guards also protect against grazing by small animals such as voles and rabbits. There are other methods of protection (eg spiral guards) that prevent grazing and bark-ringing but do not provide any shelter to the growing tree (BS 4043[353]).

Shrubs are available either bare root or containerised, the latter being suitable for year round planting (BS 5236[354]).

Intruder protection

A number of plant species (eg of thorn-bearing shrubs or climbers) can form intruder-deterring barriers or hedges round buildings in conjunction with suitable fencing. The general idea is that the vegetation softens the fortress-like appearance of high security fencing such as that described in Chapter 5.3.

Several species of shrubs have thorns on the stem and branches. Traditionally these species were used in farm hedges to retain stock within fields. Dense planting can also deter intruders. However, unless supplemented with appropriate security fencing, the thorny shrubs cannot by themselves form an impenetrable barrier.

There are a number of examples of suitable shrubs.
- Hawthorn
- Blackthorn
- Berberis
- Chenomeles
- Holly
- Pyracantha
- Sea buckthorn
- Gorse
- Roses

Ground cover

Ground cover planting is typically employed for large areas of open ground. By careful selection of species and planting densities it is possible to create a deep carpet of foliage that prevents weed growth; it therefore does not require a great deal of maintenance. A wide variety of plant types and species can be used to create ground cover but typical examples used are:
- *Berberis wilsoniae*
- *Ceanothus repens*
- *Cotoneaster prostrata*
- *Euonymus fortunei*
- *Hebe* spp
- *Hedera* spp
- *Hypericum calycinum*
- *Lonicera pileata*
- *Pachysandra* spp
- *Potentilla* spp
- *Rosa* spp
- *Rubus calcynoides*
- *Sarcococca* spp
- *Viburnum* spp
- *Vinca* spp

These shrubs are planted at relatively high densities so that they will close the canopy within one or two growing seasons and thereby suppress weed growth (BS 3936-10[355]) (Figure 6.15).

Figure 6.15 Ground cover planting. This particular example needs regular pruning to ensure that it does not get out of hand

6.2 Trees, plants and grass

Mulching and anti-desiccant spraying

Mulching is a technique of spreading material over the soil surface to prevent water evaporation from the soil and to inhibit weed growth that competes with shrubs for water and nutrients. It can therefore reduce the amount of watering and weeding necessary both to establish and maintain landscape planting.

A wide variety of materials have been used as mulch. Chipped bark, wet straw and rotting leaves have been used, but polyethylene sheeting is now common, and even old carpets and slate. The plastics sheeting is not visually attractive and should be removed when plants have established. Bark and other mulch constituents can be topped up us necessary to maintain the required depth as the bottom layer will rot away into the soil (Figure 6.16).

The new Botanic Garden of Wales makes extensive use of small waste slate, of which there are enormous quantities in the country, for mulches. Rainwater penetration is assured, yet light exclusion is of a very high order.

Turf

Various grades of turf are available from specialist suppliers. The great advantage of laying turf compared with grass seeding is the instant finish quality provided by turf. It is important that the topsoil base for the turf is prepared properly and is appropriately consolidated.

Turf should be laid within 18 hours of delivery during the summer, and 24 hours during spring and autumn. It is normally laid to stretcher bond from planks laid on the suitably prepared surfaces. Joints should be close-butted and may be pegged or pinned. The whole surface is then top dressed, brushed and rolled with a heavy garden roller to ensure a uniform finish. Liberal watering over the following few weeks is important. Where laid to sloping surfaces the turves may require extra support by means of cables or netting (BS 3969[356]).

Seeding

Seeding grass areas is less expensive than turfing but requires that the seeded areas be kept free from foot and vehicular traffic for the first growing season. Wild flower seed can be included in the seed mix to create wild flower meadows; however, the mowing regime must be adjusted to allow the flowering species to flourish.

Seed mixes are normally selected for particular areas with the primary use of that area in mind. Seed for hardwearing areas will differ from that for lawns. It may also be feasible to choose particular mixes for steeply sloping areas, many of which will prevent erosion. Alternatively, reinforcing mesh may be buried just below the surface. The advice of a nurseryman should be sought. It is important to avoid using old seed.

Hydraulic seeding

For extensive grassed areas on steep slopes, it may be useful to consider employing specialist contractors to spray seed, stabiliser and nutrient emulsions, though these tend to be used for example on motorway embankments and large scale land reclamation schemes rather than on smaller scale developments. Hydraulic seeding is not suitable for areas where intensive use is anticipated.

Grassed hardstandings

Perforated pavings designed to provide the appearance of grass, but to give access for fire engines, need to be laid on a consolidated base – fire engines are amongst those vehicles imposing the heaviest wheel loads.

'Grass-crete' pavings are rarely used now. Plastics mesh and grid reinforcement can allow grass areas to withstand occasional vehicular access. Appropriate consolidation of the subsoil and topsoil is essential to prevent subsidence.

Figure 6.16 Wood and bark chippings used as a mulch to suppress weed growth

Edgings

Trimming the edges of grassed areas tends to be labour intensive, and this can sometimes be avoided altogether if the edge of the grassed area is either flush with the surrounding paving, or is armoured appropriately, so that the mower effectively removes the growth without scalping the turf.

Where lawn areas abut buildings or freestanding walls it is good practice to incorporate a 300 mm mowing strip. This is a grass free area, usually gravel or unit paving which allows grass cutting machinery to ride over the edge of the grass and trim all the grass area.

Climbing plants

It is possible that climbing plants can have detrimental effects on buildings (Figure 6.17). There is some evidence that Virginia creeper, *Vitis quinquefolia*, which clings by sucker pads on tendrils, does not in general damage building materials. However, the common ivy, *Hedera helix*, which clings by short adventitious roots on the stems, can penetrate masonry, eaves and roof tiling or slating. Cutting the bases of stems does not invariably kill the plant if there is sufficient moisture to sustain growth remaining in the materials which make up the wall. In removing growths of ivy from affected walls and roofs, the plants have to be removed with care if damage to the fabric is not to ensue. With respect to roofs, any attempt to pull off the growths wholesale will also tend to pull off the roof covering.

Apart from the obvious effect of blocking gutters and perhaps harbouring pests, climbing plants which cling by suckers probably do little harm to the fabric. The Royal Botanic Gardens, Kew, believe that it would be most unlikely for Virginia creeper to root above ground level. Vigorous climbing roses will occasionally enter roof voids through gaps in open eaves, but the shoots will barely survive if no light reaches them.

Main performance requirements and defects

Soil analyses

Before planting it is essential to assess the nature of the topsoil. The selection and specification of plant material will depend on the quality of the growing medium.

If there are no apparent issues with ground contamination it is usual to test for the following:
- alkalinity/acidity (pH)
- organic material
- soil type (earth, sand, clay, silt etc, and proportions of these)
- nitrogen, phosphorus and potassium (NPK) content
- grit and stone content
- vegetation content

It is recommended that a number of samples are taken from a variety of locations as soil quality can vary greatly over a site.

Transplanting of shrubs and trees

The correct preparation of the planting medium is important for both trees and shrubs. Pits for root-balled trees need to be 500 mm wider than the root ball and prepared with free-draining gravel at the base. Once the tree is in position the pit should be backfilled with appropriate material; this is usually a mixture of sand, fertiliser, organic material and top soil excavated from the pit. The backfilling medium should be well firmed to evenly compact the soil and the tree should be well watered.

Small bare root trees and shrubs can be notch planted (planted into the vee-shaped declivity formed by driving a spade into the soil and working it to and fro) into existing soil if it is of sufficiently good quality.

Transplanting large trees is a complex operation and unless undertaken by a specialist landscape contractor is unlikely to be successful; the root system is likely to suffer significant damage from which the tree may not recover.

Larger trees and those in important locations are often supported below ground by guying the root-ball with cables to large timber baulks, traditionally railway sleepers.

Protection

The protection needed for newly planted trees is entirely dependent on the environmental circumstances. Tree guards are only required if there is a specific threat to the survival of the newly planted tree. If site perimeter fencing is erected which is stock and rabbit proof, there will be no necessity to protect the individual trees from grazing. Similarly, tree shelters are only required if trees to be protected are planted in relatively exposed positions.

In urban locations, formal tree planting is often protected with 1.8 m high metal tree guards to prevent vandalism and damage. Tree ties are usually fixed to the tree guards.

Figure 6.17 Climbing plants should be cut back and trimmed to ensure that they do not have any detrimental effect on the building fabric

6.2 Trees, plants and grass

Staking
Stakes need to be substantial in size, normally around 75–100 mm diameter.

Stakes are frequently left in position too long. Once the tree is self-supporting (usually after 3–4 years) the stake should be removed. The stake tie can often be trapped within the tree trunk growth forming a permanent wound that can easily become a site for infections and rot to establish.

Guying
Steel wire is commonly used, and it should be adequately protected (eg galvanised). As with staking, the trunk should not be damaged by the tie or strap used for guying.

Underground guying is not removed and rarely becomes a constraint on growth.

Guarding
Young trees and shrubs are vulnerable to attack, especially by bark-ringing by rabbits. The common method of protection is by spiral wound plastics sheaths round the trunk which are able to expand with the growth of the tree. Alternatively, translucent stiff plastics cylinders can be used, with the tree growing inside the cylinder.

Weed growth around the bases of young trees and shrubs may be inhibited by suitable mulching. Using herbicides in this situation is to be avoided.

Health and safety
Use of pesticides, herbicides etc
It is very important that chemicals used in conjunction with landscaping work do not have a detrimental effect on the environment beyond the immediate purpose in hand. Herbicide and pesticide use must only be undertaken by trained personnel with appropriate qualifications.

Accidents
Accidents such as those involving falls or tripping on level ground and surfaces in garden areas and around the home (eg on driveways and footpaths) are about three times as many as those occurring between different levels (eg on steps). Some idea of overall numbers of accidents can be gained from the fact that reported falls on external steps alone amount to over 30,000 per annum[357]. (Similar data for non-domestic situations are not available.) Proper maintenance of the immediate surroundings of buildings, as well as their initial design, will play key roles if accident rates are to be kept down.

Maintenance
Appropriate maintenance is essential for any planting scheme to survive and flourish. The level of maintenance required is dependent on the nature of the planting. Herbaceous borders and formal lawns will require rather more input than shelter belt plantings or wild flower meadows. Inadequate maintenance is often apparent with landscape schemes; weeds can choke and kill shrub plantings and lack of watering in the early stages will prevent plants from becoming properly established.

Water conservation
Efficient landscaping can reduce the amount of water required for irrigation and the watering of plants. Presently, the external use of water for gardening is between 2% and 3% of the total water consumption for domestic properties. Many commercial and public buildings have extensive landscaped areas. Low transpiration plants native to the Mediterranean can be used for ornamental gardening. Efficient watering systems which monitor wind speed and air temperature can control irrigation rates so that water is more fully utilised and evaporation is reduced. See also *Technical bulletin on watering*[358].

The installation of irrigation systems is still relatively rare in the UK. However the provision of watering tubes for larger trees allows water to reach the areas for maximum root growth.

Planted areas
For shrubs, apart from correct selection of the plant material, proper preparation of the planting area and the level of maintenance in the early stages of growth are the key factors in durability. In order to facilitate maintenance and successful establishment, the following points should be considered as part of the initial specification:
- introduction of suitable land drainage
- ground preparation
- planting density
- protection of planting in vulnerable areas
- appropriate irrigation and maintenance systems

For grassed areas, apart from correct selection of the seed mix, proper preparation of the area to be seeded is important for durability. The main points to consider in specifications are:
- removal of existing root systems liable to regrow
- introduction of suitable land drainage
- introduction of suitable irrigation systems
- restoration of topsoil to reduced subsoil levels
- adequacy of scarification and recompaction of subsoils
- suitability and adequacy of herbicides
- seeding density
- rolling
- protection of newly seeded areas

For specifying topsoil, see BS 3883[359].

6 Hard and soft landscaping

Figure 6.18 Formal hedging requires regular trimming – often an underestimated long term commitment, though clearly not with this example

Surgery and pruning of trees

It is important that any works on existing trees are undertaken by qualified and experienced arboriculturists or tree surgeons, and that they are carried out in accordance with BS 3998[108].

Drastic pollarding or lopping of trees to reduce height is a common mistake. This will frequently kill the tree or result in future growth that makes the overall tree shape unattractive. Crown thinning should be specified as a percentage of the existing crown and should not exceed 50%. Crown raising can allow more light in to ground level if growth of other plants at lower levels is restricted.

It is no longer recommended practice to paint cuts after surgery or pruning; it is considered better to allow a tree's own wound healing processes to provide protection.

Shrub pruning

Shrub and ground cover planting may require occasional pruning to maintain appropriate shape to the plants. Keeping the plants to a specific size and shape, as with formal hedges, will need frequent pruning (Figure 6.18). Ground cover shrubs and boundary hedges may need to be trimmed twice a year, depending on species. If this is neglected, a more drastic operation may be called for which might have unfortunate results (Figure 6.19).

Grassed areas

Grassed areas can be difficult to mow or may become dumping areas for rubbish. Where new areas of grass are to be planted, an assessment will need to be made of the cost of maintenance including regular mowing during the growing season and the replacement or re-seeding of worn areas. Routes worn down by pedestrians may indicate that paths have been sited inappropriately.

Grass is rarely successful in areas of high shade (eg under tree canopies or close to the north face of taller buildings). Thin growth which leaves patches of bare earth and extensive moss growth are symptoms of this problem. Poor drainage and water logging will prevent grass establishment. Specific symptoms linked to mineral deficiencies within the growing medium may require expert interpretation. Some grassed areas may require a top-dressing of NPK fertiliser to ensure continued growth.

Durability

With appropriate maintenance, landscape schemes can last indefinitely – trees for example can live upwards of 200 years. However, many shrub species will require replacement after 15–20 years as they become over-mature. A maintenance plan should include for a long term ongoing programme of tree and plant replacement.

Stakes

Trees should not be staked with stakes treated with preservative unless it has been established that the preservative will not injure the tree. English oak or larch stakes are normally classed as durable timber without the need for preservative treatment.

Work on site

Workmanship

The key issues are:
- careful plant handling and storage. Plants must not be allowed to dry out or be damaged prior to or during planting
- appropriate soil improvement and avoidance of soil compaction
- firming of soil around plant roots
- appropriate watering and maintenance regimes after planting
- creating a level and even soil tilth before grass turfing or seeding

Inspection

The main elements to assess are:
◊ condition of plant material
◊ moisture content of topsoil
◊ firmness of planting
◊ smoothness and levelness of grassed areas, whether turfed or seeded

The main potential problems are:
◊ unsuitable seeding mixes on areas subject to hard wear
◊ unsuitable seeding mixes for shaded areas
◊ no lawn edgings leading to mowers scalping the grass
◊ over-matured shrubs

Figure 6.19 This Leylandii windbreak was allowed to get out of hand, and the severe trimming has had an unfortunate affect on appearance from which it will not fully recover

References and further reading

Each numbered reference below is shown only under the chapter in which it first appears in the text. Items may appear in further reading lists before being cited as references.

In many cases, existing buildings will have complied with legislation and British Standards in force at the time of construction. Therefore many references will be older editions that would have specified requirements and reflect conditions current at the time.

Publications produced by the former Department of the Environment, Transport and the Regions (DETR), and of the Ministry of Agriculture Fisheries and Food (MAFF), are now the responsibility of their successor departments – the Department of the Environment, Food and Rural Affairs (DEFRA) or the Department of Transport, Local Government and the Regions (DTLR) – or to the Department of Trade and Industry (DTI).

Other organisations for which abbreviations are shown are the British Standards Institution (BSI), the Construction Industry Research and Information Association (CIRIA), and the Interdepartmental Committee on the Redevelopment of Contaminated Land (ICRCL)

Preface
[1] **Health and Safety Commission.** *Managing construction for health and safety.* Construction (Design and Management) Regulations 1994, Approved code of practice. Sudbury, HSE Books, 1995

Chapter 0
[2] **Lillywhite M S T and Webster C J D.** Investigations of drain blockages and their implications for design. *The Public Health Engineer,* April 1979
[3] **Harrison H W.** Quality in new-build housing. *BRE Information Paper* IP 3/93. Garston, Construction Research Communications Ltd, 1993
[4] **DETR.** *English house condition survey 1996.* London, The Stationery Office, 1998
[5] **Scottish Homes.** *Scottish house condition survey 1996.* Survey report. Edinburgh, Scottish Homes, 1997
[6] **Welsh Office.** *Welsh house condition survey 1993.* Cardiff, Welsh Office, 1993
[7] **Northern Ireland Housing Executive.** *Northern Ireland house condition survey 1996.* Belfast, Northern Ireland Housing Executive, 1997
[8] **Construction Quality Forum.** *Database analysis.* Report. Garston, BRE, 1994
[9] **Building Research Station.** *Principles of modern building,* Volume 1: Walls, partitions and chimneys. London, The Stationery Office, 1959
[10] **BRE.** *BRE housing design handbook. Energy and internal layout.* BRE Report. Garston, Construction Research Communications Ltd, 1994
[11] **BRE.** *Assessing traditional housing for rehabilitation.* BRE Report. Garston, Construction Research Communications Ltd, 1990
[12] **Bonshor R B and Bonshor L L.** *Cracking in buildings.* BRE Report. Garston, Construction Research Communications Ltd, 1996
[13] **Willis, Professor R.** Architecture of Worcester Cathedral. *Journal of the Institute of Architects*
[14] **Willis, Professor R.** *The Builder,* 2 March 1861
[15] **Phillips D.** *Excavations at York Minster,* Volume II. London, The Stationery Office, 1985
[16] **Fielden, B.** *York Minster: what is being done.* York, York Minster Appeal Fund, 1968
[17] **Anon.** *Handbook to the cathedrals of England. St Paul's.* London, John Murray, 1879
[18] **Lea F M and Watkins C M.** The durability of reinforced concrete in sea-water. Twentieth report. *National Building Studies Research Paper* No 30. Garston, BRE, 1960
[19] **Billington N S and Roberts B M.** *Building services engineering.* Pergamon, Oxford, 1982
[20] **Harrison H W and Trotman P M.** *BRE building elements. Building services. Performance, diagnosis, maintenance, repair and the avoidance of defects.* Garston, Construction Research Communications Ltd, 2000
[21] **Ward W H.** Soil movement and weather. *Procs of 3rd International Conference on Soil Mechanics and Foundation Engineering, Switzerland, 16–27 August 1953,* vol 1, session 4, pp 477–82
[22] **Crilly M S and Chandler R J.** A method of determining the state of desiccation in clay soils. *BRE Information Paper* IP 4/93 Garston, Construction Research Communications Ltd, 1993
[23] **Bonshor R B and Bonshor L L.** 75 years of building research 1921–1996. *BRE Occasional Paper* OP2. Garston, BRE,1996
[24] **Charles J A, Driscoll R M C, Powell J J M and Tedd P.** Seventy-five years of building research: geotechnical aspects. *Procs of the Institution of Civil Engineers, Geotechnical Engineering 119, July 1996*
[25] **Ministry of Housing and Local Government.** The Building Regulations. Statutory Instrument 1965 No 1373. London, The Stationery Office, 1965
[26] **BSI.** Code of practice for building drainage. *British Standard* BS 8301:1985. London, BSI, 1985
[27] **BSI.** Drain and sewer systems outside buildings. *British Standard* BS EN 752:1996–98. London, BSI, 1996–98

[28] **CIRIA.** Sustainable urban drainage systems: design manual for Scotland and Northern Ireland. *Report* C521. London, CIRIA, 2000

[29] **CIRIA.** Sustainable urban drainage systems: design manual for England and Wales. *Report* C522. London, CIRIA, 2000

[30] **Toyne S.** Is your home on an environment blacklist? and New danger hits 500,000 house prices. *The Sunday Times*, 17 September 2000

[31] **Mathiason N.** Only the cash is drying up. *The Observer*, 5 November 2000

[32] **BRE.** *75 years of building innovation*. Garston, Building Research Establishment, 1996

Chapter 1.1

[33] **BRE.** Concrete in aggressive ground, Parts 1-4. *BRE Special Digest* SD 1. Garston, Construction Research Communications Ltd, 2001

[34] **BRE.** Assessment of damage in low-rise buildings with particular reference to progressive foundation movement. *BRE Digest* 251. Garston, Construction Research Communications Ltd, 1981

[35] **Burford D.** Heave of tunnels beneath the Shell Centre, London, 1959–1986. *Géotechnique*, 1988, **38** (1) 135–7

[36] **Tomlinson M J.** *Foundation design and construction* (5th edition). Harlow, Longman Scientific and Technical, 1991

[37] **BSI.** Code of practice for site investigations. *British Standard* BS 5930: 1999. London, BSI, 1999

[38] **DoE.** *Review of foundation conditions in Great Britain*. London, The Stationery Office, 1995

[39] **DoE and The Welsh Office.** *The Building Regulations 1991 Approved Document A* (1992 edition). London, The Stationery Office, 1995

[40] **Moore J F A and Jones C W.** In situ deformation of Bunter sandstone. *BRE Current Paper* CP 20/75. Garston, BRE, 1975

[41] **Driscoll R M C.** A review of British experience of expansive clay problems. *Procs of 5th International Conference on Expansive Soils, Adelaide, 1983*

[42] **Driscoll R M C and Crilly M S.** *Subsidence damage to domestic buildings: lessons learned and questions remaining*. FBE Report. Garston, Construction Research Communications Ltd, 2000

[43] **BRE.** Low-rise buildings on shrinkable clay soils: Part 1 (new edition). *BRE Digest* 240. Garston, Construction Research Communications Ltd, 1993

[44] **Collins R J.** Case studies of floor heave due to microbiological activity in pyritic shales. *Procs of Symposium on Microbiology in Civil Engineering, Silsoe College, Bedford, 3–5 September 1990*

[45] **Ministry of Housing and Local Government.** The Building Regulations 1965. Statutory Instrument 1965 No 1373. London, The Stationery Office, 1965

[46] **Marsland A and Butcher A P.** In situ tests on highly weathered chalk near Luton, England. *Procs of International Symposium on Soil and Rock Investigation by In-situ Testing, 1983*

[47] **Thorburn S and Cooke R W.** Observations of settlement of a two storey school building underlain by peat. *Procs of Conference on Marginal and Derelict Land, Glasgow, 1986*

[48] **Driscoll R M C, Powell J J M and Uglow I M.** Geotechnical properties of an alluvial site of light-weight buildings. *Procs of Conference on Marginal and Derelict Land, Glasgow, 1986*

[49] **Scott K F.** Private correspondence with H W Harrison, 1995

[50] **Skempton A W.** Notes on the compressibility of clays. *Geological Society of London Quarterly Journal* (July 28 1944), **100**

[51] **King R J.** *Handbook of the eastern cathedrals*. London, John Murray, 1862

[52] *Scots Magazine*, February 1749

[53] **Barron J.** Northern highlands. *Inverness Courier*, 1816

[54] *The Highland News* (Inverness), 5 April 1952

[55] **BSI.** Design provisions for earthquake resistance of structures. Eurocode 8. *British Standard* DD ENV 1998:1996–99. London, BSI, 1996–99

[56] **White R B.** *Prefabrication – a history of its development in Great Britain*. London, The Stationery Office, 1965

[57] **Institution of Civil Engineers.** *Mining subsidence*. ICE, London, 1959

[58] **BSI.** Methods of test for soils for civil engineering purposes. Classification tests. *British Standard* BS 1377-2:1990. London, BSI, 1990

[59] **Moss R M and Matthews S L.** In-service structural monitoring: a state of the art report. *Structural Engineer* (17 January 1995), **73** (2) 23–31

[60] **Driscoll R M C.** Measured shift work. *Building* (20 October 1989), **254** (42)

[61] **BSI.** Code of practice for safety precautions in the construction of large diameter boreholes for piling and other purposes. *British Standard* BS 5573:1978. London, BSI 1978

[62] **BRE.** Site investigation for low-rise building: desk studies. *BRE Digest* 318. Garston, Construction Research Communications Ltd, 1987

[63] **BRE.** Site investigation for low-rise building: trial pits. *BRE Digest* 381. Garston, Construction Research Communications Ltd, 1993

[64] **BRE.** Site investigation for low-rise building: soil description. *BRE Digest* 383. Garston, Construction Research Communications Ltd, 1993

[65] **Abbiss C P and Viggiani G.** Surface wave damping measurements of the ground with a correlator. *Procs of 13th International Conference on Soil Mechanics and Foundation Engineering, New Delhi, 1994*

Further reading

Burland J B and Burbidge M C. Settlement of foundations on sand and gravel. *Procs of the Institution of Civil Engineers, December 1985*, Part 1, vol 78, pp 1325–81

BRE. Foundations on shrinkable clay: avoiding damage due to trees. *BRE Defect Action Sheet (Design)* DAS 96. Garston, Construction Research Communications Ltd, 1987

Boden J B and Driscoll R M C. The design and performance of house foundations on shrinkable/expansible clays. *Municipal Engineer* (1987), **4** (4) 181–213

DoE. *Landslides*. London, The Stationery Office, 1990

DoE. *Mining instability*. London, The Stationery Office, 1992

DoE. *Natural underground cavities*. London, The Stationery Office, 1993

DoE. *Seismic risk*. London, The Stationery Office, 1993

BRE. Site investigation for low-rise building: the walk-over survey. *BRE Digest* 348. Garston, Construction Research Communications Ltd, 1989

BRE. Soils and foundations: 1. *BRE Digest* 63. Garston, Construction Research Communications Ltd, 1976

BRE. Soils and foundations: 2. *BRE Digest* 64. Garston, Construction Research Communications Ltd, 1976

BRE. Soils and foundations: 3. *BRE Digest* 67. Garston, Construction Research Communications Ltd, 1976

Ove Arup and Partners. *Earthquake hazard and risk in the United Kingdom*. London, Ove Arup and Partners, 1993

Chapter 1.2

[66] **Pye P W and Harrison H W.** *BRE building elements. Floors and flooring. Performance, diagnosis, maintenance, repair and the avoidance of defects*. Garston, Construction Research Communications Ltd, 1997

[67] **Cairney T.** *Contaminated land. Problems and solutions*. Blackie, London, 1993

[68] **BSI.** Code of practice for foundations. *British Standard* BS 8004:1986. London, BSI 1986

References and further reading

[69] **BRE.** Fill. Part 1: Classification and load carrying characteristics. *BRE Digest* 274. Garston, Construction Research Communications Ltd, 1983 (revised 1991)

[70] **BRE.** Fill. Part 2: Site investigation, ground improvement and foundation design. *BRE Digest* 275. Garston, Construction Research Communications Ltd, 1983 (minor revisions 1992)

[71] **Charles J A and Burford D.** Settlement and groundwater in opencast mining backfills. *Procs of 9th European Conference on Soil Mechanics and Foundation Engineering, Dublin, 1987*, vol 1, 289–92

[72] **Environment Committee.** First Report to the House of Commons. Contaminated Land, Volume 1. London, The Stationery Office, 1990

[73] **DoE (ICRCL).** Guidance on the assessment and redevelopment of contaminated land. Report No 59/83 (2nd edition). London, DoE, 1983

[74] **BSI.** Code of practice for the identification of potentially contaminated land and its investigation. *British Standard* DD 175:1988. London, BSI, 1988

[75] **Leach B A and Goodger H K.** Building on derelict land. CIRIA Special Publication 78. London, CIRIA, 1991

[76] **DoE (ICRCL).** Notes on the fire hazards of contaminated land. No 61/84 (2nd edition). London, DoE, 1986

[77] **Paul V.** Performance of building materials in contaminated land. BRE Report. Garston, Construction Research Communications Ltd, 1994

[78] **BRE.** Slurry trench cut-off walls to contain contamination. *BRE Digest* 395. Garston, Construction Research Communications Ltd, 1994

[79] **Trenter N A and Charles J A.** A model specification for engineered fills for building purposes. *Procs of the Institution of Civil Engineers, Geotechnical Engineering 119, July 1996*

[80] **BRE.** Fill and hardcore. *BRE Digest* 222, Garston, Construction Research Communications Ltd, 1979

[81] **Nixon P J.** Floor heave in buildings due to the use of pyritic shales as fill material. *Chemistry and Industry*, 4 March 1978

[82] **Charles J A and Watts K S.** Building on fill: geotechnical aspects (2nd edition). [BRE Report.] Garston, Construction Research Communications Ltd, 1993 (2001)

[83] **Charles J A.** The causes, magnitudes and control of ground movements in fills. *Procs of 4th International Conference on Ground Movements and Structures, UWIST, Cardiff, 8–11 July 1991*. London, Pentech Press, 1992

[84] **Burford D and Charles J A.** Long term performance of houses built on opencast ironstone mining backfill at Corby, 1975–1990. *Procs of 4th International Conference on Ground Movements and Structures, UWIST, Cardiff, 8–11 July 1991*, vol 4, pp 54–67. London, Pentech Press, 1992

[85] **BSI.** Methods of test for soils for civil engineering purposes. Chemical and electro-chemical tests. *British Standard* BS:1377-3: 1990. London, BSI, 1990

[86] **British Drilling Association.** *Guidance notes for the safe drilling of landfills and contaminated land*. Brentwood, BDA, 1992

[87] **DoE (ICRCL).** Notes on the redevelopment of scrap yards and similar sites. No 42/80 (2nd edition). London, DoE, 1983

Further reading

BRE. *Procs of Symposium, Founding buildings on difficult ground, 28 October 1987*. Garston, BRE, 1987

Charles J A and Watts K S. The assessment of the collapse potential of fills and its significance for building on fill. *Procs of the Institution of Civil Engineers, Geotechnical Engineering 119, July 1996*

BSI. Code of practice for strengthened/reinforced soils and other fills. *British Standard* BS 8006:1995. London, BSI, 1995

BSI. Structural design of low-rise buildings. Code of practice for stability, site investigation, foundations and ground floor slabs for housing. *British Standard* BS 8103-1:1995. London, BSI, 1995

CIRIA. Building on derelict land. *CIRIA Special Publication* 78. London, CIRIA, 1991

DoE (ICRCL). Notes on the redevelopment of sewage works and farms. No 23/79 (2nd edition). London, DoE, 1983

DoE (ICRCL). Notes on the redevelopment of gas works sites. No. 18/79 (5th edition). London, DoE, 1986

DoE (ICRCL). Notes on the development and after-use of landfill sites. No. 17/78 (8th edition). London, DoE, 1990

DoE (ICRCL). Asbestos on contaminated sites. No. 64/85 (2nd edition). London, DoE, 1986

Bowley M J. *Sulphate and acid attack on concrete in the ground: recommended procedures for soil analysis*. BRE Report. Garston, Construction Research Communications Ltd, 1995

Johnson R. *Protective measures for housing on gas-contaminated land*. BRE Report. Garston, Construction Research Communications Ltd, 2001

Chapter 1.3

[88] **BSI.** Code of practice for drainage of roofs and paved areas. *British Standard* BS 6367:1983. London, BSI, 1983

[89] **BSI.** Code of practice for building drainage. *British Standard* BS 8301:1985. London, BSI, 1985

[90] **BRE.** Soakaway design. *BRE Digest* 365. Garston, Construction Research Communications Ltd, 1991

[91] **BSI.** Drain and sewer systems outside buildings. Hydraulic design and environmental considerations. *British Standard* BS EN 752-4:1998. London, BSI, 1998

[92] **Graves H M and Phillipson M C.** *Potential implications of climate change in the built environment*. FBE Report. Garston, Construction Research Communications Ltd, 2000

[93] **Scottish Executive.** The Building Standards (Scotland) Regulations 1990. Edinburgh, The Stationery Office, 1990

[94] **Charles J A and Burford D.** The effect of a rise of water table on the settlement of opencast mining backfill. *BRE Information Paper* IP 15/85. Garston, Construction Research Communications Ltd, 1985

[95] **Charles J A and Burford D.** Settlement and ground-water in opencast mining backfill. *Procs of 9th European Conference on Soil Mechanics and Foundation Engineering, Dublin, 1987*, vol 1, pp 289–92

[96] **Charles J A, Burford D and Hughes D B.** Settlement of opencast coal mining backfill at Horsley, 1973–1992. *Procs of Conference, Engineered Fills, Newcastle upon Tyne, September 1993*. London, Thomas Telford, 1993

Further reading

BRE. Drying out buildings. *BRE Digest* 163. Garston, Construction Research Communications Ltd, 1974

BRE. *Design guidance on flood damage to dwellings*. Edinburgh, The Stationery Office, 1996

Chapter 1.4

[97] **Harrison H W.** *BRE building elements. Roofs and roofing. Performance, diagnosis, maintenance, repair and the avoidance of defects*. Garston, Construction Research Communications Ltd, 1996

[98] **Anderson B R, Clark A J, Baldwin R and Milbank N O.** BREDEM: the BRE Domestic Energy Model. *BRE Information Paper* IP 16/85. Garston, Construction Research Communications Ltd, 1985

[99] **BRE.** Climate and site development. Part 3: improving microclimate through design. *BRE Digest* 350. Garston, Construction Research Communications Ltd, 1990

[100] **BRE.** Wind environment around tall buildings. *BRE Digest* 141. Garston, Construction Research Communications Ltd, 1972

[101] **Harrison H W and De Vekey R C.** *BRE building elements. Walls, windows and doors. Performance, diagnosis, maintenance, repair and the avoidance of defects.* Garston, Construction Research Communications Ltd, 1998

[102] **BRE.** Climate and site development. Part 2: influence of microclimate. *BRE Digest* 350. Garston, Construction Research Communications Ltd, 1990

[103] **BRE.** Estimating daylight in buildings: Part 1. *BRE Digest* 309. Garston, Construction Research Communications Ltd, 1986

[104] **BRE.** Estimating daylight in buildings: Part 2. *BRE Digest* 310. Garston, Construction Research Communications Ltd, 1986

Further reading

O'Rourke T. *Planning for passive solar design.* Garston, BRECSU

Graves H M and Phillipson M C. *Potential implications of climate change in the built environment.* FBE Report. Garston, Construction Research Communications Ltd, 2000

Chapter 1.5

[105] **BRE.** Low-rise building foundations: the influence of trees in clay soils. *BRE Digest* 298 (new edition 1999). Garston, Construction Research Communications Ltd, 1999

[106] **Chandler R J and Gutierrez C I.** The filter paper method of suction measurement. *Géotechnique* (1986), **36** (2) 265–8

[107] **Institution of Structural Engineers.** *Subsidence of low rise buildings.* London, ISE, 2000

[108] **BSI.** Recommendations for tree work. *British Standard* BS 3998:1989. London, BSI, 1989

[109] **National House-Building Council.** *Standards.* Amersham, NHBC, 1992

[110] **Cheney J E.** 25 years' heave of a building constructed on clay, after tree removal. *Ground Engineering* (1988), **21** (5) 13–27

[111] **BSI.** Guide for trees in relation to construction. *British Standard* BS 5837: 1991. London, BSI, 1991

Further reading

Singh C J (compiler). The effect of trees and vegetation on damage to and settlement of structures. *BRE Library Bibliography* 264. Garston, BRE, 1994

BRE. Foundations on shrinkable clay: avoiding damage due to trees. *BRE Defect Action Sheet (Design)* DAS 96. Garston, Construction Research Communications Ltd, 1987

Cheney J E. Long term heave of a building founded on clay soil after tree removal. *Procs of Conference on Geotechnical Instrumentation in Civil Engineering Projects, Nottingham, 1989*

BRE. Low-rise buildings on shrinkable clay soils: Part 1. *BRE Digest* 240. Garston, Construction Research Communications Ltd, 1993

BRE. Low-rise buildings on shrinkable clay soils: Part 2. *BRE Digest* 241. Garston, Construction Research Communications Ltd, 1976

BRE. Low-rise buildings on shrinkable clay soils: Part 3. *BRE Digest* 242. Garston, Construction Research Communications Ltd, 1993

BRE. Assessment of damage in low-rise buildings. *BRE Digest* 251. Garston, Construction Research Communications Ltd, 1976

Chapter 2.1

[112] **Jackson B H.** *Recollections of Thomas Graham Jackson.* London, OUP, 1950

[113] **BRE.** Site investigation for low-rise building: desk studies. *BRE Digest* 318. Garston, Construction Research Communications Ltd, 1987

[114] **Roscoe G H and Driscoll R.** *A review of routine foundation design practice.* BRE Report. Garston, Construction Research Communications Ltd, 1987

[115] **Charles J A, Burford D and Watts K S.** Field studies of the effectiveness of 'dynamic consolidation'. *Procs of 10th International Conference on Soil Mechanics and Foundation Engineering, Stockholm, 1981,* vol 3, pp 617–22

[116] **Watts K S, Saadi A, Woods L A and Johnson D.** Preliminary report on a field trial to assess the design and performance of vibro ground treatment with reinforced strip foundations. *Procs of the 2nd International Conference on Polluted and Marginal Land, Brunel University, June 1992*

[117] **Watts K S, Charles J A and Butcher A P.** Ground improvement for low-rise housing using vibro at a site in Manchester. *Municipal Engineer* (1989), **6** (3) 145–57

[118] **Charles J A, Burford D and Watts K S.** Improving the load carrying characteristics of uncompacted fill by pre-loading. *Municipal Engineer* (1986), **3** (1) 1–19

[119] **Charles J A, Burford D and Watts K S.** Preloading uncompacted fills. *BRE Information Paper* IP 16/86. Garston, Construction Research Communications Ltd, 1986

[120] **Terzaghi K and Peck R B.** *Soil mechanics in engineering practice* (2nd edition). Wiley, New York, 1967

[121] **BSI.** Eurocode 7. Geotechnical design. *British Standard* DD ENV 1997:1995 and 2000. London, BSI, 1995 and 2000

[122] **Charles J A.** The depth of influence of loaded areas. *Géotechnique*, 1994

[123] **Burland J B and Wroth C P.** Settlement of buildings and associated damage. *BRE Current Paper* CP 33/75. Garston, BRE, 1975

[124] **Skempton A W and Macdonald D H.** The allowable settlement of buildings. *Procs of the Institution of Civil Engineers, 1955,* vol 5, no 3, part 3

[125] **National House-Building Council.** Building near trees. *Standards,* Chapter 4.2. Amersham, NHBC, 1995

[126] **Simpson B, Blower T, Craig R N and Wilkinson W B.** The engineering implications of rising ground-water levels in the deep aquifer beneath London. *CIRIA Special Publication* 69. London, CIRIA, 1989

[127] **Knipe C V, Lloyd J W, Lerner D N and Greswell R.** Rising ground-water levels in Birmingham and the engineering implications. *CIRIA Special Publication* 92. London, CIRIA, 1993

[128] **Younger P L.** Possible environmental impact of the closure of two collieries in County Durham. *Journal of the Institute of Water and Environment Management* (1993), **7** 521–31

[129] **BRE.** Simple measuring and monitoring of movement in low-rise buildings. Part 2: settlement, heave and out-of-plumb. *BRE Digest* 344 (revised 1995). Garston, Construction Research Communications Ltd, 1989 (1995)

[130] **Robson P.** *Structural appraisal of traditional buildings.* London, Gower Technical, 1990

[131] **BRE.** Why do buildings crack? *BRE Digest* 361. Garston, Construction Research Communications Ltd, 1991

[132] **Melville and Gordon.** *The repair and maintenance of houses.* London, Estates Gazette Ltd, 1973

[133] **Burford D, Crilly M S and Handley V.** Monitoring and repair of a building damaged by ground movements using the Hoopsafe system. *Procs of Conference on Structural Faults and Repair, 1999*

[134] **Anon.** The day our house fell down. London, *The Times*, 12 October 1999

[135] **Hunt R, Dyer R H and Driscoll R.** *Foundation movement and remedial underpinning in low-rise buildings.* BRE Report. Garston, Construction Research Communications Ltd, 1991

[136] **BRE.** Mini-piling for low-rise buildings. *BRE Digest* 313. Garston, Construction Research Communications Ltd, 1986

[137] **BRE.** Rising damp in walls: diagnosis and treatment. *BRE Digest* 245. Garston, Construction Research Communications Ltd, 1986

[138] **Elson W K.** Design of laterally loaded piles. *CIRIA Report* 103. London, CIRIA, 1984

[139] **Chandler R J, Crilly M S and Montgomery-Smith G.** 1992. A low-cost method of assessing clay desiccation for low-rise buildings. *Procs of the Institution of Civil Engineers, Civil Engineering*, vol 92, pp 82–9

[140] **Crilly M S and Driscoll R M C.** The behaviour of lightly loaded piles in swelling ground and implications for their design. *Procs of the Institution of Civil Engineers, Geotechnical Engineering* 143, January 2000, pp 3–16

Further reading

BSI. Structural design of low-rise buildings. Code of practice for stability, site investigation, foundations and ground floor slabs for housing. *British Standard* BS 8103-1:1995. London, BSI, 1995

BSI. Code of practice for foundations. *British Standard* BS 8004:1986. London, BSI, 1986

BRE. Simple measuring and monitoring of movement in low-rise buildings. Part 1: cracks. *BRE Digest* 343. Garston, Construction Research Communications Ltd, 1989

BRE. Assessment of damage in low-rise buildings with particular reference to progressive foundation movement. *BRE Digest* 251 (revised 1995). Garston, Construction Research Communications Ltd, 1981 (1995)

BRE. Estimation of thermal and moisture movements and stresses: Part 1. *BRE Digest* 227. Garston, Construction Research Communications Ltd, 1979

BRE. Estimation of thermal and moisture movements and stresses: Part 2. *BRE Digest* 228. Garston, Construction Research Communications Ltd, 1976

BRE. Estimation of thermal and moisture movements and stresses: Part 3. *BRE Digest* 229. Garston, Construction Research Communications Ltd, 1979

BRE. Underpinning. *BRE Digest* 352. Garston, Construction Research Communications Ltd, 1993

BRE. Damage to structures from ground-borne vibration. *BRE Digest* 353. Garston, Construction Research Communications Ltd, 1990

Watts K S. *Specifying vibro stone columns.* BRE Report. Garston, Construction Research Communications Ltd, 2000

Chapter 2.2
Further reading

BRE. Repairing brick and block masonry. *BRE Digest* 359. Garston, Construction Research Communications Ltd, 1991

BRE. Low-rise buildings on shrinkable clay soils: Part 1. *BRE Digest* 240. Garston, Construction Research Communications Ltd, 1993

BRE. Low-rise buildings on shrinkable clay soils: Part 2. *BRE Digest* 241. Garston, Construction Research Communications Ltd, 1980 (minor revisions 1990)

BRE. Low-rise buildings on shrinkable clay soils: Part 3. *BRE Digest* 242. Garston, Construction Research Communications Ltd, 1980

BRE. Safety of large masonry walls. *BRE Digest* 281. Garston, Construction Research Communications Ltd, 1984

BRE. Building mortar. *BRE Digest* 362. Garston, Construction Research Communications Ltd, 1991

BRE. External masonry walls: eroding mortars – repoint or rebuild? *BRE Defect Action Sheet (Design)* DAS 70. Garston, Construction Research Communications Ltd, 1986

BRE. External masonry walls: repointing – specification. *BRE Defect Action Sheet (Design)* DAS 71. Garston, Construction Research Communications Ltd, 1986

Chapter 2.3

[141] **BSI.** Concrete. (Part 1) Guide to specifying concrete. (Part 2) Methods for specifying concrete mixes. *British Standard* BS 5328-1 and 2:1997. London, BSI, 1997

[142] **BSI.** Steel fabric for the reinforcement of concrete. *British Standard* BS 4483:1998. London, BSI, 1998

[143] **Thomas M D A and Matthews J D.** Performance of fly ash concrete in UK structures. *ACI Materials Journal* (1993), **90** (6) 586–93

[144] **Cooke R W and Thorburn S.** Observations of settlement and foundation loading on four and five storey housing blocks on alluvial soils near the River Clyde. *Procs of Symposium on Soil–Structure Interaction, Institution of Structural Engineers, London, 1984*, pp 9–14

[145] **Halliwell M A and Crammond N J.** *Avoiding the thaumasite form of sulfate attack: two-year report.* BRE Report. Garston, Construction Research Communications Ltd, 2000

[146] **Walton P L.** Effects of alkali-silica reaction on concrete foundations. *BRE Information Paper* IP 16/93. Garston, Construction Research Communications Ltd, 1993

Further reading

BSI. Structural use of concrete. Code of practice for design and construction. *British Standard* BS 8110-1:1997. London, BSI, 1997

BRE. Concrete. Part 1: materials. *BRE Digest* 325. Garston, Construction Research Communications Ltd, 1987

BRE. Concrete. Part 2: specification, design and quality control. *BRE Digest* 326. Garston, Construction Research Communications Ltd, 1987

BRE. Sulfate and acid resistance of concrete in the ground. *BRE Digest* 363 (new edition 1996). Garston, Construction Research Communications Ltd, 1991 (1996)

BRE. Alkali-silica reaction in concrete. *BRE Digest* 330 (in 4 Parts). Garston, Construction Research Communications Ltd, 1999

BRE. Simple foundations for low-rise housing. Site investigation. *BRE Good Building Guide* GBG 39 Part 1. Garston, Construction Research Communications Ltd, 2000

BRE. Simple foundations for low-rise housing. 'Rule-of-thumb' design. *BRE Good Building Guide* GBG 39 Part 2. Garston, Construction Research Communications Ltd, 2001

Chapter 2.4

[147] **Ward W H and Green H.** House foundations: the short-bored pile. *Municipal Services Congress, Institution of Civil Engineers, 6 November 1952*

[148] **BRE.** Choosing piles for new construction. *BRE Digest* 315. Garston, Construction Research Communications Ltd, 1986

[149] **Price G, Wardle I F and Jennings D.** The use of slope measuring devices for monitoring bending strains in piles. *Procs of 21st Conference of the British Society of Strain Measurement, Cambridge, 2–5 September 1985*

[150] **Price G and Wardle I F.** Recent developments in pile and soil instrumentation systems. *Procs of the International Symposium on Field Measurements in Geomechanics, Zurich, September 1983.* Rotterdam, A A Balkema, 1984, pp 533–42

[151] **BSI.** Code of practice for safety precautions in the construction of large diameter boreholes for piling and other purposes. *British Standard* BS 5573:1978. London, BSI, 1978

[152] **BRE.** Concrete in sulphate-bearing soils and groundwaters. *BRE Digest* 250 (minor revisions 1986). Garston, Construction Research Communications Ltd, 1981 (1986)

[153] **Quarry Products Association.** *Practical application of the DETR Expert Group report on the thaumasite form of sulfate attack on buried concrete.* London, QPA, 1999

[154] **BRE.** Timbers: their natural durability and resistance to preservative treatment. *BRE Digest* 296. Garston, Construction Research Communications Ltd, 1985

[155] **BSI.** Structural use of timber. Code of practice for the preservative treatment of structural timber. *British Standard* BS 5268-5:1989 London, BSI, 1989

[156] **BSI.** Code of practice for preservation of timber. *British Standard* BS 5589:1989. London, BSI, 1989

Further reading
BRE. The durability of steel in concrete: Part 1. Mechanism of protection and corrosion. *BRE Digest* 263. Garston, Construction Research Communications Ltd, 1982
BRE. The durability of steel in concrete: Part 2. Diagnosis and assessment of corrosion-cracked concrete. *BRE Digest* 264. Garston, Construction Research Communications Ltd, 1982

Chapter 3
[157] **Covington S A.** Basements in housing: a feasibility study. *BRE Current Paper* CP 4/78. Garston, BRE, 1978
[158] **British Cement Association.** *Basements, land use and energy conservation with market and construction survey.* Crowthorne, BCA, 1998. (See also http://www.basements.org.uk)

Chapter 3.1
[159] **Scivyer C R and Jaggs M P R.** *A BRE guide to radon remedial measures in existing dwellings. Dwellings with cellars and basements.* BRE Report. Garston, Construction Research Communications Ltd, 1998
[160] **Tedd P and Charles J A.** Interaction of a propped retaining wall with the stiff clay in which it is embedded. *Procs of Soil–Structure Interaction Symposium, Institution of Structural Engineers, November 1984,* pp 47–49
[161] **BSI.** Code of practice for use of masonry. Materials and components, design and workmanship. *British Standard* BS 5628-3:2001. London, BSI, 2001
[162] **BSI.** Code of practice for design of concrete structures for retaining aqueous liquids. *British Standard* BS 8007:1987. London, BSI, 1987

Further reading
Burford D. Heave of tunnels beneath the Shell Centre, 1959–1986. *Géotechnique* (1988), **38** (1)

Chapter 3.2
[163] **BSI.** Code of practice for protection of structures against water from the ground. *British Standard* BS 8102:1990. London, BSI, 1990
[164] **BRE** Damp proofing existing basements. *BRE Good Building Guide* GBG 3. Garston, Construction Research Communications Ltd, 1993
[165] **BRE.** Treating dampness in basements. *BRE Good Repair Guide* GRG 23. Garston, Construction Research Communications Ltd, 1999
[166] **BRE.** Repairing flood damage: immediate action. *BRE Good Repair Guide* GRG 11 Part 1. Garston, Construction Research Communications Ltd, 1997
[167] **BRE.** Repairing flood damage: ground floors and basements. *BRE Good Repair Guide* GRG 11 Part 2. Garston, Construction Research Communications Ltd, 1997
[168] **Bravery A F, Berry R W, Carey J K and Cooper D E.** *Recognising wood rot and insect damage in buildings.* BRE Report. Garston, Construction Research Communications Ltd, 1992
[169] **Berry R W.** *Remedial treatment of wood rot and insect attack in buildings.* BRE Report. Garston, Construction Research Communications Ltd, 1994
[170] **BRE.** Repairing flood damage: services, secondary elements, finishes, fittings. *BRE Good Repair Guide* GRG 11 Part 4. Garston, Construction Research Communications Ltd, 1997

Chapter 3.3
[171] **British Cement Association.** *Basements for dwellings. Approved Document. The Building Regulations 1991.* Crowthorne, BCA, 1997
[172] **Anderson B R.** U-values for basements. *BRE Information Paper* IP 14/94. Garston, Construction Research Communications Ltd, 1994
[173] **BSI.** Code of practice for control of condensation in buildings. *British Standard* BS 5250:1989. London, BSI, 1989
[174] **BRE.** Drying out buildings. *BRE Digest* 163. Garston, Construction Research Communications Ltd, 1974
[175] **BRE.** Rising damp in walls: diagnosis and treatment. *BRE Digest* 245 (minor revision 1986). Garston, Construction Research Communications Ltd, 1984 (1986)
[176] **BRE.** Remedies for condensation and mould in traditional housing. Audiovisual package. Garston, BRE, 1986
[177] **DoE and The Welsh Office.** *The Building Regulations 1991 Approved Document B.* London, The Stationery Office, 1992
[178] **BSI.** Fire precautions in the design, construction and use of buildings. Code of practice for residential buildings. *British Standard* BS 5588-1:1990. London, BSI, 1990
[179] The Housing Act 1957. London, The Stationery Office

Chapter 4.1
[180] **Ministry of Public Building and Works.** Co-ordination of underground services on building sites. The Common trench. *R&D Bulletin* 1. London, The Stationery Office, 1973
[181] **BSI.** Specification for identification of pipelines and services. *British Standard* BS 1710:1984. London, BSI, 1984
[182] **DoE.** *English house condition survey 1991.* London, The Stationery Office, 1993
[183] The Water Act 1981. London, The Stationery Office, 1981
[184] The Water (Scotland) Act 1980. Edinburgh, The Stationery Office, 1980
[185] Model Water Byelaws, 1986. London, The Stationery Office, 1986
[186] Water Supply/Water Fittings Regulations 1999. London, The Stationery Office, 1999
[187] **BSI.** Water supply. Requirements for systems and components outside buildings. *British Standard* BS EN 805:2000. London, BSI, 2000
[188] **BSI.** Specification for design, installation, testing and maintenance of services supplying water for domestic use within buildings and their curtilages. *British Standard* BS 6700:1997. London, BSI, 1997

Chapter 4.2
[189] **DoE and The Welsh Office.** *The Building Regulations 1991. Approved Document H.* London, DETR, 1992
[190] **Scottish Executive.** *Technical Standards. For compliance with the Building Standards (Scotland) Regulations 1990, Part M.* Edinburgh, The Stationery Office, 1990
[191] **DoE (Northern Ireland).** *The Building Regulations (Northern Ireland) 1994, Technical Booklet N.* Belfast, The Stationery Office, 1994
[192] **BSI.** Gravity drainage systems inside buildings. British Standard BS EN 12056-1 to 5:2000. London, BSI, 2000
[193] **BSI.** Drain and sewer systems outside buildings. *British Standard* BS EN 752: 1996–98. London, BSI, 1996–98
[194] **Balmer and Mears.** Surveying Severn Trent's sewers. *Procs of the Institution of Civil Engineers Conference on Restoration of Sewage Systems, 1981*
[195] Sewage (Scotland) Act 1968. Edinburgh, The Stationery Office, 1968
[196] **Building Research Station.** Drainage pipelines – 1. *BRS Digest* 130. Garston, Construction Research Communications Ltd, 1971
[197] **BSI.** Drain and sewer systems outside buildings. Generalities and definitions. *British Standard* BS EN 752-1:1996. London, BSI, 1996
[198] **BSI.** Drain and sewer systems outside buildings. Performance requirements. *British Standard* BS EN 752-2:1997. London, BSI, 1997
[199] **BSI.** Drain and sewer systems outside buildings. Planning. *British Standard* BS EN 752-3:1997. London, BSI, 1997
[200] **BSI.** Drain and sewer systems outside buildings. Rehabilitation. *British Standard* BS EN 752-5:1998. London, BSI, 1998
[201] **BSI.** Drain and sewer systems outside buildings. Pumping installations. *British Standard* BS EN 752-6:1998. London, BSI, 1998

References and further reading

[202] **BSI.** Drain and sewer systems outside buildings. Maintenance and operations. *British Standard* BS EN 752-7:1998. London, BSI, 1998

[203] **BSI.** Construction and testing of drains and sewers. *British Standard* BS EN 1610: 1998. London, BSI, 1998

[204] **BSI.** Pressure sewerage systems outside buildings. *British Standard* BS EN 1671:1997. London, BSI, 1997

[205] **McKay W B.** Building construction, Volume 2. London, Longmans Green, 1944

[206] **BSI.** Precast concrete pipes, fittings and ancillary products. Specification for inspection chambers. *British Standard* BS 5911:1982–94. London, BSI, 1982–94

[207] **BSI.** Precast concrete pipes, fittings and ancillary products. Specification for unreinforced and reinforced manholes and soakaways of circular cross section. *British Standard* BS 5911-200:1989. London, BSI, 1989

[208] **BSI.** Specification for plastics inspection chambers for drains. *British Standard* BS 7158:1989. London, BSI, 1989

[209] **BSI.** Gullies for buildings. Requirements. *British Standard* BS EN 1253-1:1999. London, BSI, 1999

[210] **BSI.** Gullies for buildings. Test methods. *British Standard* BS EN 1253-2: 1999. London, BSI, 1999

[211] **DETR.** *The Building Regulations 1991. Approved Document H. Draft.* London, DETR, 26 July 2000. (Also available on www.construction.detr.gov.uk/consult/parth/index.htm)

[212] **BSI.** Code of practice for design and installation of small sewage treatment works and cesspools. *British Standard* BS 6297: 1983. London, BSI, 1983

[213] **BSI.** Small wastewater treatment systems for up to 50 PT. Prefabricated septic tanks. *British Standard* BS EN 12566-1:2000. London, BSI, 2000

[214] **Griggs J and Grant N.** Reed beds: application and specification. *BRE Good Building Guide* GBG 42 Part 1. Garston, Construction Research Communications Ltd, 2000

[215] **Griggs J and Grant N.** Reed beds: design, construction and maintenance. *BRE Good Building Guide* GBG 42 Part 2. Garston, Construction Research Communications Ltd, 2000

[216] **Grant N and Griggs J.** Reed beds for the treatment of domestic wastewater. BRE Report. Garston, Construction Research Communications Ltd, 2001

[217] **BSI.** Procedure for type testing of small biological domestic wastewater treatment plants. *British Standard* BS 7781: 1994. London, BSI, 1994

[218] **BSI.** Wastewater treatment plants. *British Standard* BS EN 12255:1999–2001. London, BSI, 1999–2001

[219] **BSI.** Sewerage. Guide to pumping stations and pumping mains. *British Standard* BS 8005-2:1987. London, BSI, 1987

[220] **BSI.** Wastewater lifting plants for buildings and sites. Principles of construction and testing. Lifting plants for wastewater containing faecal matter. *British Standard* BS EN 12050-1:2001. London, BSI, 2001

[221] **BSI.** Wastewater lifting plants for buildings and sites. Principles of construction and testing. Lifting plants for faecal-free wastewater. *British Standard* BS EN 12050-2: 2001. London, BSI, 2001

[222] **BSI.** Wastewater lifting plants for buildings and sites. Principles of construction and testing. Lifting plants for wastewater containing faecal matter for limited applications. *British Standard* BS EN 12050-3:2001. London, BSI, 2001

[223] **Carroll R and Britten J.** Non-sewered sanitation for housing and institutional buildings. *Building Issues* (1992), **4** (1). Lund Centre for Habitat Studies, Lund University, 1992

[224] **BSI.** Specification for unplasticized polyvinyl chloride (PVC-U) pipes and plastics fittings of nominal sizes 110 and 160 for below ground gravity drainage and sewerage. *British Standard* BS 4660:1989. London, BSI, 1989

[225] **BSI.** Specification for unplasticized PVC pipe and fittings for gravity sewers. *British Standard* BS 5481:1977. London, BSI, 1977

[226] **BSI.** Adhesives. Freeze–thaw stability. *British Standard* BS EN 1239:1998. London, BSI, 1998

[227] **BSI.** Plastics piping systems for soil and waste (low and high temperature) within the building structure. Acrylonitrile-butadiene-styrene (ABS). Specifications for pipes, fittings and the system. *British Standard* BS EN 1455-1:2000. London, BSI, 2000

[228] **BSI.** Plastics piping systems for soil and waste discharge (low and high temperature) within the building structure. Polypropylene (PP). Specifications for pipes, fittings and the system. *British Standard* BS EN 1451-1:2000. London, BSI, 2000

[229] **BSI.** Plastics piping systems for non-pressure underground drainage and sewerage. Polypropylene (PP). Specifications for pipes, fittings and the system. *British Standard* BS EN 1852-1:1998. London, BSI, 1998

[230] **BSI.** Plastics piping systems for non-pressure drainage and sewerage. Glass-reinforced thermosetting plastics (GRP) based on unsaturated polyester resin (UP). Fittings. *British Standard* BS EN 1636:2001. London, BSI, 2001

[231] **BSI.** Specification for thermoplastics waste pipe and fittings. *British Standard* BS 5255:1989. London, BSI, 1989

[232] **Wise A F E.** Regulating the flow. *Building Services*, October 1985

[233] **BSI.** Plastics piping systems for non-pressure underground drainage and sewerage. Polyethylene (PE). Part 1. Specification for pipes, fittings and the system. 96/126602 DC, *British Standard* BS EN 12666-1 (draft for comment). London, BSI

[234] **BSI.** Specification for pitch-impregnated fibre pipes and fittings for below and above ground drainage. *British Standard* BS 2760:1973. London, BSI, 1973

[235] **BSI.** Specification for asbestos-cement pipes, joints and fittings for sewerage and drainage. *British Standard* BS 3656:1981. London, BSI, 1981

[236] **BSI.** Precast concrete pipes, fittings and ancillary products. Specification for unreinforced and reinforced pipes and fittings with flexible joints. *British Standard* BS 5911-100:1988. London, BSI, 1988

[237] **BSI.** Precast concrete pipes, fittings and ancillary products. Specification for glass composite concrete (GCC) pipes and fittings with flexible joints. *British Standard* BS 5911-101:1988. London, BSI, 1988

[238] **BSI.** Fibre-cement pipes for sewers and drains. Pipes, joints and fittings for gravity systems. *British Standard* BS EN 588-1: 1997. London, BSI, 1997

[239] **BSI.** Specification for vitrified clay pipes, fittings and ducts, also flexible mechanical joints for use solely with surface water pipes and fittings. *British Standard* BS 65:1991. London, BSI, 1991

[240] **BSI.** Vitrified clay pipes and fittings and pipe joints for drains and sewers. Requirements. *British Standard* BS EN 295-1: 1991. London, BSI, 1991

[241] **BSI.** Vitrified clay pipes and fittings and pipe joints for drains and sewers. Quality control and sampling. *British Standard* BS EN 295-2:1991. London, BSI, 1991

[242] **BSI.** Vitrified clay pipes and fittings and pipe joints for drains and sewers. (Part 3) Test methods. (Part 4) Requirements for special fittings, adaptors and compatible accessories. *British Standard* BS EN 295-3 and 4:1991 and 1995. London, BSI, 1991 and 1995

[243] **BSI.** Vitrified clay pipes and fittings and pipe joints for drains and sewers. Requirements for vitrified clay manholes. *British Standard* BS EN 295-6:1996. London, BSI, 1996

[244] **BSI.** Vitrified clay pipes and fittings and pipe joints for drains and sewers. Requirements for vitrified clay pipes and joints for pipe jacking. *British Standard* BS EN 295-7:1996. London, BSI, 1996

[245] **BSI.** Specification for grey iron pipes and fittings. *British Standard* BS 4622:1970. London, BSI, 1970

[246] **BSI.** Specification for ductile iron pipes and fittings. *British Standard* BS 4772:1988. London, BSI, 1988

[247] **BSI.** Specification for manhole covers, road gully gratings and frames for drainage purposes. Cast iron and cast steel. *British Standard* BS 497-1:1976. London, BSI, 1976

[248] **BSI.** Ductile iron pipes, fittings, accessories and their joints for sewerage applications. Requirements and test methods. *British Standard* BS EN 598:1995. London, BSI, 1995

[249] **BSI.** Specification for centrifugally cast (spun) iron pressure pipes for water, gas and sewage. *British Standard* BS 1211:1958. London, BSI, 1958

[250] **BSI.** Specification for cast iron spigot and socket pipes (vertically cast) and spigot and socket fittings. Fittings. *British Standard* BS 78-2:1965. London, BSI, 1965

[251] **BSI.** Specification for aggregates from natural sources for concrete. *British Standard* BS 882:1992. London, BSI, 1992

[252] **Lillywhite M S T and Webster C J D.** Investigations of drain blockages and their implications for design. *The Public Health Engineer*, April 1979

[253] **Anon.** Recycled bedding for clay pipes. *Building Control*, September 2000

[254] **BRE.** Recycled aggregates. *BRE Digest* 433. Garston, Construction Research Communications Ltd, 1998

[255] **BRE.** Access to domestic underground drainage systems. *BRE Digest* 292. Garston, Construction Research Communications Ltd, 1984

[256] **Edgell G J.** Water testing brickwork manholes. *British Ceramic Research Association Technical Note* 373. Stoke-on-Trent, BCRA, 1986

[257] The Public Health Act 1936. London, The Stationery Office, 1936

[258] Water Industry Act 1991. London, The Stationery Office, 1991

[259] Building Act 1984. London, The Stationery Office, 1984

[260] **Hall J and Griggs J.** Rats in drains. *BRE Information Paper* IP 6/90. Garston, Construction Research Communications Ltd, 1990

[261] **BSI.** Concrete unreinforced tubes and fittings with ogee joints for surface water drainage. *British Standard* BS 4101:1967. London, BSI, 1967

[262] **BRE.** Plastics drainage pipes: storage and handling. *BRE Defect Action Sheet (Site)* DAS 40. Garston, Construction Research Communications Ltd, 1983

[263] **BRE.** Clay-ware drainage pipes: storage and handling. *BRE Defect Action Sheet (Site)* DAS 49. Garston, Construction Research Communications Ltd, 1984

[264] **BSI.** Workmanship on building sites. Code of practice for below ground drainage. *British Standard* BS 8000-14:1989. London, BSI, 1989

[265] **BRE.** Plastics drainage pipes: laying, jointing and backfilling. *BRE Defect Action Sheet (Site)* DAS 39. Garston, Construction Research Communications Ltd, 1983

[266] **BRE.** Flexibly jointed clay-ware drainage pipes: jointing and backfilling. *BRE Defect Action Sheet (Site)* DAS 50. Garston, Construction Research Communications Ltd, 1984

[267] **BSI.** Sewerage. Guide to rehabilitation of sewers. *British Standard* BS 8005-5:1990. London, BSI, 1990

[268] **BRE.** Domestic foul drainage systems: avoiding blockages – specification. *BRE Defect Action Sheet (Design)* DAS 89. Garston, Construction Research Communications Ltd, 1986

[269] **BRE.** Domestic foul drainage systems: avoiding blockages – installation. *BRE Defect Action Sheet (Site)* DAS 90. Garston, Construction Research Communications Ltd, 1986

Further reading

Young O C. Pipelaying principles. *National Building Studies Special Report* 35. London, The Stationery Office, 1964

Young O C. Loading charts for the design of rigid buried pipelines. *National Building Studies Special Report* 37. London, The Stationery Office, 1966

Anon. The watertightness of brickwork manholes. *National Builder*, December 1982

Griggs J and Grant N. Reed beds: (Part 1) application and specification; (Part 2) design, construction and maintenance. *BRE Good Building Guide* GBG 42 Parts 1 and 2. Garston, Construction Research Communications Ltd, 2000

BSI. Sewerage. Guide to new sewerage construction. *British Standard* BS 8005-1:1987 London, BSI, 1987

BSI. Sewerage. Guide to planning and construction of sewers in tunnel. *British Standard* BS 8005-3:1989. London, BSI, 1989

BSI. Sewerage. Guide to design and construction of outfalls. *British Standard* BS 8005-4:1987 London, BSI, 1987

Garner J F. *The law of sewers and drains under the Public Health Acts.* London, Shaw and Sons, 1981

Payne R. *Drain maintenance: estate management.* Harlow, Construction Press, 1982

Payne J A and Butler D. Septic tanks and small sewage treatment works. A guide to current practice and common problems. *CIRIA Technical Note* 164. London, CIRIA, 1993

Water Regulations Advisory Scheme. Reclaimed water systems. Information about installing, modifying or maintaining reclaimed water systems. *WRAS Information and Guidance Note* 9-02-04. Oakdale (Gwent), WRAS, 1999

Water Regulations Advisory Scheme. Reclaimed water systems. Marking and identification of pipework for reclaimed (greywater) systems. *WRAS Information and Guidance Note* 9-02-05. Oakdale (Gwent), WRAS, 1999

Chapter 4.3

[270] **BSI.** Specification for clayware field drain pipes and junctions. *British Standard* BS 1196:1989. London, BSI, 1989

[271] **BSI.** Specification for plastics pipes and fittings for use as subsoil field drains. *British Standard* BS 4962:1989. London, BSI, 1989

[272] **BSI.** Precast concrete pipes, fittings and ancillary products. Specification for ogee pipes and fittings (including perforated). *British Standard* BS 5911-110:1992. London, BSI, 1992

[273] **BSI.** Precast concrete pipes, fittings and ancillary products. Specification for porous pipes. *British Standard* BS 5911-114:1992. London, BSI, 1992

[274] **BSI.** Precast concrete pipes, fittings and ancillary products. Specification for road gullies and gully cover slabs. *British Standard* BS 5911-230:1994. London, BSI, 1994

[275] **Hall M J, Hockin D L and Ellis J B.** *The design of flood storage reservoirs.* Oxford, Butterworth-Heinemann/CIRIA, 1993

[276] **DoE.** Reservoirs Act 1975. London, The Stationery Office, 1975

[277] **BSI.** Workmanship on building sites. Code of practice for below ground drainage. British Standard BS 8000-14:1989. London, BSI 1989

Chapter 4.4

[278] **BSI.** Requirements for electrical installations. IEE Wiring Regulations. Sixteenth edition. *British Standard* BS 7671:2001. London, BSI, 2001

[279] **Health & Safety Executive.** Avoiding danger from buried electricity cables. *HSE Leaflet* GS 33. Sudbury, HSE Books, 1985

Chapter 5.1

[280] **BSI.** Code of practice for earth retaining structures. *British Standard* BS 8002:1994. London, BSI, 1994

[281] **Hope S, Ho K, Price G et al.** Stansted Airport vertical and lateral load tests on bored pile elements to be used in a retaining wall. *Conference, Instrumentation in Geotechnical Engineering*, 3–5 April 1989

[282] **BSI.** Code of practice for earthworks. *British Standard* BS 6031:1981. London, BSI, 1981

[283] **BRE.** Building brickwork or blockwork retaining walls. *BRE Good Building Guide* GBG 27. Garston, Construction Research Communications Ltd, 1996

References and further reading

[284] **Sills G C, Burland J B and Czechowski M K.** Behaviour of an anchored diaphragm wall in stiff clay. *BRE Current Paper* CP 38/78. Garston, BRE, 1978. (Also published in proceedings of 9th International Conference on Soil Mechanics and Foundation Engineering, Tokyo, 1977, vol 2, pp 147–54)
[285] **BRE.** Building simple plan brick or blockwork freestanding walls. *BRE Good Building Guide* GBG 14. Garston, Construction Research Communications Ltd, 1994
[286] **BRE.** Freestanding brick walls – repairs to copings and cappings. *BRE Good Building Guide* GBG 17. Garston, Construction Research Communications Ltd, 1993
[287] **BRE.** Damp-proof courses. *BRE Digest* 380. Garston, Construction Research Communications Ltd, 1993
[288] Building and buildings. The Noise Insulation Regulations 1975. Statutory Instrument 1975 No 1763. London, The Stationery Office, 1975
[289] **Department of Transport and the Welsh Office.** *Calculation of road traffic noise.* London, The Stationery Office, 1975 and 1988
[290] **BSI.** Specification for dimensions of bricks of special shapes and sizes. *British Standard* BS 4729:1990. London, BSI, 1990
[291] **BSI.** Sills and copings. Specification for copings of precast concrete, cast stone, clayware, slate and natural stone. *British Standard* BS 5642-2:1983. London, BSI, 1983
[292] **BRE.** Building reinforced, diaphragm and wide plan freestanding walls. *BRE Good Building Guide* GBG 19. Garston, Construction Research Communications Ltd, 1994
[293] **BSI.** Workmanship on building sites. Code of practice for masonry. *British Standard* BS 8000-3:2001. London, BSI, 2001

Further reading
BSI. Code of practice for ground anchorages. *British Standard* BS 8081:1989. London, BSI, 1989
BRE. Choosing external rendering. *BRE Good Building Guide* GBG 18. Garston, Construction Research Communications Ltd, 1994

Chapter 5.2
[294] **BRE.** Freestanding masonry boundary walls: stability and movement. *BRE Defect Action Sheet (Design)* DAS 129. Garston, Construction Research Communications Ltd, 1989
[295] **Korff J O A.** *Design of free-standing walls.* Windsor, Brick Development Association, 1984
[296] **BRE.** Surveying brick or blockwork freestanding walls. *BRE Good Building Guide* GBG 13. Garston, Construction Research Communications, 1992

Chapter 5.3
[297] **BSI.** Specification for pedestrian restraint systems in metal. *British Standard* BS 7818:1995. London, BSI, 1995
[298] **BSI.** Fences. Specification for open mesh steel panel fences. *British Standard* BS 1722-14:2001. London, BSI, 2001
[299] **BSI.** Fences. Specification for chain link fences. *British Standard* BS 1722-1:1999. London, BSI, 1999
[300] **BSI.** Fences. Specification for rectangular wire mesh and hexagonal wire netting fences. *British Standard* BS 1722-2:1989. London, BSI, 1989
[301] **BSI.** Fences. Specification for strained wire fences. *British Standard* BS 1722-3. London, BSI, 1986
[302] **BSI.** Fences. Specification for anti-intruder fences in chain link and welded mesh. *British Standard* BS 1722-10:1990. London, BSI, 1990
[303] **BSI.** Fences. Specification for woven wood and lap boarded fences. *British Standard* BS 1722-11:1992. London, BSI, 1992
[304] **BSI.** Fences. Specification for cleft chestnut pale fences. *British Standard* BS 1722-4:1986. London, BSI, 1986
[305] **BSI.** Fences. Specification for close-boarded fences. *British Standard* BS 1722-5:1986. London, BSI, 1986
[306] **BSI.** Fences. Specification for wooden palisade fences. *British Standard* BS 1722-6:1986. London, BSI, 1986
[307] **BSI.** Fences. Specification for wooden post and rail fences. *British Standard* BS 1722-7:1999. London, BSI, 1999
[308] **BSI.** Fences. Specification for mild steel (low carbon steel) continuous bar fences and hurdles. *British Standard* BS 1722-8:1990. London, BSI, 1990
[309] **BSI.** Fences. Specification for mild steel (low carbon steel) fences with round or square verticals and flat horizontals. *British Standard* BS 1722-9:1992. London, BSI, 1992
[310] **BSI.** Fences. Specification for steel palisade fences. *British Standard* BS 1722-12:1990. London, BSI, 1990
[311] **BSI.** Metallic and other inorganic coatings. Thermal spraying. Zinc, aluminium and their alloys. *British Standard* BS EN 22063:1994. London, BSI, 1994
[312] **BSI.** Specification for hot dip galvanized coatings on iron and steel articles. *British Standard* BS 729:1971. London, BSI, 1971
[313] **BSI.** Specification for steel wire for general fencing purposes. *British Standard* BS 4102:1998. London, BSI, 1998
[314] **BSI.** Specification for testing zinc coatings on steel wire and for quality requirements. *British Standard* BS 443:1982. London, BSI, 1982
[315] **BSI.** Fences. Specification for organic powder coatings to be used as a plastics finish to components and mesh. *British Standard* BS 1722-16:1992. London, BSI, 1992
[316] **Occasione J F et al.** Atmospheric corrosion investigation of aluminium-coated, zinc-coated and copper bearing steel wire and wire products. *American Society for Testing and Materials Special Technical Publication* 585A. Ann Arbor (USA), ASTM, 1984
[317] **Smith G A and Orsler R J.** *The biological natural durability of timber in ground contact.* BRE Report. Garston, Construction Research Communications Ltd, 1996
[318] **Smith G A and Orsler R J.** *Long-term field trials on preserved timber in ground contact* (revised to 1993). BRE Report. Garston, Construction Research Communications Ltd, 1995
[319] **BSI.** Copper/chromium/arsenic preparations for wood preservation. *British Standard* BS 4072:1999. London, BSI, 1999
[320] **BSI.** Wood preservation using coal tar creosotes. Specification for preservative. *British Standard* BS 144-1:1990. London, BSI, 1990
[321] **MAFF.** *United Kingdom atmospheric corrosivity values.* London, MAFF, 1986. (Available from DEFRA, Cartographic Branch, Lion House, Willowburn Estate, Alnwick, Northumberland NE66 2PF)
[322] **BSI.** Code of practice for performance and loading criteria for profiled sheeting in building. *British Standard* BS 5427:1976. London, BSI, 1976
[323] **BSI.** Hot dip galvanized coatings on fabricated iron and steel articles. Specifications and test methods. *British Standard* BS EN ISO 1461:1999. London, BSI 1999
[324] **BSI.** Specification for zinc-rich priming paint (organic media). *British Standard* BS 4652:1995. London, BSI, 1995
[325] **BSI.** Specification for water-borne priming paints for woodwork. *British Standard* BS 5082:1993. London, BSI, 1993
[326] **BSI.** Specification for solvent-borne priming paints for woodwork. *British Standard* BS 5358:1993. London, BSI, 1993
[327] **BSI.** Specification for undercoat and finishing paints. *British Standard* BS 7664:2000. London, BSI, 2000
[328] **BSI.** Code of practice for painting of buildings. *British Standard* BS 6150:1991. London, BSI, 1991
[329] **BSI.** Specification for lead-based priming paints. *British Standard* BS 2523:1966. London, BSI, 1966

Further reading
Suttie E D. Durability of timber in ground contact. *BRE Information Paper* IP 14/01. Garston, Construction Research Communications Ltd, 2001

BSI. Field test method for determining the relative protective effectiveness of a wood preservative in ground contact. *British Standard* BS 7282:1990. London, BSI, 1990

BSI. Durability of wood and wood-based products. Natural durability of solid wood. Guide to the principles of testing and classification of natural durability of wood. *British Standard* BS EN 350-1:1994. London, BSI, 1994

BSI. Durability of wood and wood-based products. Natural durability of solid wood. Guide to natural durability and treatability of selected wood species of importance in Europe. *British Standard* BS EN 350-2:1994. London, BSI, 1994

BSI. Fences. Chain link fences for tennis court surrounds. *British Standard* BS 1722-13:1978. London, BSI, 1978

BSI. Gaps, gates and stiles. Specification. *British Standard* BS 5709:2001. London, BSI, 2001

Chapter 5.4

[330] BSI. Road lighting. Guide to the general principles. *British Standard* BS 5489-1:1992. London, BSI, 1992

[331] BSI. Road lighting. Code of practice for lighting for subsidiary roads and associated pedestrian areas. *British Standard* BS 5489-3:1992. London, BSI, 1992

[332] **Chartered Institute of Building Services Engineers.** Lighting the environment. LG 06. London, CIBSE, 1995

[333] BSI. Intruder alarm systems. *British Standard* BS 4737:1977–88. London, BSI, 1977–88

Further reading

BSI. Glossary of building and civil engineering terms. Services. Lighting. *British Standard* BS 6100-3.4:1985. London, BSI, 1985

Chapter 6.1

[334] BSI. Stabilized material for civil engineering purposes. Methods of test for cement-stabilized and lime-stabilized materials. *British Standard* BS 1924-2:1990. London, BSI 1990

[335] BSI. Precast concrete flags, kerbs, channels, edgings and quadrants. Precast, unreinforced concrete paving flags and complementary fittings. Requirements and test methods. *British Standard* BS 7263-1:2001. London, BSI, 2001

[336] BSI. Specification for dressed natural stone kerbs, channels, quadrants and setts. *British Standard* BS 435:1975. London, BSI, 1975

[337] BSI. Precast, unreinforced concrete paving blocks. Requirements and test methods. *British Standard* BS 6717:2001. London, BSI, 2001

[338] BSI. Pavements constructed with clay, natural stone or concrete pavers (6 Parts). *British Standard* BS 7533:1997–2001. London, BSI, 1997–2001

[339] BSI. Hot-applied joint sealants for concrete pavements (3 Parts). *British Standard* BS 2499:1992–93. London, BSI, 1992–93

[340] BSI. Coated macadam (asphalt concrete) for roads and other paved areas. Specification for constituent materials and for mixtures. *British Standard* BS 4987-1:2001. London, BSI, 2001

[341] BSI. Bitumens for building and civil engineering. Specification for bitumens for roads and other paved areas. *British Standard* BS 3690-1:1989. London, BSI, 1989

[342] BSI. Specification. Dense tar surfacing for roads and other paved areas. *British Standard* BS 5273:1975. London, BSI, 1975

[343] BSI. Hot rolled asphalt for roads and other paved areas. Specification for constituent materials and asphalt mixtures. *British Standard* BS 594-1:1992. London, BSI, 1992

[344] BSI. Hot rolled asphalt for roads and other paved areas. Specification for the transport, laying and compaction of rolled asphalt. *British Standard* BS 594-2:1992. London, BSI, 1992

[345] BSI. Specification for mastic asphalt (limestone fine aggregate) for roads, footways and pavings in buildings. *British Standard* BS 1447:1988. London, BSI, 1988

[346] **Yates T J S and Richardson D.** Flooring, paving and setts. Requirements for safety in use. *BRE Information Paper* IP 10/00. Garston, Construction Research Communications Ltd, 2000

[347] BSI. Screeds, bases and in-situ floorings. Code of practice for polymer modified cementitious wearing surfaces. *British Standard* BS 8204-3:1993. London, BSI, 1993

[348] **Pascoe T and Bartlett P.** *Making crime our business. A crime audit guide for Registered Social Landlords.* BRE Report. Garston, Construction Research Communications Ltd, 2000

[349] **Schaffer R J.** The weathering of natural building stones. *National Building Studies Special Report* 18. London, Department of Scientific and Industrial Research, 1932. (Reprinted in facsimile form by BRE)

Further reading

BSI. Methods for measuring the skid resistance of pavement surfaces (2 Parts). *British Standard* BS 7941-2:1999 and 2000. London, BSI, 1999 and 2000

Collins R. Assessment of concrete slabs. *International Journal of Pavement Engineering and Asphalt Technology*, **2** (1) 75–84

Chapter 6.2

[350] **The Horticultural Trades Association.** *The national plant specification.* Reading, HTA, 2001

[351] **The Landscape Institute.** *List of trees and shrubs for landscape planting.* London, The Landscape Institute, 1989

[352] BSI. Nursery stock. Specification for trees and shrubs. *British Standard* BS 3936-1:1992. London, BSI, 1992

[353] BSI. Recommendations for transplanting root-balled trees. *British Standard* BS 4043:1989. London, BSI, 1989

[354] BSI. Recommendations for cultivation and planting of trees in the advanced nursery stock category. *British Standard* BS 5236:1975. London, BSI, 1975

[355] BSI. Nursery stock. Specification for ground cover plants. *British Standard* BS 3936-10:1989. London, BSI, 1989

[356] BSI. Recommendations for turf for general purposes. *British Standard* BS 3969:1998. London, BSI, 1998

[357] **DTI (Consumer Safety Unit).** *Accident rates from home and leisure.* London, DTI, 1990

[358] **Silsoe College.** *Technical bulletin on watering.* London, The Landscape Institute, 1996

[359] BSI. Specification for topsoil. *British Standard* BS 3882:1994. London, BSI 1994

Further reading

Lovejoy D. *Landscape handbook.* London, E & FN Spon, 1997

BSI. Code of practice for general landscape operations (excluding hard surfaces). *British Standard* BS 4428:1989. London, BSI, 1989

Rorison I H and Hunt R. *Amenity grassland – an ecological perspective.* Chichester, Wiley, 1980

Index

Foundations, basements and external works is systematically structured to enable the reader seeking particular information to identify quickly the parts of the book relevant to his or her search. The broad structure of chapter and sub-chapter titles will be seen in the contents list on page iii.

Chapter 0, the introduction to *Foundations, basements and external works*, describes the development of these services since late Victorian times, and typical aspects and problems that have arisen, and continue to arise.

Chapter 1 and its sub-chapters explain the issues and characteristics of sites, how they affect the design, construction and use of building fabric below ground level, and the areas within the curtilages of buildings.

Chapters 2–6 describe foundations; below-ground accommodation; public utilities provided for buildings; site perimeter and other retaining structures, and external lighting and security; and landscaping.

In the other books in the BRE Building Elements series, the authors have used standard headings (or adaptations of the standard headings) to help readers identify particular areas of interest.

With *Foundations, basements and external works*, some chapters have necessarily been treated differently with headings reflecting the variety of technical criteria.

Main contents of sub-chapters in Chapters 2–6
Characteristic details
Section headings provide basic descriptions, types and details of below-ground and external elements.

Main performance requirements and defects (not necessarily in the order shown)
The following broadly drawn sections describe the influences more frequently met with foundations, basements and external works, and the consequences they have for building users:
- strength and stability
- commissioning, testing and monitoring
- unwanted side effects
- health and safety
- security
- durability
- maintenance

Work on site
The most frequently used section headings are:
- workmanship
- supervision of critical features
- inspection (in panel)

Using the Index
The Index excludes words and expressions that are already presented in the list of contents or the list of section headings. Therefore the reader should undertake his or her search in the following order:
1. list of contents (pages iii and iv)
2. list of section headings (previous column)
3. index (starts next page)

Words and expressions which appear in the list of contents or in section headings **and** in other contexts are shown in the index with page numbers for the other contexts.

Page references for captions to illustrations are shown in bold.

Index

Access for firefighting vehicles, 226
Access roads, lighting for, 220
Access to basements, 150
Accidents outside buildings, 237
Accumulated temperature differences, 63
Acid attack, 118
Additives, concrete, 115
Aerial photography, 17, 70, 75, 90
Air, anabatic flow of, 68
    katabatic flow of, 68
Air temperatures, 63, 67
Alkali silica reaction, 92, 115, 119
Allowable bearing pressures, 30
Alluvial clay, 24, 32
Alluvium, 28, 31
Amenity lighting, 221, 223
Anabatic flow of air, 68
Anchored walls, 198
Anchors, rock, 199
    soil, 199
Anti-intruder fences, 213
Archaeological investigations, 28
Asbestos cement pipes, 172, 179
Asphalt, 229
Atmospheric corrosivity rates, 218
Augers, 123
    continuous flight, 77, 123, 128
    hand, **41**
    mechanical, **100**
Back-falls, 180
Backdrops, 167
Backfill compaction, 182
Backfilling, trench, 182
Backfilling old mine shafts, 51
Backfilling retaining walls, 202
Bacteria, pathogenic, 47
Bark-ringing of trees, 234
Basement floors, thermal performance of, 151
Basement structures, distress in, 137
Basement usage, 139
Basements, 8, 12
    access to, 150
    dampness in, 141
    door frames in, 146
    fire protection of, 154
    flotation of, 137
    horizontal loads on, 136
    services in, 146
    U-values of, 152
    ventilation of, 152
    vertical loads on, 136
    window frames in, 146
Beams, ground, 114
Bearing capacity, 84
Bedding material, recycled, 176
Beddings, granular, 174
Bell pits, 33
Benching, **166**
Bentonite slurry, **133**
Bio-remediation, 56
Blackwater, vi
Blasting, quarry, 86
Blockages, clearing drain, 183
    drain, 176, 180, 181, 184

Bollards, 232
'Boning' rods, 162
Bored piles, contiguous, 134
Bored piling machines, 123
Borehole depth, **41**
Boreholes, 39, 41, 91
Boring rig, percussion, **41**
Boulder clays, 30, 32
Boundary walls, 3, **209**
Box foundations, 114
Brick stitching, 97
Bricks, calcium silicate, 110
    clay, 110
Brickwork under-sailing, 93
Brownfield sites, 12, 43, **48**
Brushwood beneath footings, 109
Buoyancy forces (in ground), 31
Buoyant structures, 78
Cable bedding, 193
Cable TV, 193
Cables, colour coding of electricity, 157
    dangers from electricity, 193
Caissons, 113
Calcium silicate bricks, 110
Cantilevered retaining walls, 198
Cappings for retaining walls, 205
Caps, pile, 122
Car parking, 226
Carbon dioxide in landfill gas, 47
Carbonation of concrete, 92, 218
Casagrande standpipe, 38
Cast iron pipes, 159, 172, 173, 179
Catch pits, 187
Cavities, drained, 143
CCTV, 220, 223
CCTV inspection of a drain, **183**
Ceilings, cellar, 132
Cellar floors, 132
Cellar structures, distress in, 137
Cellar walls, 132
Cement mortar, 109
Cesspits, 168
Cesspools, 168
Chain link netting, 213, 216
Chalk, 27
Channels (in paving), 228
Chemical attack, 18, 92, 93
Chemical stabilisation, 84
Chemical treatments of clay soils, 98
Chimney breasts, 146
Chimney (geology), **17**
Clay bricks, 110
Clay field test, 25
Clay pipes, 178
    fired, 172
Clay soils, chemical treatments of, 98
Clays, 24, 25, **26**, 30, 32
    plastic limits of, 71
    shrinkable, 70
Cleft chestnut pale fencing, 214
Climate, UK, 62
Climbing plants, 236
Clinker, 229
Close boarded fencing, 214
Coal mines, grouting in old, 50

Cobbles, 9, 227, 228
Cohesionless soils, 27
Cohesive soils, 23, 25
Cohesive soils properties, 29, 30
Collapse compression, 49, 52, 61, 82, 93
Collapses of cathedral towers, 6
Colliery workings, old, **21**
Colour coding of electricity cables, 157
Colour coding of pipes, 157
Common trenches, 157
Compaction, 31
    backfill, 182
    dynamic, 80
    ground, **32**
    rapid impact, 81
    vibro, 12
Compaction of paving bases, 232
Compaction testing, 55
Compactors, 200
Composting toilets, 171
Compressible soil layers, 73
Concrete additives, 115
Concrete carbonation, 92
Concrete in contaminated land, 119
Concrete mixes, 115
Concrete pavings, in situ, 228
Concrete piles, 124, 127
Concrete pipes, 172, 178
Concrete raft, **115**
Concrete reinforcement, 115
Concrete, underpinning with mass, 98
Condensation, 152
Cone penetration tests, 29, 37
Conservation, water, 237
Consolidation of cohesive soils, 25
Constant head test, 61
Constructions, stepped, 151
Contaminants, types of, 45
Contaminated land, concrete in, 119
Contamination, salt, 148
Contamination containment, **48**
Contiguous bored piles, 134
Continuous bar fencing, 214
Continuous flight augered piles, 126
Continuous flight augers, 77, 123, 128
Controlled fill, 49
Copings for freestanding walls, 210
Copings for retaining walls, 205
Copper pipes, 160
Corrosion, microbial, 119
    reinforcement, 119
Corrosion of steel fixings, 92
Corrosion of steel piles, 127
Corrosive soils, 159
Corrosivity rates, atmospheric, 218
Corseting, 97
Crack patterns, **85**
Crack width monitoring, 95
Cracking, 3, 18, 34, 35, 90, 92, 107, 111, 147, 210
    environmental stress, 160
Cracks classifications, 91
Cradles, 174
Creep, 91
    slope, 20, 90

Index

Creosote, 219
Crib walls, 199, 204
Cribwork, 197
Crime prevention, 220
Cross-falls, gradients of, 188
Crown holes, **17**
Crown raising (trees), 238
Crown thinning (trees), 238
Cryptoflorescence, 110
Crypts, 8, 132
Cut-and-fill construction, 20
Cut-off walls, 48
Cutters, under-reaming, 123
Damage investigation, 90
Damp, penetrating, 130
    rising, 130
Dampness, 3, 138
    remedial work for, 143
Dampness in basements, 141
Dampness in retaining walls, 203
Dampproofing, 3, 141
Dampproofing partition walls, 145
Daylight availability, 69
Daylighting, 156
Decay, timber, 146
Deformable packing, **100**
Degradation, microbial, 160
Demec points, 95
Densifying fill, 61
Density index, 23
Desiccation, 12, 71, 89, 104
Desiccation profiles, 106
Desludging, 181
Detectors, intruder, 221
Detention tanks, 188
Diagnosis of foundation damage, 92
Diaphragm walls, 133
Differential movements, 3, 86
Displacement piles, 121
Displacement transducers, 96
Ditches, surface water, 186
Door frames in basements, 146
Downdrag, 20
DPCs, defective, 145
Drain, CCTV inspection of a, **183**
Drain blockages, 2, 180, 181, 184
    clearing, 183
Drain gradients, 176
Drain lengths, 163, 164
Drain repairs, 183
Drainage blankets, 186
Drainage for retaining walls, 202
Drainage grilles, gaps in, 231
Drainage layouts, 163
Drainage methods to control porewater
        pressures, 61
Drainage mounds, 188
Drainage of paved surfaces, 188, 230
Drainage practice, differences in, 164
Drainage sumps, 143
Drained cavities, 143
Drains, blockages in, 176
    French, 186
    hydraulic design for, 176
    land, 187

Drains, (cont)
    lining sleeves for, 183
    owners' responsibilities for, 181
    protection of, 182
    redundant, 178
    repair techniques for, 183
    trenchless technology for, 183
    under-, 186
    vertical, 83
    wastewater, 10, 162
Drains surcharging, 57
Driven tube window samplers, 91
Driving rain, 67
Dry lining, ventilated, 145
Drying out, 147
Ductile iron pipes, 173
Durability of masonry, 3
Durability of piles, 127
Dwellings, unfit, 4
Dynamic compaction, 80
Earth, reinforced, 197
Earth sheltered buildings, 130
Earthquakes, 35
Edgings (paving), 228
Efflorescence, 110, 147
Electrical inspection certificates, 148
Electrical resistance meter, 141
Electricity cables, colour coding of, 157
        dangers from, 194
Electricity mains, 193
Electrolevels, 38, **96**, 126
Electronic tilt meters, 96
Embedded walls, 198
End-bearing piles, 121
Engineered fills, 49, 53
Environmental stress cracking, 160
Epoxy resin coating of paved surfaces, 231
Ettringite, 53, 111, 118
Expansive clay, 25
Factors of safety, 85
Factors of safety with retaining walls, 201
Fencing, 9
Fills, 82, 88, 93
    densifying, 61
    performance specification for, 107
    trench, 11, 113
Fire protection of basements, 154
Fire resistance of walls and floors, 154
Fired clay pipes, 172
Firefighting shafts, 154
Firefighting vehicles access, 226
Fires, underground, 48
Flags (paving), 227, 228
'Floaters', 23
Flood plains, 59
Flooded areas, treatment of, 147, 148
Flooding, 14, 57, 59, 138
Flooding frequencies, 57
Floodlighting, security, 223
Floor slabs, underpinning unstable, 105
Floor slab piling, 102
Floors, cellar, 132
    fire resistance of, 154
    thermal performance of basement, 151
Flotation of basements, 137

Fluvio-glacial clays, 32
Footings, concrete strip, 112
Footpaths, ramps to, 231
'Footprint', vi
Foundations, changes in levels of, 116
    hollow box, 136
    medieval, **5**
    old, 20
    raft, 21
Freezing water mains, 159
French drains, 186
Friction piles, 121
Frost, 110, 117
Frost action, 231
Frost attack, 92
Frost heave, **53**, 89, 93, 119
Frost hollows, 64, 68
Frost-susceptible soils, 117
Fuel storage in cellars, 129
Fungal growth, **153**
Gabion walls, 197, 199, 204, 205
Galvanised steel fencing, 218
Gamma radiation, 29
Garages, 226
Gas mains, 194
Gas supplies, 11
Gas works sites, 46
Gates, **215**
    wrought iron, 215
Geo-membranes, 49, 55
Geological maps, 18
Geology, 18
Geophones, 39
Geotextile membranes, 187, 202
Gradients, drain, 176
    pipe, 166
Gradients of cross-falls, 188
Granular beddings, 174
Granular soils properties, 29
Graphitization, 159
'Grass-crete', 235
Gravel, 27, 229
Gravimetric method of moisture
        measurement, 142
Gravity walls, 198
Grease traps, 180
Greenfield sites, 12
Greywater, vi
Greywater reuse, 169
Ground, made, 209
Ground anchors, 133
Ground beams, 114
Ground compaction, **32**
Ground cover, 234
Ground heave, 126
Ground improvement, 79, 80
Ground investigation costs, 16
Ground pollution, 44
Groundborne vibration, 35, 86
Groundwater levels, 93
    rising, 88
Grouting, 97
Grouting in old coal mines, 50
Grouting slip surfaces, 198
Guard rails, 212, 216

Index

Guards for trees, 237
    spiral, 234
Gullies, 168, 188, 189
Guying trees, 237
Ha-has, 196
Hand augers, **41**
Hard water, 158
Hardcore settlement, 51
Hardstandings, grassed, 235
Head, 18, 31
Heat, venting, 154
Heat island effect, 62
Heat islands, 68
Heave, 18, 34, 53, 73, 75, 76, 89, 90, 111, 113, 136
    designing underpinning to prevent, 106
    frost, **53**, 89, 119
    ground, 126
Hedges for deterring intruders, 234
Helmets for piles, 128
Hoggin, 229
Hollow box foundations, 136
Horizontal loads on basements, 136
Hurdles, 214
Hydraulic seeding (grass), 235
Hydrocarbons in contaminated land, 47, 119
Hygrometers, 148
Hygroscopic salts, 138, 152
Inclinometers, 38, 96
Industrial pollution, 69
Industrial sites, former, 46
Inspection certificates, electrical, 148
Inspection chambers, 162, 167
Insulation, thermal, 87, 152
Insurance risks, 34
Intercepting traps, 10, 164, **177**
Intruder detectors, 221
Intruders, hedges for deterring, 234
Inundation, 61
Investigation of damaged buildings, 87, 90
Ion exchange, 198
Iron gates, wrought, **215**
Iron pipes, cast, 159, 172, 173, 179
    ductile, 173
Jacking of rafts, 114
Jacks, 104
Katabatic flow of air, 68
Kerbs, 228
Knuckle bends, 2
Land drains, 187
Land, recycled, 15
Landfill, 46
Landfill gas, 47
    removing, 54
Landslips, 14, 18, 19, 35, 90
Lap boarded fencing, 214
Lateral forces on foundations, 106
Lead pipes, 13, 160
Libore secant piling, 135
Light wells, 138, 156
Lightweight plasters, 152
Lime mortar, 109
Limit state design, 85
Lining sleeves for drains, 183

Liquefaction, 35
    spontaneous, 117
Liquid limits, 71
Loaded areas, depth of influence, 85
Loads on basements, horizontal, 136
    vertical, 136
Longwall mining, 33, 34
Macadam, 227
    stone, 9
Made ground, 17, 43, 44, 45, 209
Maguire's Rule, 10, 165
Mains, electricity, 193
    gas, 194
Mains water, 13
Manhole, vortex, 168
Manhole covers, 180
Manholes, 162, 167
    watertight, 177
Maps, 17, 18
Mascar bowl, 165, 168
Masonry, moisture content of, 141
Masonry durability, 3
Masonry footings, 109
Mastic asphalt tanking, 144
Materials, thermal effects on building, 91
Means of escape, 154
Medieval foundations, **5**
Membranes, geotextile, 187, 202
    liquid applied, 145
    self-adhesive, 144
Microbial corrosion, 119
Microbial degradation, 160
Mine shafts, backfilling old, 51
Mine workings, 88
Mines, 20, 32, 33, 34
Mini-piles, 122
Mini-piling, 101
    design of remedial, 103
Mining, 18
Mining practice, 33
Mining subsidence, 36, 116
Model Water Byelaws, 158
Moisture changes, 91
Moisture content of masonry, 141
Moisture content profiles, soil, **72**
Moisture measurement, Gravimetric method of, 142
Monitoring foundations, 94
    techniques for, 94
Monitoring pile performance, 126
Monitoring unstable slopes, 37
Mortar, cement, 109
    lime, 109
Mould growth, 138
Movement, differential, 86
Mowing strip (grass), 236
Mudstone, 29
Mulching, 235
Needle piling, 101
Netting, chain link, 213, 216
    wire, 213
Noise, pile driving, 126
Noise attenuation by vegetation, 69
Noise barriers, 204
Non-displacement piles, 122

Non-sewered sanitation, 170
Open mesh steel panel fencing, 215
Overburden, removal of, 89
Packing, deformable, **100**
Paint systems for fencing, 218
Pali radice, 102
Parking, car, 226
Partition walls, dampproofing, 145
Passive infrared sensors, 224
Paved surfaces, drainage of, 188, 230
    epoxy resin coating of, 231
    pressure hosing of, 231
    slipperiness of, 230
Pavers, 227, 228
Paving bases, compaction of, 232
Pavings, 9
    in situ concrete, 228
Pavings round trees, 73
Peak particle velocity, 86
Peat, 30, 44
Pedestrian areas, lighting for, 220
Pedestrian subways, 223
Penetrating damp, 1309
Percolating groundwater, 33, 88
Percolation values, 192
Percussion boring rig, **41**
Permissible stress, 85
Photography, aerial, 17, 70, 75, 90
Pier-and-beam underpinning, 99
Piers, 113
Piezometers, 38
Pile-and-beam underpinning, 99
Piled raft underpinning, 99
Piled retaining walls, 198
Piles, 77, 80
    contiguous bored, 134
    reinforced concrete, 7
    root, 102
Piling, floor slab, 102
    Libore secant, 135
    mini-, 101
    needle, 101
    raked, 101
    secant, 134
    shaft resistance with, 125
    sheet, 197
    skin friction with, 126
    Stent wall secant, 135
Piling machines, bored, 123
Pipe bedding, 182
Pipe cradles, 174
Pipe gradients, 166
Pipe joints, 180, 182
    flexible, 167
    push-fit, 179
Pipe sizes, 166
Pipes, 165
    asbestos cement, 172, 179
    bedding flexible, 174
    bedding rigid, 173
    cast iron, 159, 172, 173, 179
    clay, 178
    colour coding of, 157
    concrete, 172, 178
    copper, 160

Index

Pipes, (cont)
    cover for flexible, 175
    cover for rigid, 175
    ductile iron, 173
    fired clay, 172
    flexible, 165
    lead, 160
    loads on, 173
    pitch fibre, **171**, 173, 179
    plastics, 160, 171, 179
    rigid, 165
    rocker, 175
    steel, 160
Pipes near foundations, 175
Pipes under roads, 173
Pitch fibre pipes, **171**, 173, 179
Plasters, lightweight, 152
Plastic limits of clays, 71
Plastics pipes, 160, 171, 179
Plump holes, 33
Poker vibrators, 82
Pollarding, 75, 238
Pollutants in rainwater run-off, 190
Pollution, 177
    ground, 44
    industrial, 69
Porewater pressures, 19, 35
    drainage methods to control, 61
Posts, 216, 217
Pre-loading, 83
Precision optical levels, 96
Precision water gauges, 96
Pressure hosing of paved surfaces, 231
Primers, 219
'Privies', 13
Propped walls
Protection for trees, 236
Pruning shrubs, 238
Pruning trees, 238
Pulverised fuel ash, 115, 118
Pumps, sewage, 170
Push-fit pipe joints, 179
Quadrants (paving), 228
Quarries, 20
Quarry blasting, 86
Quarry workings
Quicksand, 60
Radiation, solar, 64
Radon, 14, 153
Radon levels, reducing, 154, 155
Raft foundations, 21, 117
Rafts, 114
    concrete, **115**
    jacking of, 114
Rails, guard, 212, 216
Rain, driving, 67
Rainfall intensity, 57, **58**, 59
Rainwater disposal, 59
Rainwater run-off, 57
    pollutants in, 190
Raked piling, 101
Raking shoring, 133
Ramps to footpaths, 231
Rapid impact compaction, 81
Re-lamping, 224

Re-strike times, 224
Recycled land, 15
Reed beds, 169, 170
Refuse tips, 44
Reinforced concrete piles, 7
Reinforced earth, 197
Reinforced walls, 198
Reinforcement, concrete, 115
Reinforcement corrosion, 119
Reinforcement of fills, 53
Renders, cementitious, 144
Repairing superstructures, 97
Replacement piles, 122
Resistance meter, electrical, 141
Rising damp, 130
Rock, 22
Rock anchors, 199
Rock properties, 29
Rocker pipes, 175
Rodding eyes, 162, 182
Rods, 'boning', 162
Rooms, windowless, 156
Root-balled trees, 233
Root piles, 102
Rot, timber, 148
Rowe cell, 25
Run-up times, 224
Safety, factors of, 85
Salt contamination, 148
Salt crystallisation, 118
Salts, hygroscopic, 138, 152
Sand, 27
Sand-wicks, 186
Sanitation, non-sewered, 170
Secant piling, 134
    Libore, 135
    Stent wall, 135
Seeding grass, 235
Seeding (grass), hydraulic, 235
Seepage, 60
Seismic forces, 35
Sensors, passive infrared, 224
Septic tanks, 169
Septicity, 177
Service entry points, 193
Service trenches, 116
Services in basements, 146
Settlement, 3, 52
    differential, 86
    limits for maximum, 86
    tolerable, 85
Settlement of hardcore, 51
Setts, 9, 227, 228
Sewage pumps, 170
Sewage treatment works, 169
    redundant, 46
Sewers, private, 162
Shading effect of trees, 69
Shaft resistance (piling), 125
Shales, 27
Shear failure, 88
Shear strength of soil, 28
Sheet pile walls, 198
Sheet piling, 197
Sheeted steel fencing, 215

Shell piles, 122
Shelter belts, 66
Shelter from wind, 65
Short bored piles, 11, 122
Shrinkable clays, 25, 70
Shrinkable clays locations, **26**
Shrubs, pruning, 238
Sieving soil, 23
Silt, 23
Silt traps, 187
Silty clay, 24
Site investigations, 16, 38, 39
Skid resistance, 232
Skin friction of soil, 77, 84, 126
Slag, 55
Slip indicators, 37
Slip plane layers, 73
Slip surfaces, grouting, 198
Slipperiness of paved surfaces, 230
Slope creep, 20, 90
Slope stabilisation, 74
Slope stability, 196
Slopes, monitoring unstable, 37
    regrading, 196
Sloping sites, 21
Slurry, bentonite, **133**
Smoke, venting, 154
Soakage tests, 189
Soft water, 158
Soil, skin friction of, 77, 84
    top-, 237
Soil analyses, 236
Soil anchors, 199
Soil classification, 22
Soil layers, compressible, 73
Soil mechanics, 7
Soil moisture content profiles, **72**
Soil nature, 23
Soil particle size, 23, 24
Soil sieving, 23
Soil stabilisation, 97
Soil strength measuring, 36
Soil testing, in situ, 37
Soil water suction, 71
Soils, cohesionless, 27
    cohesive, 23, 25
    corrosive, 159
    frost-susceptible, 117
Solar access, 67, 68
Solar gain, 64
Solar radiation, 64
Solution features, 20
Stabilisation, chemical, 84
    slope, 74
    soil, 97
Stability, slope, 194
Staking trees, 233, 237
Standard penetration test, 29, 37
Steel fixings, corrosion of, 92
Steel palisade fencing, 215, 217
Steel piles, 123
    corrosion of, 127
Steel pipes, 160
Steel wire fencing, 213, 216
Stent wall secant piling, 135

Step irons, **166**
Stepped constructions, 151
Stitching, brick, 97
Stone macadam, 9
Stools (sacrificial props), 99
Storm frequencies, 57
Strained wire fencing, 213
Stress cracking, environmental, 160
Strip footings, concrete, 112
Strip foundations, testing performance of, 117
Structured fill, 49
Structures, buoyant, 78
Subsidence, 14, 33, 34, 36, 52, 87
    mining, 116
Subsidence damage, action against, 74
Subsurface erosion, 19
Subways, pedestrian, 223
Suction, soil water, 71
Sulfate attack, 53, 89, 110, 111, 117, 118, 127
Sumps, drainage, 143
Superstructures, repairing, 97
Surcharge, vi
Surcharging, 83
Surcharging (drains), 57, 163, 177, 184, 190
Surcharging (poor ground), **83**
Surface erosion, 19
Surface wave velocity, 39
Sustainable urban drainage systems, 185, 188
Swales, 186
Swallow holes, 17, 20, 28, 32, 45
Tanking, 139, 141
    mastic asphalt, 144
Tanks, detention, 188
Tarmacadam, 229
Tell-tales, 95
Temperature differences, accumulated, 63
Temperatures, air, 67
Testing performance of strip foundations, 117
Testing underpinning, 107
Tests, soakage, 189
Thaumasite, 111, 118
Thermal bridging, 152
Thermal effects on building materials, 91
Thermal insulation, 87, 152
Thermal performance of basement floors, 151
Timber decay, 146
Timber fencing, 213
Timber in the ground, 217
Timber piles, 123, 127
Timber rot, 148
Timbers, built-in, 146
Toilets, composting, 171
Tolerable settlement, 85
Topsoil, 237
Traffic loadings, protection from, 175, 176
Transmission poles, 194
Transplanting trees, 233
Traps, grease, 180
    intercepting, 10, 164, **177**
    silt, 187

Tree pruning, 75
Tree reduction, 75
Tree removal, 70, 75
Tree removal precautions, 72
Tree root barriers, 76
Tree root pruning, 76
Trees, 94
    damage by, 72
    pavings round, 73
    shading effect of, 69
Trench backfilling, 182
Trench fill, 11, 113
Trenches, 182
    common, 157
    service, 116
    backfilling, 175
Trenchless technology (drains), 183
Trial pits, 39, 40
Trial pits for soakaways, 191, 192
Tumbling bay, 167
Turf, 235
U-values of basements, 152
UK climate, 62
Under-reaming cutters, 123
Under-sailing of brickwork, 93
Underdrains, 186
Underground house, **151**
Underpinning, 7, 79, 97, 98
    design of, 104
    partial, 107
    pier-and-beam, 99
    pile-and-beam, 99
    piled raft, 99
    testing, 107
Underpinning design to prevent heave, 106
Underpinning unstable floor slabs, 105
Underpinning with mass concrete, 98
Uplift forces on foundations, 106
Vandalism, 231
Vane shear test, 37
Variability of ground, 22
Variable head test, 61
Varnishes, 219
Varved clays, 24, 30
Vehicle impacts, 208
Ventilated dry lining, 145
Ventilation of basements, 152
Venting heat and smoke, 154
Vermin, 177
Vertical drains, 83
Vertical loads on basements, 136
Vibrating wire strain gauges, 38
Vibration, 19, 35
    groundborne, 86
Vibrators, poker, 82
Vibro, 82
Vibro compaction, 12
Vibro treatment, 81
Vortex manhole, 168
Vortexes, 167
'Waisting' of piles, 124, 128
Walls, 9
    cellar, 132
    cut-off, 48
    diaphragm, 133

Walls, (cont)
    embedded, 198
    fire resistance of, 154
    reinforced, 198
Waste, household, 46
Wastewater, vi
Wastewater drainage, 13
Wastewater drains, 10
Wastewater lifting plants, 170
Water, hard, 158
    mains, 13
    percolation of, 88
    soft, 158
    suction of soil, 71
Water Byelaws, Model, 158
Water conditioners, 159
Water conservation, 237
Water flow through ground, 60
Water gauges, precision, 96
Water levels, portable, 90
Water mains, 10, 158
    freezing, 159
Water Regulations, 158
Water supply, 10
Waterstopping, cementitious crystallisation, 140
    post-injected, 140
Waterstops, 139
    flexible, 140
    water-swellable, 140
Weathering, 19
Weep holes, 202
Wells, 10, 20
Whips (trees), 233
Wind, shelter from, 65
Wind control, 66
Wind sheltering, 66
Wind speed near buildings, 65
Wind speed zones, **207**
Wind speeds, basic hourly, **63**
Wind turbulence, 64
Window frames in basements, 146
Windowless rooms, 156
Wire coatings, 216
Wire fencing, 213
Wire netting, 213
Wire strengths, 216
Wood block pavings, 228
Wood fencing, 213
Wood stains, 219
Wooden palisade fencing, 214
Wooden post-and-rail fencing, 214
Work in confined spaces, 14
Woven wood fencing, 214
Wrought iron gates, **215**